Polymers and the Environment

Polymers and the Environment

Editors

Jesús-María García-Martínez
Emilia P. Collar

MDPI • Basel • Beijing • Wuhan • Barcelona • Belgrade • Manchester • Tokyo • Cluj • Tianjin

Editors
Jesús-María García-Martínez
Polymer Engineering Group
(GIP)-Institute of Polymer
Science and Technology
(ICTP)
CSIC
Madrid
Spain

Emilia P. Collar
Polymer Engineering Group
(GIP)-Institute of Polymer
Science and Technology
(ICTP)
CSIC
Madrid
Spain

Editorial Office
MDPI
St. Alban-Anlage 66
4052 Basel, Switzerland

This is a reprint of articles from the Special Issue published online in the open access journal *Polymers* (ISSN 2073-4360) (available at: www.mdpi.com/journal/polymers/special_issues/polymer_environment).

For citation purposes, cite each article independently as indicated on the article page online and as indicated below:

LastName, A.A.; LastName, B.B.; LastName, C.C. Article Title. *Journal Name* **Year**, *Volume Number*, Page Range.

ISBN 978-3-0365-7723-4 (Hbk)
ISBN 978-3-0365-7722-7 (PDF)

© 2023 by the authors. Articles in this book are Open Access and distributed under the Creative Commons Attribution (CC BY) license, which allows users to download, copy and build upon published articles, as long as the author and publisher are properly credited, which ensures maximum dissemination and a wider impact of our publications.

The book as a whole is distributed by MDPI under the terms and conditions of the Creative Commons license CC BY-NC-ND.

Contents

About the Editors vii

Preface to "Polymers and the Environment" ix

Jesús-María García-Martínez and Emilia P. Collar
Polymers and the Environment: Some Current Feature Trends
Reprinted from: *Polymers* 2023, 15, 2093, doi:10.3390/polym15092093 1

Debora Puglia, Francesca Luzi and Luigi Torre
Preparation and Applications of Green Thermoplastic and Thermosetting Nanocomposites Based on Nanolignin
Reprinted from: *Polymers* 2022, 14, 5470, doi:10.3390/polym14245470 5

Giulia Infurna, Gabriele Caruso and Nadka Tz. Dintcheva
Sustainable Materials Containing Biochar Particles: A Review
Reprinted from: *Polymers* 2023, 15, 343, doi:10.3390/polym15020343 21

Andrea Genovese, Flavio Farroni and Aleksandr Sakhnevych
Fractional Calculus Approach to Reproduce Material Viscoelastic Behavior, including the Time–Temperature Superposition Phenomenon
Reprinted from: *Polymers* 2022, 14, 4412, doi:10.3390/polym14204412 39

Noel León Albiter, Orlando Santana Pérez, Magali Klotz, Kishore Ganesan, Félix Carrasco and Sylvie Dagréou et al.
Implications of the Circular Economy in the Context of Plastic Recycling: The Case Study of Opaque PET
Reprinted from: *Polymers* 2022, 14, 4639, doi:10.3390/polym14214639 55

Mohor Mihelčič, Alen Oseli, Miroslav Huskić and Lidija Slemenik Perše
Influence of Stabilization Additive on Rheological, Thermal and Mechanical Properties of Recycled Polypropylene
Reprinted from: *Polymers* 2022, 14, 5438, doi:10.3390/polym14245438 71

Kit O'Rourke, Christian Wurzer, James Murray, Adrian Doyle, Keith Doyle and Chris Griffin et al.
Diverted from Landfill: Reuse of Single-Use Plastic Packaging Waste
Reprinted from: *Polymers* 2022, 14, 5485, doi:10.3390/polym14245485 87

Rafael E. Hincapie, Ante Borovina, Torsten Clemens, Eugen Hoffmann, Muhammad Tahir and Leena Nurmi et al.
Optimizing Polymer Costs and Efficiency in Alkali–Polymer Oilfield Applications
Reprinted from: *Polymers* 2022, 14, 5508, doi:10.3390/polym14245508 107

Ángel Agüero, Esther Corral Perianes, Sara Soledad Abarca de las Muelas, Diego Lascano, María del Mar de la Fuente García-Soto and Mercedes Ana Peltzer et al.
Plasticized Mechanical Recycled PLA Films Reinforced with Microbial Cellulose Particles Obtained from Kombucha Fermented in Yerba Mate Waste
Reprinted from: *Polymers* 2023, 15, 285, doi:10.3390/polym15020285 131

Ewa Liwarska-Bizukojc
Effect of Innovative Bio-Based Plastics on Early Growth of Higher Plants
Reprinted from: *Polymers* 2023, 15, 438, doi:10.3390/polym15020438 151

Marica Bianchi, Andrea Dorigato, Marco Morreale and Alessandro Pegoretti
Evaluation of the Physical and Shape Memory Properties of Fully Biodegradable Poly(lactic acid) (PLA)/Poly(butylene adipate terephthalate) (PBAT) Blends
Reprinted from: *Polymers* **2023**, *15*, 881, doi:10.3390/polym15040881 . . . . **163**

About the Editors

Jesús-María García-Martínez

Jesús-María García-Martínez holds a Ph.D. in Chemistry (Chemical Engineering) from Universidad Complutense de Madrid (1995), two M.Sc. level degrees, and has received more than 70 highly specialized courses. He is a Tenured Scientist at the Institute of Polymer Science and Technology (ICTP) of the Spanish National Research Council (CSIC). Since 1992, within the Polymer Engineering Group (GIP), he has co-authored more than 180 scientific and/or technical works in topics related to polymer engineering, chemical modification of polymers, heterogeneous materials based on polymers, interphases, and interfaces, polymer composites, polyblends and alloys, organic–inorganic hybrids materials, polymer recycling, quality, and so on. Furthermore, he has participated in more than 32 research and industrial projects (national and international programs) and is co-author of one currently active industrial patent on polymer recycling. From 2000 to 2005, he also assumed the position of Quality Director for the ICTP (CSIC) ISO 17025 Accreditation Project of the ICTP Laboratories. Additionally, Dr. García-Martínez is actively reviewing tasks for WOS, and SCOPUS-indexed journals, with more than 370 reports in the last years, being awarded twice with the Publons Reviewer Awards (2018 2019) and once with the POLYMERS Outstanding Reviewer Award (2019). Editor of more than 60 papers, he has been a guest Editor for many special issues in Q1 journals. Since 2016, he has served as Head of the Department of Chemistry and Properties of Polymer Materials within the ICTP (CSIC).

Emilia P. Collar

Ph.D. in Industrial Chemistry (U. Complutense, 1986). Since 1990 she has been permanent member of staff (Tenured Scientist) at the Consejo Superior de Investigaciones Científicas (CSIC), after two years (1986–1988) as CSIC's postdoc fellow and one (1989) as Chemical Engineering Assistant Teacher at the Universidad Complutense de Madrid. In the Polymer Science and Technology Institute (ICTP/CSIC), she works at the Polymer Engineering Group (GIP), founded in 1982 by Prof. O. Laguna, being the GIP's Head since 1999. Between 1990 and 2005, she supervised 6 doctoral theses and 5 post-graduate ones, under 14 research public funds projects, jointly to 18 research private funded contracts issuing 53 technical reports for different companies. Furthermore, from 2001 to 2005, she performed the positions of Technical Director of the Physical and Mechanical Properties Laboratory and Deputy-Technical Director of the Thermal Properties Laboratory under the successful ISO 17025 Accreditation Project for the CSIC/ICTP's Laboratories, ACiTP, commanded by the ICTP's Head. Author and co-author of more than 20 chapters on books, two currently active industrial patents on polymer recycling, and more than 150 papers mainly on SCI Journals, her research work lines deal with polymers and environment under the general frame of heterogeneous materials based on polymers. From 2006 to date, she has participated in three public Spanish-funded research projects and one EU project under its 7th Framework Program.

Preface to "Polymers and the Environment"

The aim of this reprint devoted to the topic "Polymers and the Environment" is to pursue environmentally friendly objectives for polymer-based materials under a two-fold perspective, applied and academic.

So, it is perhaps noteworthy to mention that, as far as in the early 1980s, the first global environmental crisis occurred with an emphasis on the role of plastics in big cities' massive solid-waste streams. It was apparent then (and now) that the best environmental management practices required solid scientific and technical knowledge (usually under technical standards). Once at the end of their useful life, these plastics become involved in their materials (polymers and additives) into a circular economy strategy conjugated with the non-steady too scenarios of the other key sectors of the economy, industry, society, and policy.

Thus, the above-mentioned two-fold perspective, applied and academic, is used to link tandem polymers and the environment has led, 40 years later, to a wide polymer research field devoted to continuously improving the environmental performance of polymer and polymer-based materials.

This strategy comprises all the steps in the polymer management chain, from the raw materials to the polymers themselves, many of which come from classical and/or renewable sources (the so-called bioplastics).

Additionally, there is a need to improve the processability, the ultimate properties, and the performance by employing friendly environment additives; the recyclability of the materials; and the development of innovative and disruptive processes allowing better mechanical and/or energy recovery, including chemical recycling.

Therefore, this reprint includes approaches related to this frontrunner polymer science and technology area. As the reader will appreciate, all the works compiled in this volume fully match all the philosophies mentioned under these lines.

Jesús-María García-Martínez and Emilia P. Collar
Editors

Editorial

Polymers and the Environment: Some Current Feature Trends

Jesús-María García-Martínez *[ ] and Emilia P. Collar *[ ]

Polymer Engineering Group (GIP), Polymer Science and Technology Institute (ICTP), Spanish National Research Council (CSIC), C/Juan de la Cierva, 3, 28006 Madrid, Spain
* Correspondence: jesus.maria@ictp.csic.es (J.-M.G.-M.); ecollar@ictp.csic.es (E.P.C.)

Citation: García-Martínez, J.-M.; Collar, E.P. Polymers and the Environment: Some Current Feature Trends. *Polymers* **2023**, *15*, 2093. https://doi.org/10.3390/polym15092093

Received: 19 April 2023
Accepted: 25 April 2023
Published: 27 April 2023

Copyright: © 2023 by the authors. Licensee MDPI, Basel, Switzerland. This article is an open access article distributed under the terms and conditions of the Creative Commons Attribution (CC BY) license (https://creativecommons.org/licenses/by/4.0/).

In the early 1980s, the first global environmental crisis occurred with an emphasis on the role of plastics in big cities' massive solid waste streams. It was apparent then (and now) that the best environmental management practices required solid scientific and technical knowledge (usually under technical standards). Once at the end of their useful life, these plastics become involved in their materials (polymers and additives) into a circular economy strategy conjugated with the non-steady too scenarios of the other key sectors of the economy, industry, society, and policy. Thus, a twofold perspective, applied and academic, to link tandem polymers and the environment has led, forty years later, to a wide polymer research field devoted to continuously improving the environmental performance of polymer and polymer-based materials. This strategy comprises all the steps in the polymer management chain, from the raw materials to the polymers themselves, many of which come from classical and/or renewable sources (the so-called bioplastics). There is a need to improve the processability, ultimate properties, and performance through friendly environment additives; the recyclability of the materials; and innovative processes that will allow for better mechanical and/or energy recovery, including chemical recycling. Therefore, this Special Issue includes a number of interesting works related to this frontrunner polymer R&D area [1,2]. The articles compiled in this volume fully match all the philosophies mentioned above.

Torre, Puglia, and Luzi have contributed a fascinating review paper on the preparation and applications of green thermoplastic and thermosetting nanocomposites based on nanolignin [3]. In this review article, the authors pay attention to lignin's valorization from plenty of lignocellulosic wastes. Lignin is a natural polymer with a cross-linked structure, valuable antiradical activity, and unique thermal and UV absorption properties besides being biodegradable, making it very valuable to be used either as a nanofiller or raw material to synthesize eco-friendly polymeric matrices. In the resume, this interesting review summarizes frontrunner synthesis methods for bio-based and/or biodegradable thermoplastic and thermosetting nanocomposites jointly with applications of lignin nanoparticles as reinforcements in nanocomposites.

The other review in this volume, authored by Infurna, Caruso, and Dintcheva [4], is devoted to using biochar particles as reinforcement in sustainable materials. The article is focused on how the conversion of polymer waste, food waste, and biomasses let obtain a solid phase named char/biochar particles (in addition to fuels and syngas) to be used in a wide variety of eco-friendly applications based on its chemical composition, porosity, and absorption ability depending of the thermochemical decomposition method used. Furthermore, the authors mention the actual uses of these particles as fertilizers for soil retirement and water treatment, introducing new services for formulating sustainable polymer and biopolymer-based composites, or as fillers in asphalts. Finally, the authors describe the advantages and disadvantages of using these biochar particles in polymer composites.

The article by Farroni et al. [5] is devoted to the fractional calculus approach to reproduce the material viscoelastic behavior based on an exceptionally reliable model parameterization procedure using the poles–zeros formulation. In the article, the authors state that the design of modern products and processes cannot leave out the usage of viscoelastic materials that provide extreme design freedoms at relatively low cost, paying attention to the possibility of reuse and recycling of these types of materials. The authors conclude that a limited number of the experimental curves could feed the identification methodology and predict the complete viscoelastic material behavior.

Based on the case study of a recycled opaque polyethylene terephthalate (r-O-PET), Arbiter, Santana, Maspoch, et al. report the implications of the circular economy in the context of plastic recycling [6]. The authors have evaluated r-O-PET with TiO_2 as reinforcement for recycled polypropylene matrices (r-PP), applying the life cycle assessment to different scenarios. By considering two different recycled blends and two virgin raw plastics as reference materials, when comparing the environmental performance of the proposed treatments, they found a significant reduction of environmental costs when substituting virgin materials PP and PA66 for the blends evaluated in this study.

Perše et al. contributed to this issue with a valuable study on the influence of stabilizers on the rheological, thermal, and mechanical properties of recycled polypropylene [7]. In their article, the authors have shown that after twenty recycling steps by successive reprocessing of polypropylene, the well-known fact that the main degradation processes of polypropylene are based on oxidation and chain scission, which the addition of such stabilizer can efficiently delay. Hence, they find that a tiny proportion of the additive dramatically improves the properties of the reprocessed polypropylene at least after twenty reprocessing rounds, acting as a hardener by promoting cross-linking reactions between polymer chains.

Dipa et al. investigated the reuse of packaging single-use plastic wastes recovered from landfill in the current context of a circular economy [8]. In essence, the authors study LDPE-based packaging films mostly end up in landfill after single-use because they are not commonly recycled despite their flexible nature, low strength, and low cost due to the expensive step of separation and sorting the different plastic waste streams that limit a broader use in recycling routes. Their research investigates the properties of PE-based waste mixed plastics with traces of PP, concluding that the final properties and absence of degradation provide a scientific insight into adopting these materials to return to the material markets.

An article on optimizing the polymer costs and efficiency in alkali–polymer oilfield applications is the proposal published by Hincapie et al. [9]. Indeed, the authors have performed a fascinating study by presenting various evaluations critical to prior field applications. For this purpose, the authors combine different laboratory approaches to optimize the usage of polymers in combination with alkalis to improve project economics. In essence, the authors optimize the performance of alkali–polymer floods by employing lower polymer viscosities during the injection but increasing polymer viscosities in the reservoir owing to the "thermal aging" of the polymers at a high pH. They evaluated alkali–polymers in the 8-Torton Horizon reservoir of the Matzen field in Austria in the Schoenkirchen area. Finally, the authors conclude that alkali–polymer injection leads to substantial incremental oil production of reactive oils. Therefore, alkali–polymer flood displacement efficiency must be evaluated by incorporating the aging of the polymer solutions, reaching significant cost savings in the process.

Arrieta et al. use yerba mate waste to obtain a kombucha beverage that produces microbial cellulose as a helpful byproduct to reinforce a mechanically recycled poly(lactic acid) (r-PLA) matrix [10] in their article titled Plasticized Mechanical Recycled PLA Films Reinforced with Microbial Cellulose Particles Obtained from Kombucha Fermented in Yerba Mate Waste. The authors also use microbial cellulosic particles obtained from the fresh yerba mate for comparison. To simulate the revalorization of the industrial PLA products rejected during the production line, PLA was subjected to three extrusion cycles and further addition of plasticizer. The transparent films showed flexible and good structural and mechanical performance, jointly with antioxidant properties revealed by contact with fatty food models, implying the potential of this recycled food waste in food packaging and agricultural films.

Considering that plastic particles are widespread in the environment, including terrestrial ecosystems, and they may disturb the physicochemical properties of soil and plant growth, Ewa Liwarska-Bizukojc has studied the effect of different biobased polymers on vegetal plants evolution [11]. In her investigation, she considered five innovative biobased plastics of varying chemical composition and application on the early growth of higher plants (sorghum, cress, and mustard), being found that at the early stages, the growth of monocotyledonous plants seemed not to be affected at all by the presence of each bioplastic studied. The plastics studied were based on polylactic acid (PLA), poly-hydroxybutyrate–valerate (PHBV), and polybutylene succinate (PBS). Simultaneously, she found that some bio-based plastics inhibited the root growth and stimulation of shoot growth of dicotyledonous plants.

Finally, Dorigato, Morreale, and coworkers present an evaluation of the physical and shape memory properties of fully biodegradable poly-lactic acid (PLA)/poly-butylene adipate terephthalate (PBAT) blends, taking in mind that this type of materials in combination has potential application in the packaging field [12]. One study aim is to provide information about the thermomechanical behavior of PBAT poorly characterized in the literature, mainly concerning the effect of the addition of PBTA on the shape memory properties of PLA. A microstructural investigation was conducted, evidencing the immiscibility and the low interfacial adhesion between the PLA and PBAT phases. The authors conclude the presence of PBAT hinders the shape memory of PLA and explain this fact in terms of the incompatibility of both polymers. Further, the authors claim that their future works will explore the shape memory effect of the samples once compatibilized with suitable additives.

To conclude, as the editors, we can say that the topic "Polymers and the Environment" arises as an essential framework in the field of Polymer Science and Technology in the present and near future. For this reason, a second Special Issue on the topic, which will be published by POLYMERS in 2023, is in progress, and new contributions will be welcomed.

Conflicts of Interest: The authors declare no conflict of interest.

References

1. García-Martínez, J.M.; Collar, E.F. Recycling of Thermoplastics. In *Handbook of Thermoplastics*, 2nd ed.; Olabisi, O. Adewale, K., Taylor and Francis Group, Eds.; CRC Press: Boca Raton, FL, USA, 2016; Chapter 27; pp. 919–940.
2. Collar, E.P.; García-Martínez, J.M. Environment, Health and Safety: Regulatory and Legislative Issues. In *Handbook of Thermoplastics*, 2nd ed.; Olabisi, O., Adewale, K., Taylor and Francis Group, Eds.; CRC Press: Boca Raton FL, USA, 2016; Chapter 28; pp. 939–954.
3. Puglia, D.; Luzi, F.; Torre, L. Preparation and Applications of Green Thermoplastic and Thermosetting Nanocomposites Based on Nanolignin. *Polymers* 2022, 14, 5470. [CrossRef] [PubMed]
4. Infurna, G.; Caruso, G.; Dintcheva, N.T. Sustainable Materials Containing Biochar Particles: A Review. *Polymers* 2023, 15, 343. [CrossRef] [PubMed]
5. Genovese, A.; Farroni, F.; Sakhnevych, A. Fractional Calculus Approach to Reproduce Material Viscoelastic Behavior, including the Time–Temperature Superposition Phenomenon. *Polymers* 2022, 14, 4412. [CrossRef] [PubMed]
6. León Albiter, N.; Santana Pérez, O.; Klotz, M.; Ganesan, K.; Carrasco, F.; Dagréou, S.; Maspoch, M.L.; Valderrama, C. Implications of the Circular Economy in the Context of Plastic Recycling: The Case Study of Opaque PET. *Polymers* 2022, 14, 4639. [CrossRef] [PubMed]
7. Mihelčič, M.; Oseli, A.; Huskić, M.; Slemenik Perše, L. Influence of Stabilization Additive on Rheological, Thermal and Mechanical Properties of Recycled Polypropylene. *Polymers* 2022, 14, 5438. [CrossRef] [PubMed]
8. O'Rourke, K.; Wurzer, C.; Murray, J.; Doyle, A.; Doyle, K.; Griffin, C.; Christensen, B.; Brádaigh, C.M.Ó.; Ray, D. Diverted from Landfill: Reuse of Single-Use Plastic Packaging Waste. *Polymers* 2022, 14, 5485. [CrossRef] [PubMed]
9. Hincapie, R.E.; Borovina, A.; Clemens, T.; Hoffmann, E.; Tahir, M.; Nurmi, L.; Hanski, S.; Wegner, J.; Janczak, A. Optimizing Polymer Costs and Efficiency in Alkali–Polymer Oilfield Applications. *Polymers* 2022, 14, 5508. [CrossRef] [PubMed]
10. Agüero, Á.; Corral Perianes, E.; Abarca de las Muelas, S.S.; Lascano, D.; de la Fuente García-Soto, M.d.M.; Peltzer, M.A.; Balart, R.; Arrieta, M.P. Plasticized Mechanical Recycled PLA Films Reinforced with Microbial Cellulose Particles Obtained from Kombucha Fermented in Yerba Mate Waste. *Polymers* 2023, 15, 285. [CrossRef] [PubMed]

11. Liwarska-Bizukojc, E. Effect of Innovative Bio-Based Plastics on Early Growth of Higher Plants. *Polymers* **2023**, *15*, 438. [CrossRef] [PubMed]
12. Bianchi, M.; Dorigato, A.; Morreale, M.; Pegoretti, A. Evaluation of the Physical and Shape Memory Properties of Fully Biodegradable Poly(lactic acid) (PLA)/Poly(butylene adipate terephthalate) (PBAT) Blends. *Polymers* **2023**, *15*, 881. [CrossRef] [PubMed]

Disclaimer/Publisher's Note: The statements, opinions and data contained in all publications are solely those of the individual author(s) and contributor(s) and not of MDPI and/or the editor(s). MDPI and/or the editor(s) disclaim responsibility for any injury to people or property resulting from any ideas, methods, instructions or products referred to in the content.

Review

Preparation and Applications of Green Thermoplastic and Thermosetting Nanocomposites Based on Nanolignin

Debora Puglia [1,*], Francesca Luzi [2] and Luigi Torre [1]

1. Department of Civil and Environmental Engineering, University of Perugia, 05100 Terni, Italy
2. Department of Materials, Environmental Sciences and Urban Planning (SIMAU), Polytechnic University of Marche, 60131 Ancona, Italy
* Correspondence: debora.puglia@unipg.it; Tel.: +39-0744-492916

Abstract: The development of bio-based materials is of great importance in the present environmental circumstances; hence, research has greatly advanced in the valorization of lignin from lignocellulosic wastes. Lignin is a natural polymer with a crosslinked structure, valuable antiradical activity, unique thermal- and UV-absorption properties, and biodegradability, which justify its use in several prospective and useful application sectors. The active functionalities of lignin promote its use as a valuable material to be adopted in the composite and nanocomposites arenas, being useful and suitable for consideration both for the synthesis of matrices and as a nanofiller. The aim of this review is to summarize, after a brief introduction on the need for alternative green solutions to petroleum-based plastics, the synthesis methods for bio-based and/or biodegradable thermoplastic and thermosetting nanocomposites, along with the application of lignin nanoparticles in all green polymeric matrices, thus generating responsiveness towards the sustainable use of this valuable product in the environment.

Keywords: bio-based; thermosetting; thermoplastic; nanocomposites; lignin nanoparticles

1. Introduction

The rising global demand for fossil resources for nonenergy purposes, as in the case of plastics production, has intensely motivated research to find alternative solutions to petrochemical plastics; however, progress has still not reached a commercially viable scale. The demand for cost-effective, ecofriendly materials has also increased to reduce waste management and pollution issues, thus academic/industry interest in sustainable bio-based materials has accordingly exploded in recent years. Even if numerous synthetic biopolymers have been used to this purpose, the need to suit different applications has allowed for progress even in the field of natural biopolymers. According to this, significant progress in lignin valorization and the use of sustainable and natural resources has been accomplished in the last years, in particular with regard to bio-based and biodegradable natural polymers based on this source, which are facing increasing consideration as they are environmentally green and economically reasonable. Replacing petroleum-based-derived materials with sustainable and environmentally friendly materials has been also considered as a crucial activity in the current period, so consideration has been given to the progress of lignin bio-based and/or biodegradable materials with thermomechanical performance that can compete or even surpass the petroleum-based products presently used. Lignin is considered an excellent substitute feedstock for the preparation of chemical products and polymers, even if one of the main difficulties still remaining is the lack of a well-defined structure and the partial flexibility linked to its origin, including extraction fragmentation procedures [1]. Although lignin is presently often considered as a filler or additive, it is hardly appreciated as a natural resource for chemical production. Nevertheless, it may be an outstanding candidate for chemical reactions due to its extraordinary reactivity (i.e., the presence of abundant aliphatic and phenolic hydroxyl groups) for the preparation of bio-based materials.

Following its fast development, efforts have incessantly been made to advance its compatibility with other additives in multicomponent systems, as in the case of its mixing with thermoplastics and thermosetting green matrices [2]. On the other hand, in parallel to its application at the macroscale as an additive in bio-based polymeric materials, research has advanced to solve these limits and has opened a different perspective towards the use of lignin-based nanomaterials as functional fillers in bio-based matrices [3]. To realize the potential of this material, one of the possible routes to follow includes lignin use in nanocomposite assemblies, where synergic interactions are extremely advantageous [4]. To this end, we review both existing possibilities, i.e., renewable thermoset and thermoplastic polymers based on lignin and the use of nanolignin as the active ingredient in these specific matrices, with particular attention to their application in niche sectors.

2. Synthesis of Bio-Based Polymeric Matrices from Lignin

Lignin is characterized by a complicated three-dimensional structure, obtained by polymerizing three phenylpropane units that arise from three aromatic alcohols: p-coumaryl alcohol, coniferyl alcohol, and sinapyl alcohol (Figure 1a) [5].

Its complexity can be controlled by accurate deconstruction, so the synthesis of polymers starting from aromatic compounds ideally obtained from lignin has attracted a growing curiosity. A large quantity of polymers obtained from lignin derivatives originate from different products (Figure 1b), gained by applying various depolymerization techniques [6,7]: as a function of the phenol substrates, various chemical changes and polymerization routes can be settled, leading to (semi)aromatic polymeric systems covering a widespread range of diverse thermal and mechanical characteristics.

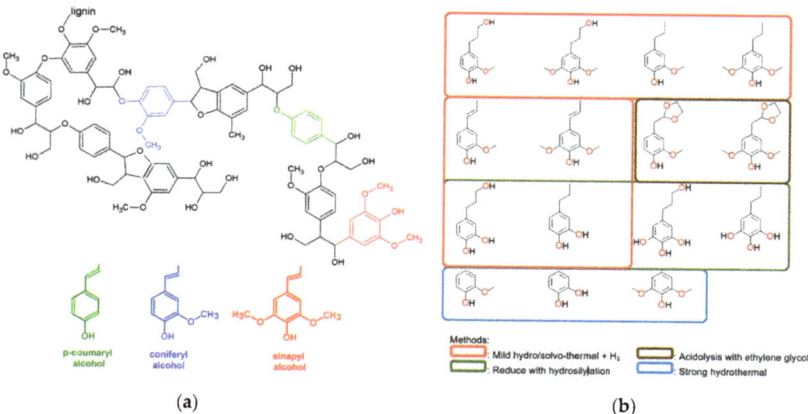

Figure 1. (**a**) Typical fragments of lignin structure with its main monolignols and (**b**) the potential phenolic products from lignin degradation. Reprinted with permission from Refs. [8,9]. Copyright 2019, Springer Nature Switzerland AG & Wiley.

Different approaches can be considered to realize high-performance polymers from lignin. The different adopted depolymerization methods have great prospective to get several monomer precursors, e.g., as alcohols, aldehydes, and acids, from lignin (Figure 2a). Vanillin is one of the most explored precursors. Other derived lignin compounds are ferulic acid, coumaric acid, sinapic acid, vanillic acid, syringic acid, caffeic acid, cinnamic acid, syringaldehyde, guaiacol, syringol, phenol, cresol, and catechol. The different molecules can be arranged as blocks for various polymers (Figure 2b), e.g., polyesters, polycarbonates, phenolic resins, polyurethanes, and epoxy resins. Many reviews extensively refer to the synthesis of lignin-derived compounds and their use in the realization of bio-based polymers [11–16]: here, we will exclude the case of lignin polymer blends, while the possibility of graft polymerization or lignin conversion of monomers to polymers will be considered.

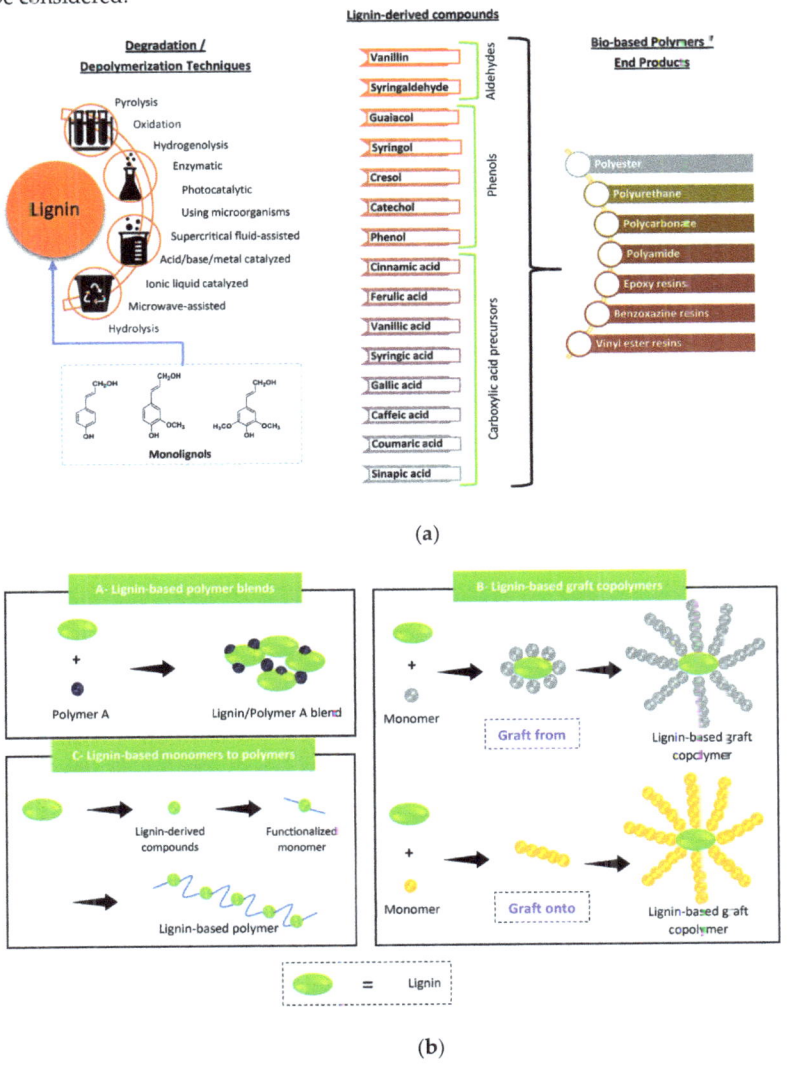

Figure 2. (a) Lignin depolymerization or degradation procedures and its derived monomers and polymers; (b) Different methods for lignin-based polymer synthesis. Reprinted with permission from Ref. [10]. Copyright 2021, Royal Society of Chemistry.

2.1. Synthesis Routes of Bio-Based Thermoplastics from Lignin-Based Molecules

The use of valorized lignin increases the potential to realize biodegradable biopolymers, such as polyhydroxyalkanoates, polyhydroxybutyrates, polylactic acid, or nonbiodegradable matrices, as in the case of polyurethane, polyolefins, polyamides, etc. [17] (Figure 3).

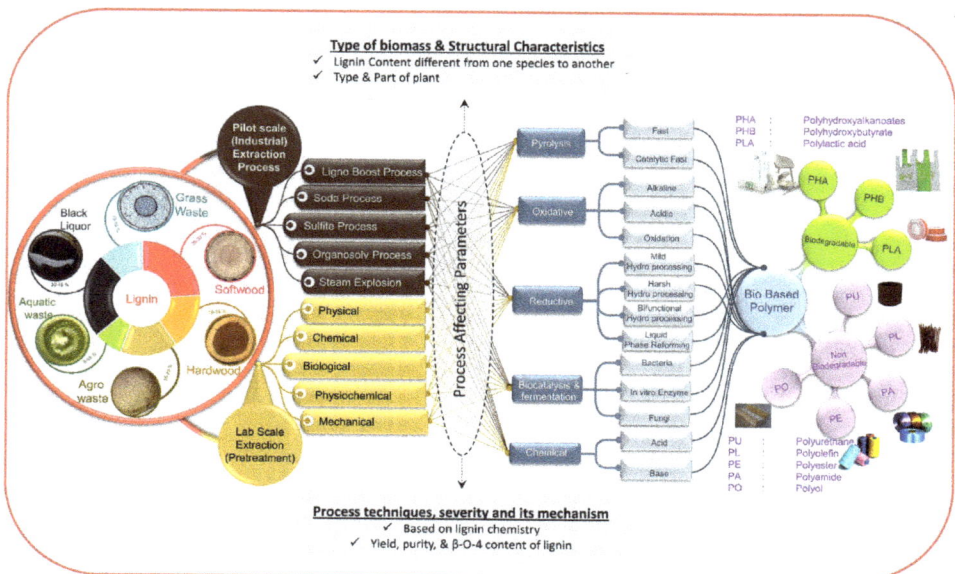

Figure 3. Schematic representation of lignin valorization on biopolymer production. Reprinted with permission from [17]. Copyright 2019, Elsevier Ltd.

The first case of polymerized vanillic acid to make polyesters was defined in 1955 [18]. Carboxylate was obtained by the conversion of vanillic acid and by the esterification process of a phenolic segment with ethylene dihalides. After that, the carboxylate was chemically esterified with ethylene glycol and reduced to linear polyester, showing a glass transition temperature of 80 °C and a melting temperature of 210 °C [18]. In 1981, a similar strategy was established by Lange and coauthors to produce vanillic and syringic acid-based polymers. In the second path, the phenolic moiety of vanillic acid interacted with ethylene oxide [19]. In the refining technique, lignin has a complex structure that makes it hard to be directly adapted into high-value materials: nevertheless, with the advanced clarification of the lignin structure and its microbial metabolism, it has been possible to alter the lignin structure to give, for example, high-value-added products by means of biological procedures, as in the case of PHAs. In nature, numerous bacteria have settled various metabolic paths to convert lignin to PHAs with short-, medium-, or long-chain structures; specifically, lignin derivatives can be processed to acetyl-CoA for PHA. Currently, PHAs from lignin or lignin-related aromatic compounds have been obtained by selecting many bacteria, such as *Oceanimonas doudoroffii* (PHA from sinapinic acid and syringic acid), *Cupriavidus basilensis* B-8, *Pseudomonas putida* A514, and *Pseudomonas putida* KT2440. The produced PHA can be adapted to convert into varied chemical precursors, such as alkenoic acids and hydrocarbons, which indicate that lignin can be adapted to become fuel-range hydrocarbons, chemical precursors, and biomaterials [20].

Lignin has been also explored as a natural resource for the synthesis of polyurethanes because it retains hydroxyl groups on its surface [21]. It fits well even into PU chemistries, since it acts both as crosslinking agent, due to the accessibility of numerous hydroxyl groups on each molecule and as a hard segment due to the aromatic nature. The last years have also viewed a reliable tendency that seeks to take advantage of the vast availability of renewable feedstocks, such as lignin derivatives, terpenes, vegetable oils, and polyols as precursors for the synthesis of nonisocyanates polyurethanes (NIPUs). Many approaches have been explored to use the aforesaid renewable resources to synthetize NIPUs combined with the prerequisites of green chemistry [22].

Meng et al. studied a different method that utilizes lignin extracted from the cosolvent lignocellulosic fractionation (CELF) of poplar wood to realize bio-NIPUs [23]. In this method, hardwood poplar is initially reduced in different fractions via a CELF treatment with the aim of recovering a lignin stream rich in phenolic content. The CELF lignin was then aminated by a Mannich reaction [23] and finally reacted with bicyclic carbonates to produce an innovative NIPU. The authors suggested the utilization of the CELF reaction to obtain a lignin rich in phenolic OH groups to raise its reactivity in amination. In addition, the authors studied, via the Mannich reaction, the amination of CELF lignin by using diethylenetriamine and formaldehyde in acidic environments. Successively, as an alternative to the reaction of hydroxyl groups of lignin with isocyanate groups to gain the traditional PU linkage (pathway I, Figure 4), the amine group in the CELF lignin reacted with cyclic carbonate originated from carbonation of epoxides to give lignin-based NIPU (pathway II, Figure 4).

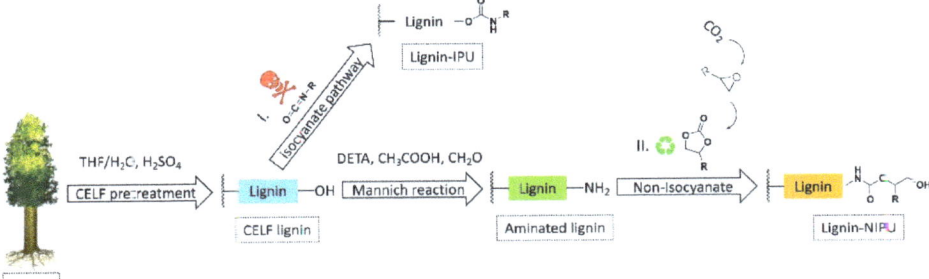

Figure 4. Synthesis procedures for polyurethanes obtained from lignin via (**I**) isocyanates pathway and (**II**) nonisocyanates way. Reprinted with permission from [23]. Copyright 2022, Elsevier Ltd.

According to this, the mechanical behavior of NIPUs based on lignin can be definitely varied from rigid to elastic by basically changing the lignin constituents of the polymeric material. The thermal characteristics of NIPUs were enhanced thanks to the addition of aminated lignin, and NIPU containing 55 and 23 wt% of lignin exhibited a high elongation at break (~140%) and tensile strength (~1.2 MPa), respectively. The obtained experimental results reveal that the reaction of cyclic carbonate with aminated lignin can be considered as a significant strategy for the synthesis of lignin-based NIPU with a relatively high lignin amount [23]. Lignin can also work as a polyol in the polyester synthesis, and hydroxyl-based chemistries can be adapted to give terminal hydroxyl groups, carboxylic acid, acyl groups, and epoxy groups. Polyester copolymers based on lignin can be obtained by the reaction of hydroxyl groups with additional reactive functional groups (diacyl chloride, dicarboxylic acid, adipic acid, and/or phthalic anhydride): in the case of branched lignin-based poly (ester-amine), reactions of triethanolamine, lignin, and adipic acid have to be considered [24]. Copolymerization of vinyl monomers and lignin is usually applied to realize lignin-based vinyl polymeric systems by different radical initiation routes [25].

Graft copolymerization, or in general, derivatization reactions, can generate new lignin containing thermoplastics by incorporating technical lignin and synthetic material. By appropriately choosing low-glass-transition temperature chemistries, lignin derivatives with variable thermomechanical properties can be obtained. On the other hand, the new progress on controlled/living polymerization and specific and effective synthetic techniques (e.g., click chemistry) propose new pathways for the design and realization of high-performance thermoplastic materials based on lignin derivatives having functional properties [26]. Even with a fruitful commercialization of many lignin-containing thermoplastics, constant efforts are still required to advance and create a new generation of lignin-based thermoplastics with precise structures, strong melt workability, good mechanical, and thermal performances.

2.2. Synthesis Routes of Bio-Based Thermosets from Lignin-Based Molecules

Thermosetting polymers can be produced from many bio-based resources, as in the case of vegetable oils, lactic acid, and citric acid, giving polymeric materials with more than 90% of a renewable amount [27]. Various bio-based thermosetting resins have been considered through the manipulation of virgin renewable feedstock; therefore, research on how to properly convert residual biomass to stimulate the production of new materials with remarkable properties is of great impact. Many functionalization approaches, including chemical or physical modifications, have been taken into account to broaden the application fields of these new resins with a singular glance to their processability and recyclability [28,29]. Referring to lignin, its aromatic structure gives molecular rigidity, and providing high-glass-transition temperatures and stiffness, as well the presence of aliphatic and aromatic hydroxyl functionalities, is a crucial characteristic for application where high thermal stability and high network levels are required [30]. Nevertheless, employing lignin as a raw material is still a noteworthy duty, being characteristically heterogeneous in its native form. Moreover, the processes considered to obtain lignin from biomass permanently alters the assembly of the lignin backbone, which is cleaved, fractionated, and assembled, making it practically nonidentifiable from its source in nature. Additionally, new functional groups are inserted through, for example, oxidation reactions, forming carboxylic acids, aldehydes, and ketones [31]. Oxypropylation, allylation, epoxidation, acetylation, and silylation are a few of the pathways for the modification of technical lignin found in the literature, which makes it compliant to be incorporated, by compatibilization, in polymeric matrices.

Two approaches are currently considered to obtain lignin-based thermosets. The first method utilizes lignin itself or incorporates lignin with other components to synthesize copolymers, with the key restriction that the reactivity of bulk lignin is lower than that of monomers. To overcome these restrictions, another method was developed that uses aromatic molecules obtained directly from lignin depolymerization. This approach enables molecular design and structural modification and can increase the performance of the resulting polymeric materials, as in the case, for example, of vanillin and other derivatives [32]. The main task in both cases is still to obtain reproducible and well-characterized fractions. These prospective polymer feedstocks, on the other hand, have their own limited challenges in terms of yields, prepolymerization reactions, and workability. The review from Feghali et al. [14] presents an overview on polymers obtained from lignin-based model compounds and depolymerized lignin (vanillin, vanillic acid, aromatic acid, quinones, and aromatic aldehydes by oxidation reaction): an extensive multiplicity of high-performance polymeric systems, such as polyurethane (PU), epoxy resin, phenol formaldehyde, and polyester, exhibiting good thermal and mechanical characteristics, can be synthesized with lignin as the macromonomer [33]. For example, Fersosian [34] obtained high-yield phenolic monomers through the selective cleaving of the β-O-4 bond of native lignin for the synthesis of lignin-based epoxy resin; however, the severe depolymerization condition, the low monomer yields, and the high separation costs limit the industrial use of this strategy (Figure 5a).

Compatibility and reactivity of lignin with compounds can be enhanced by means of chemical changes (e.g., propoxylation, phenolation, demethylation, and esterification reactions). Nevertheless, steric-hindrance influence and the limited compatibility of lignin-based epoxy resin weakened the crosslinking density, deteriorating the thermomechanical behavior of the thermosetting polymeric materials. Consequently, limitations such as the complex process, limited effect, low-lignin loading, and waste liquid recovery are unavoidable and still need to be solved. In lignin-modified phenolic resins, lignin is considered as the phenol able to react with formaldehyde in basic conditions or as an aldehyde to react with phenol in acidic conditions. Nevertheless, the replaced amount of phenol is partial due to scarce reactive sites and steric hindrance in lignin. To overcome these restrictions, lignin-derived phenols have been exploited to produce phenolic resins [35] (Figure 5b). The production of renewable, green, and sustainable phenolic resins based on lignin-derived monomers, having the potential to substitute traditional polymers, is currently under intensive study.

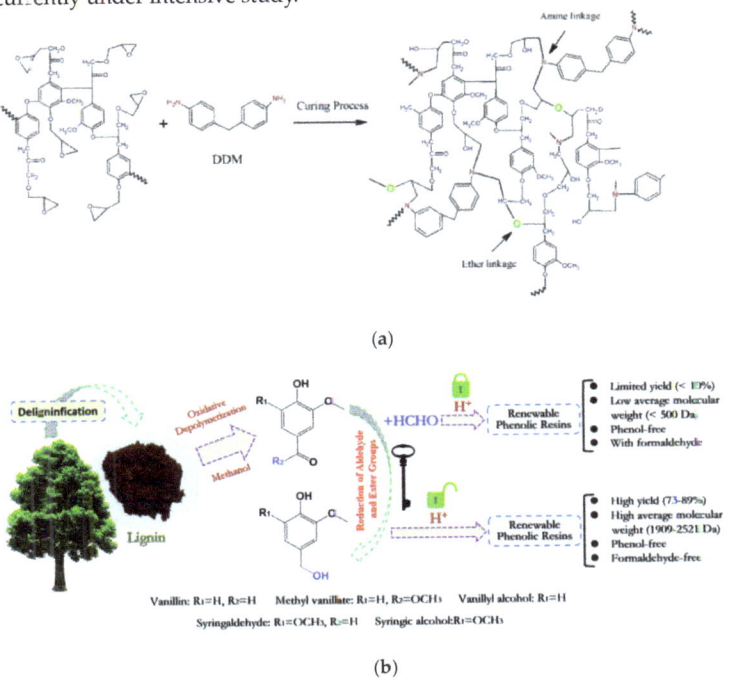

(a)

(b)

Figure 5. (a) Schematic representation and chemical structure of lignin-based epoxy resin and (b) schematic representation of the renewable phenolic resin synthesis based on lignin-derived monomers. Reprinted with permission from [34,35]. Copyright 2021, Elsevier Ltd.

3. Nanolignin as Filler in Polymeric Nanocomposites

Lignin nanoparticles have received much attention in the last years concerning the effort to utilize and apply lignin into more valued sectors [36]. In order to progress in the suitable use of lignin into different fields, it is required to ensue with chemical changes, fractionations to produce homogeneous materials, as previously described, or realize precipitated material with submicron particles for easier dispersion and enhanced features. The academic interest moved to the preparation of lignin nanoparticles (LNPs) by discovering their potential uses [37,38]. To date, the studied routes for the synthesis of LNPs are essentially chemical-based procedures which include, but are not limited to, acid-catalyzed, flash and nanoprecipitation, dialysis, solvent exchange, antisolvent process, W/O microemulsion processes, homogenization, and sonochemical synthesis. These methods have their profits and restrictions when they are utilized for LNP extractions. Therefore, the synthesis should be selected proficiently in order to yield LNPs of chosen sizes and dimensions [37,39]. Different methods, such as freeze-drying and thermal stabilization, interfacial crosslinking, polymerization and emulsion, and microbial- and enzyme-mediated, have been also considered as appropriate for the production of lignin nanoparticles. A comprehensive list of procedures that can be implemented has been reviewed in a few recent papers [40–43]. The procedures may give rise to appropriate advantages, but even distinctive faults regarding industrial use, since in some cases huge contents of solvents are necessary for the purification before precipitation, precipitation itself, and downstream processing, and in other cases a limited scalability of nanolignin production steps is manifest. Regardless the production yield, the research of suitable combination of LNPs with green matrices has progressed and prospective applications have been found and developed due to the specific characteristic of this material that can be considered for numerous potential applications (high thermal stability, manifest antioxidant properties, biodegradability, and UV-absorption features).

3.1. Nanolignin as Functional Filler in Thermoplastic Green Nanocomposites: Properties and Applications

The use of nanolignin as a reinforcing phase in macromolecules (both natural and synthetic) is a key methodology to advance in the realization of sustainable polymeric composite systems. Recently, lignin was utilized as a nanoscaled reinforcement to improve the structural characteristics of polysaccharides, proteins, natural rubber, and synthetic polymeric matrices [44,45]. In this context, the advance in the modification of lignin-based materials to give nanocomposites is, in the last decades, evident, due to the growing interest of the academic and industrial area. To provide few examples, Yang and coauthors proposed the use of lignin nanoparticles (LNP) in poly (lactic acid). They selected two different amounts (1 and 3% wt.) of nanofillers to be utilized in the polymer. Data obtained from antimicrobial analysis demonstrate the ability to hinder the growth of *Xanthomonas axonopodis pv. vesicatoria* and *Xanthomonas arboricola pv. pruni* Gram-negative bacteria over time, to positively influence the innovation, and to induce a positive effect against hazardous bacterial plant pathogens. The disintegration test under composting conditions revealed that the tested formulations reach a value up to 90% after 15 days; however, the presence of LNPs did not affect the disintegrability of different films, as shown in Figure 6a,b. The presence of LNPs did not affect the migration value, and accordingly the polymeric systems can be regarded as appropriate for the food packaging sector [46]. Data obtained from antimicrobial analysis demonstrate the ability to hinder the growth of *Xanthomonas axonopodis pv. vesicatoria* and *Xanthomonas arboricola pv. pruni* Gram-negative bacteria over time, to positively influence the innovation, and to induce a positive effect against dangerous bacterial plant pathogens. The disintegration test under composting conditions revealed that the tested formulations reach a value up to 90% after 15 days; however, the presence of LNPs did not affect the disintegrability of different films, as shown in Figure 6a,b. The presence of LNPs did not affect the migration value, and accordingly the films can be regarded as appropriate for application in the food packaging sector [46].

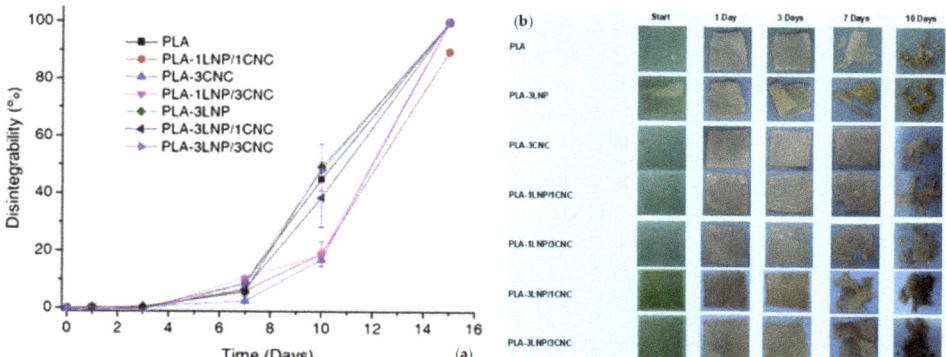

Figure 6. Disintegrability values (**a**) and visual images (**b**) of PLA and PLA binary and ternary nanocomposites at different incubation times in composting conditions. Reprinted with permission from Ref. [46]. Copyright 2016, Elsevier Ltd.

Chollet and coauthors considered the nanolignin as a new additive for flame-retardancy of poly (lactic acid) [47]. Lignin nanoparticles (LNPs) have been obtained from Kraft lignin microparticles by considering a dissolution–precipitation process. Micro- and nanolignins chemistries were altered by functionalizing the external surface with diethyl chlorophosphate (LMP-diEtP and LNP-diEtP, respectively) and diethyl (2-(triethoxysilyl)ethyl) phosphonate (LMP-SiP and LNP-SiP, respectively) to improve their flame-retardant effect in PLA. The results of inductively coupled plasma (ICP) spectrometry demonstrated that a great content of phosphorus was grafted onto the nanoparticles. Nevertheless, phosphorylated lignin nanoparticles limited PLA degradation during melt processing and the nanocomposite systems were shown to be relatively stable from the thermal point of view.

The use of lignin nanoparticles was largely applied and investigated as a method to develop new multifunctional, innovative materials in the food packaging sector. LNPs have been confirmed to provide enhanced mechanical, thermal, and antioxidant characteristics to the polymers in which they are incorporated depending on their particle size [48]. Lizundia et al. proposed the development of poly(l-lactide) (PLLA)-based nanosystems realized by the solvent casting method and combining LNPs with various metal oxide nanoparticles, such as WO_3, Ag_2O, Fe_2O_3, TiO_2, and $ZnFe_2O_4$ [49] (Figure 7a). It was found that the formulations based on nanolignin and $ZnFe_2O_4$ particles exhibited the best antioxidant behavior. Radical scavenging activity was also observed in ternary-based nanocomposites, where lignin and metal oxide nanofillers operated together synergically to boost the functional properties (Figure 7b). The antimicrobial activity of binary nanocomposites containing metal oxide NPs was correspondingly strong against PLLA, but it was only persistent for a few ternary nanocomposite films in a time result that was more obvious for *S. aureus* than for *E. coli* (Figure 7c). Lignin nanoparticles can protect towards UV light while allowing visible light to get through, and they can exceed the UV protection effect of numerous inorganic nanoparticles (Figure 7d).

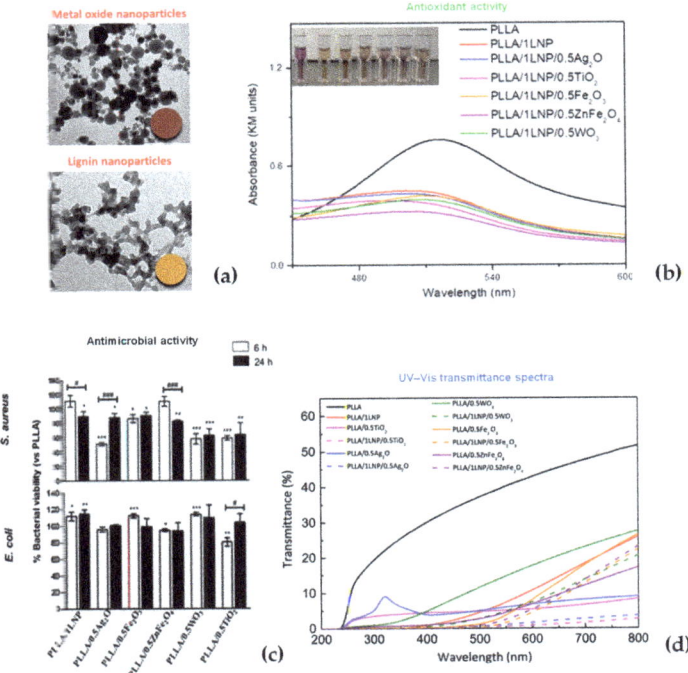

Figure 7. (a) TEM images of metal oxide (Fe_2O_3) and lignin nanoparticles (LNPs); (b) antioxidant activities of PLLA-based systems, antimicrobial activities; (c) antimicrobial activities of PLLA binary systems; (d) UV–vis spectra of PLLA binary and ternary films. Reprinted from [49]. Copyright 2020 American Chemical Society.

The central role of lignin nanoparticles as UV barrier filler was also investigated by Yang and coauthors [50]. They combined LNPs in PLA and polycaprolactone (PCL)-based formulations to increase packaging ductility and UV barrier properties. LNPs and caprolactone were first diluted in toluene to obtain a homogeneous solution and then purged with N_2 gas. The process was maintained at 120 °C for two days after the l-lactide addition. The addition of PCL determined an increase of the elongation at break up to 185%, an initial decrease of tensile strength that gradually increased to 280% after the addition of the LNP-P(LA-CL) copolymer. The toughness also rose 1.5 times above the PLA/PCL. Similar results were observed for the crystallization values and UV protection. Nanolignin–PLA/PCL-based systems can be utilized in the food packaging industry as an impact-resistant and UV-protectant material.

Cavallo et al. suggested the use of polylactic acid (PLA) films containing 1 wt% and 3 wt% of lignin nanoparticles (pristine (LNP) chemically modified with citric acid (caLNP) and acetylated (aLNP)). The different polymeric films were produced by extrusion and filming, and after that, the formulations were analyzed by determining the overall performance needed for the food packaging sector [51]. The obtained data indicated that all lignin nanoparticles induced UV-blocking, and antioxidant and antibacterial (against Gram-positive *Micrococcus luteus* and Gram-negative *Escherichia coli* bacteria) behavior to the PLA films, and a higher consequence was indeed found when increasing the filler content. Acetylation (aLNP) of the fillers moderately limited the antioxidant characteristics and the UV protection of the obtained composite systems, but it affected positively the nanoparticles distribution and aggregation, improving ductility and aesthetic quality of the films by decreasing at the same time the characteristic dark color of the lignin. Migration tests and disintegration test realized in simulated composting conditions of the nanocomposites showed that, irrespectively of their system, the realized active nanocomposites behaved likewise to neat PLA.

The use of LNPs was proposed as a valid possibility to develop promising wound dressing. Pahlevanneshan and coauthors [52] proposed the design and characterization of porous nanocomposite based on polyurethane (PU) foam synthesis. Moreover, the developed materials containing nanolignin coated with natural antimicrobial propolis for wound dressing. The antimicrobial effect was observed adding the extract to the polymeric foams, and all foams showed high biocompatibility toward L929 fibroblast cells, with the highest cell viability and cell attachment in the case of PU-LNP/propolis extract. In vivo wound-healing results, obtained by using Wistar rats' full-thickness skin wound model, showed that PU-LN/EEP has advanced wound-healing efficiency when compared to foams (Figure 8a–c) [52].

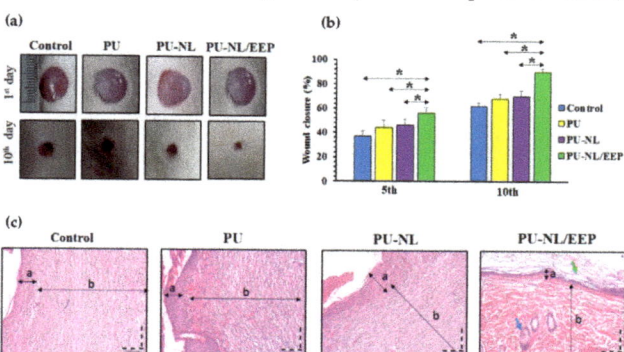

Figure 8. (a) Visual images of the wounds after 1 and 10 days of postoperation for the PU, PU-LNP, and PU-LNP/EEP groups. (b) Histograms of the wound closure for PU, PU-LNP, and PU-LNP/EEP groups. The data are expressed as mean ± standard deviation, (n = 8, *: $p < 0.05$, ns: not significant). (c) H&E-stained sections of skin specimens from the wound site of PU, PU-LNP, and PU-LNP/EEP groups. Arrows (a and b letters) designate the epidermis and dermis layers, respectively. Green and blue arrows indicate the keratin layer and sebaceous gland, respectively. Reprinted from Ref. [52]. Copyright 2021 MDPI AG.

3.2. Nanolignin as Functional Filler in Thermosetting Green Nanocomposites: Properties and Applications

Lignin can be blended, considered as a filler in a composite/nanocomposite formulation, both in its native form or chemically modified, combined in the presence of particular additives: in all cases, it has been proved that lignin can beneficially improve the overall performance of the resulting polymers. While the literature reports numerous examples of LNPs in thermosets [53–55], limited cases of nanolignin incorporation in green-based thermosetting matrices can be found. In their paper, Wang et al. [56] considered a simple and fast synthesis method to synthetize bio-based epoxy resin obtained from vanillyl alcohol; after that, vanillin-based epoxy resin (VE) was additionally reinforced by lignin-containing cellulose nanofibrils (LCNFs) with different weight contents. The authors experimentally observed that a significant improvement in the thermomechanical performance of the nanocomposites was attained with a low amount of nanofibril addition, confirming the possibility of assembling environmentally friendly and sustainable bio-based epoxy lignin nanocomposites with superior properties (Figure 9a).

A new approach was adopted to realize lignin phenol formaldehyde (LPF) resin: in the paper of Chen et al. [57], the preparation of nanolignin with a high specific surface area and porous structure was arranged and this nanofiller was then utilized as a valid phenol substitute combined with formaldehyde to produce a wood adhesive. Data showed that replacement of phenol by nanolignin could enhance the thermal characteristic of the resin, and in parallel, the modification of the curing schedule of the prepared lignin-based resin was considered.

In a quite recent paper [58], a simple foaming process to realize lignin-based polyurethane foams (LPUFs) was also considered: in that specific case, bio-based polyether polyols partially replaced petroleum-based raw components. Traces of phenolic hydroxyl groups (about 4 mmol) in lignin functioned as a direct reducing component and capping agent to silver ions by forming in situ silver nanoparticles (Ag NPs) within the LPUF skeleton. The lignin polyurethane/Ag composite foam (named as Ag NP-LPUF) was characterized by modulated thermomechanical and antibacterial properties, confirming the possibility of using these antimicrobial composite foams to encourage wound healing of full-thickness skin defects (Figure 9b).

Another example of lignin nanoparticle exploitation in a novel manner is represented by the study reported by [59], where the authors considered the realization of water-based, solvent-free, and multiresistant surface coatings: due to the presence of hydroxyl groups, the nanolignin acted as a hardener and no binder was required to realize adhesion to the substrate. In the case of the wood substrate, the particle morphology permitted proficient water repellency with a low coating weight, since the coating maintained the surface roughness of the wooden substrate while providing additional hydrophobicity.

Researchers from Washington State University, part of the NSF-supported Industry–University Cooperative Research Center for Bioplastics and Biocomposites (CB2), considered the use of a deep eutectic solvent to extract oligomeric lignin (nanoDESL) from plant biomass at a high yield and also nanosized [60]. NanoDESL shows narrower molecular size dimensions, distribution, and structural characteristics of traditional lignin. Oxypropylation of lignin was also optimized: it has been revealed that the use of polar aprotic solvents for the oxypropylation coupled with nanoDESL significantly promotes the oxypropylation reaction toward the synthesis of semiflexible PU. It was observed that the lignin-based PU containing ~20 wt% nanoDESL realized using polyol had density and compressive force comparable to the "standard" PU foam. The researchers are investigating how to enhance the reaction yield, with the goal of including 40 wt% lignin-based polyol into semiflexible foams: the potential of nanoDESL-based PU for adhesive, sealant, and coating applications is also explored by also considering their environmental toxicity and biodegradability issues. Using lignin as a source for PU synthesis not only encourages a circular economy but may also lead to the design of more ecofriendly end-of-life routes for PU plastics.

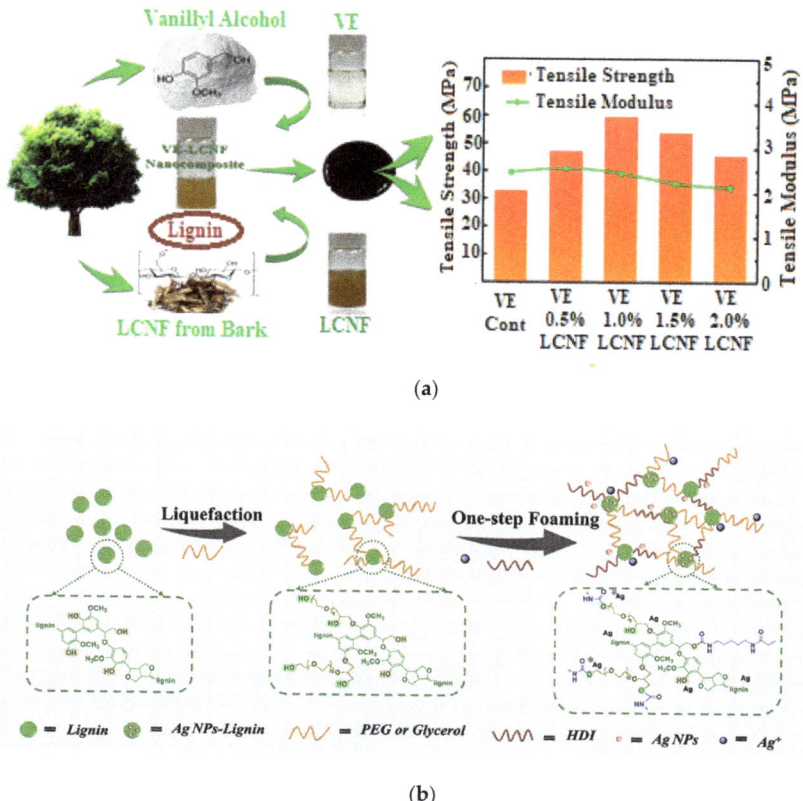

Figure 9. (a) Vanillin-based epoxy resin (VE) reinforced with lignin-containing cellulose nanofibrils (LCNFs) and the results of mechanical performance of the nanocomposites produced by considering different LCNF contents; (b) Schematic drawing of Ag NP-LPUF composite foam preparation by lignin liquefaction and one-step foaming. Reproduced with permission from [56,58]. Copyright 2020 and Copyright 2022, American Chemical Society.

All these studies provide awareness on potentialities of lignin nanosized fractions and allows for the design of reproducible and foreseeable material characteristics. It is essential to know these characteristics if we would exploit lignin as a raw material for a sustainable and innovative design. These preliminary and updated works confirm that nanolignin, if combined with green thermosetting matrices, can give fully green nanocomposites and, if effective, multifunctionality is often achieved in the presence of this nanoscaled filler.

4. Conclusions and Future Perspectives

This review, divided into two main sections, firstly provided an overview of preparation and applications of green thermoplastic and thermosetting nanocomposites based on lignin, and thereby a glance to the use of nanolignin in biopolymeric nanocomposites was also considered. Even if various structures and different properties of lignin at the nanoscale effectively can be challenging and interesting from the research point of view, the preparation and application of nanolignin-based green composites in high-value sectors are still in their infancy. Limiting factors include the achievement of uniform dispersion, and, additionally, the morphology, size, and chemistry of lignin nanoparticles, which need to be the prerequisites for the high-value-added and multifield applications of lignin. The structural and functional properties of lignin are the key points for its conversion into aromatics, polymers, and high-performance materials. Regarding this issue, we should emphasize that the production of thermoplastics or thermosetting polymers from the depolymerized lignin still involves the use of several chemicals, often causing a high environmental impact for the synthesis of bio-based and/or biodegradable polymers. Consequently, methods to chemically functionalize lignin to useful products without the use of expensive reagents or complicated synthetic routes must still be identified, with the main aim of competing with commercial commodity polymeric materials. The success of synthesizing thermoplastics and thermosetting from lignin opens up, on the other hand, new avenues to incorporate lignin as a component of value-added polymers while utilizing renewable resources. To balance the negative impact of chemically treated lignin and lignin derivatives to produce bio-based matrices, the use of lignin nanoparticles (obtained applying green process) in lignin-derived polymeric nanocomposites can be considered as a valid strategy to guarantee multifunctionality in sustainable biopolymeric matrices.

Author Contributions: Conceptualization, D.P.; methodology, D.P. and F.L; writing—original draft preparation, D.P. and F.L.; writing—review and editing, D.P. and F.L; supervision. L.T. All authors have read and agreed to the published version of the manuscript.

Funding: This research received no external funding.

Institutional Review Board Statement: Not applicable.

Data Availability Statement: Not applicable.

Conflicts of Interest: The authors declare no conflict of interest.

References

1. Laurichesse, S.; Avérous, L. Chemical modification of lignins: Towards biobased polymers. *Prog. Polym. Sci.* 2014, *39*, 1266–1290. [CrossRef]
2. Vahabi, H.; Brosse, N.; Abd Latif, N.H.; Fatriasari, W.; Solihat, N.N.; Hashim, R.; Hazwan Hussin, M.; Laoutid, F.; Saeb, M.R. 24-Nanolignin in materials science and technology—Does flame retardancy matter? *Micro Nano Technol. Biopolym. Nanomater.* 2021, 515–559. [CrossRef]
3. Liao, J.J.; Abd Latif, N.H.; Trache, D.; Brosse, N.; Hussin, M.H. Current advancement on the isolation, characterization and application of lignin. *Int. J. Biol. Macromol.* 2020, *162*, 985–1024. [CrossRef] [PubMed]
4. Lizundia, E.; Sipponen, M.H.; Garcia Greca, L.G.; Balakshin, M.; Tardy, B.L.; Rojas, O.J.; Puglia, D. Multifunctional lignin-based nanocomposites and nanohybrids. *Green Chem.* 2021, *23*, 6698–6760. [CrossRef]
5. Vásquez-Garay, F.; Carrillo-Varela, I.; Vidal, C.; Reyes-Contreras, P.; Faccini, M.; Teixeira Mendonça, R. A review on the lignin biopolymer and its integration in the elaboration of sustainable materials. *Sustainability* 2021, *13*, 2697. [CrossRef]
6. Llevot, A.; Grau, E.; Carlotti, S.; Grelier, S.; Cramail, H. From Lignin-derived Aromatic Compounds to Novel Biobased Polymers. *Macromol. Rapid Commun.* 2016, *37*, 9–28. [CrossRef]
7. Saito, T.; Brown, R.H.; Hunt, M.A.; Pickel, D.L.; Pickel, J.M.; Messman, J.M.; Naskar, A.K. Turning renewable resources into value-added polymer: Development of lignin-based thermoplastic. *Green Chem.* 2012, *14*, 3295–3303. [CrossRef]
8. Serrano, L.; Cecilia, J.A.; García-Sancho, C.; Garcia, A. Lignin Depolymerization to BTXs. *Top Curr. Chem. (Z)* 2019, *26*, 377. [CrossRef]
9. Ren, T.; Qi, W.; Su, R.; He, Z. Promising Techniques for Depolymerization of Lignin into Value-added Chemicals. *ChemCatChem* 2019, *11*, 639. [CrossRef]
10. Fadlallah, S.; Roy, P.S.; Garnier, G.; Saito, K.; Allais, F. Are lignin-derived monomers and polymers truly sustainable? An in-depth green metrics calculations approach. *Green Chem.* 2021, *23*, 1495–1535. [CrossRef]
11. Delidovich, I.; Hausoul, P.J.C.; Deng, L.; Pfützenreuter, R.; Rose, M.; Palkovits, R. Alternative Monomers Based on Lignocellulose and Their Use for Polymer Production. *Chem. Rev.* 2016, *116*, 1540–1599. [CrossRef] [PubMed]
12. Upton, B.M.; Kasko, A.M. Strategies for the Conversion of Lignin to High-Value Polymeric Materials: Review and Perspective. *Chem. Rev.* 2016, *116*, 2275–2306. [CrossRef] [PubMed]
13. Kristufek, S.L.; Wacker, K.T.; Tsao, Y.-Y.T.; Su, L.; Wooley, K.L. Monomer design strategies to create natural product-based polymer materials. *Nat. Prod. Rev.* 2017, *34*, 433–459. [CrossRef] [PubMed]
14. Feghali, E.; Torr, K.M.; van de Pas, D.J.; Ortiz, P.; Vanbroekhoven, K.; Eevers, W.; Vendamme, R. Thermosetting Polymers from Lignin Model Compounds and Depolymerized Lignins. *Top. Curr. Chem.* 2018, *32*, 376. [CrossRef] [PubMed]
15. Ganewatta, M.S.; Lokupitiya, H.N.; Tang, C. Lignin Biopolymers in the Age of Controlled Polymerization. *Polymers* 2019, *11*, 1176. [CrossRef]

16. Feghali, E.; Torr, K.M.; van de Pas, D.J.; Ortiz, P.; Vanbroekhoven, K.; Eevers, W.; Vendamme, R. Thermosetting Polymers from Lignin Model Compounds and Depolymerized Lignins. In *Lignin Chemistry. Topics in Current Chemistry Collections*; Serrano, L., Luque, R., Sels, B., Eds.; Springer: Cham, Switzerland, 2020; pp. 69–93. [CrossRef]
17. Banu, J.R.; Kavitha, S.; Yukesh Kannah, R.; Poornima Devi, T.; Gunasekaran, M.; Kim, S.-H.; Kumar, G. A review on biopolymer production via lignin valorization. *Bioresour. Technol.* **2019**, *290*, 121790. [CrossRef]
18. Bock, L.H.; Anderson, J.K. Linear polyesters derived from vanillic acid. *J. Polym. Sci.* **1955**, *17*, 553–558. [CrossRef]
19. Fache, M.; Boutevin, B.; Caillol, S. Vanillin, a key-intermediate of biobased polymers. *Eur. Polym. J.* **2015**, *68*, 488–502. [CrossRef]
20. Weng, C.; Peng, X.; Han, Y. Depolymerization and conversion of lignin to value-added bioproducts by microbial and enzymatic catalysis. *Biotechnol. Biofuels* **2021**, *14*, 84. [CrossRef]
21. Ma, X.; Chen, J.; Zhu, J.; Yan, N. Lignin-Based Polyurethane: Recent Advances and Future Perspectives. *Macromol. Rapid Commun.* **2021**, *42*, 2000492. [CrossRef]
22. Mehran, G.; Daver, F.; Ivanova, E.P.; Adhikari, B. Bio-based routes to synthesize cyclic carbonates and polyamines precursors of non-isocyanate polyurethanes: A review. *Eur. Polym. J.* **2019**, *118*, 668–684. [CrossRef]
23. Meng, X.; Zhang, S.; Scheidemantle, B.; Wang, Y.Y.; Pu, Y.; Wyman, C.E.; Cai, C.M.; Ragauskas, A.J. Preparation and characterization of aminated co-solvent enhanced lignocellulosic fractionation lignin as a renewable building block for the synthesis of non-isocyanate polyurethanes. *Ind. Crop. Prod.* **2022**, *178*, 114579. [CrossRef]
24. McDonald, A.G.; Sivasankarapillai, G. Synthesis and properties of lignin-highly branched poly(ester-amine) polymeric systems. *Biomass Bioenergy* **2011**, *35*, 919–931. [CrossRef]
25. Chung, H.; Washburn, N.R. Chemistry of lignin-based materials. *Green Mater.* **2013**, *1*, 137–160. [CrossRef]
26. Wang, C.; Kelley, S.S.; Venditti, R.A. Lignin-Based Thermoplastic Materials. *ChemSusChem* **2016**, *9*, 770. [CrossRef]
27. Ramamoorthy, S.K.; Åkesson, D.; Skrifvars, M.; Baghaei, B. Preparation and characterization of biobased thermoset polymers from renewable resources and their use in composites. Handbook of Composites from Renewable Materials. *Phys. -Chem. Mech. Charact.* **2017**, 425–457. [CrossRef]
28. Paipa-Álvarez, H.O.; Alvarado, W.P.; Delgado, B.M. Biodegradable thermosets polymers as an alternative solution to pollution generated by plastics. *J. Phys. Conf. Ser.* **2020**, *1672*, 012013. [CrossRef]
29. Liu, J.; Zhang, L.; Shun, W.; Dai, J.; Peng, Y.; Liu, X. Recent development on bio-based thermosetting resins. *J. Polym. Sci.* **2021**, *59*, 1474–1490. [CrossRef]
30. Jawerth, M.E.; Brett, C.J.; Terrier, C.; Larsson, P.T.; Lawoko, M.; Roth, S.V.; Lundmark, S.; Johansson, M. Mechanical and Morphological Properties of Lignin-Based Thermosets. *ACS Appl. Polym. Mater.* **2020**, *2*, 668–676. [CrossRef]
31. Gioia, C.; Lo Re, G.; Lawoko, M.; Berglund, L. Tunable Thermosetting Epoxies Based on Fractionated and Well-Characterized Lignins. *J. Am. Chem. Soc.* **2018**, *140*, 4054. [CrossRef]
32. Ribca, I.; Jawerth, M.E.; Brett, C.J.; Lawoko, M.; Schwartzkopf, M.; Chumakov, A.; Johansson, M. Exploring the Effects of Different Cross-Linkers on Lignin-Based Thermoset Properties and Morphologies. *ACS Sustain. Chem. Eng.* **2021**, *9*, 1692–1702. [CrossRef]
33. Huang, S.; Fu, S.; Gan, L. Lignin-Modified Thermosetting Materials. *Lignin Chem. Appl.* **2019**, 163–180. [CrossRef]
34. Ferdosian, F.; Yuan, Z.; Anderson, M.; Xu, C.C. Synthesis and characterization of hydrolysis lignin-based epoxy resins Ind. Crop. Prod. **2016**, *91*, 295–301. [CrossRef]
35. Yang, W.; Jiao, L.; Wang, X.; Wu, W.; Lian, H.; Dai, H. Formaldehyde-free self-polymerization of lignin-derived monomers for synthesis of renewable phenolic resin. *Int. J. Biol. Macromol.* **2021**, *166*, 1312–1319. [CrossRef]
36. Chen, K.; Wang, S.; Qi, Y.; Guo, H.; Guo, Y.; Li, H. State-of-the-Art: Applications and Industrialization of Lignin Micro/Nano Particles. *ChemSusChem* **2021**, *14*, 1284–1294. [CrossRef]
37. Beisl, S.; Miltner, A.; Friedl, A. Lignin from micro- to nanosize: Production methods. *Int. J. Mol. Sci.* **2017**, *18*, 1244–1274. [CrossRef]
38. Beisl, S.; Adamcyk, J.; Friedl, A. Direct Precipitation of Lignin Nanoparticles from Wheat Straw Organosolv Liquors Using a Static Mixer. *Molecules* **2020**, *25*, 1388. [CrossRef]
39. Low, L.E.; Teh, K.C.; Siva, S.P.; Chew, I.M.L.; Mwangi, W.W.; Chew, C.L.; Tey, B.T. Lignin nanoparticles: The next green nanoreinforcer with wide opportunity. *Environ. Nanotechnol. Monit. Manag.* **2020**, *2020*, 100398. [CrossRef]
40. Chauhan, P.S. Lignin nanoparticles: Eco-friendly and versatile tool for new era. *Bioresour. Technol. Rep.* **2019**, *2019*, 100374. [CrossRef]
41. Duarah, P.; Haldar, D.; Purkait, M.K. Technological advancement in the synthesis and applications of lignin-based nanoparticles derived from agro-industrial waste residues: A review. *Int. J. Biol. Macromol.* **2020**, *163*, 1828–1843. [CrossRef]
42. Luo, T.; Wang, C.; Ji, X.; Yang, G.; Chen, J.; Janaswamy, S.; Lyu, G. Preparation and characterization of size-controlled lignin nanoparticles with deep eutectic solvents by nanoprecipitation. *Molecules* **2021**, *26*, 218. [CrossRef] [PubMed]
43. Hazwan, H.M.; Appaturi, J.N.; Poh, N.E.; Abd Latif, N.H.; Brosse, N.; Ziegler-Devin, I.; Vahabi, H.; Syamani, F.A.; Fatriasari, W.; Solihat, N.N.; et al. A recent advancement on preparation, characterization and application of nanolignin. *Int. J. Biol. Macromol.* **2022**, *300*, 323–326. [CrossRef]
44. Feldman, D. Lignin nanocomposites. *J. Macromol. Sci.* **2016**, *53*, 382–387. [CrossRef]
45. Parvathy, G.; Sethulekshmi, A.S.; Jitha, S.J.; Akhila, R.; Appukuttan, S. Lignin based nano-composites: Synthesis and applications. *Process Saf. Environ. Prot.* **2021**, *145*, 395–410. [CrossRef]

46. Yang, W.; Fortunati, E.; Dominici, F.; Giovanale, G.; Mazzaglia, A.; Balestra, G.M.; Kenny, J.M.; Puglia, D. Effect of cellulose and lignin on disintegration, antimicrobial and antioxidant properties of PLA active films. *Int. J. Biolog. Macromol.* **2016**, *89*, 360–368. [CrossRef] [PubMed]
47. Chollet, B.; Lopez-Cuesta, J.M.; Laoutid, F.; Ferry, L. Lignin nanoparticles as a promising way for enhancing lignin flame retardant effect in polylactide. *Materials* **2019**, *12*, 2132. [CrossRef]
48. Ge, Y.; Wei, Q.; Li, Z. Preparation and evaluation of the free radical scavenging activities of nanoscale lignin biomaterials. *BioResour* **2014**, *9*, 6699–6706. [CrossRef]
49. Lizundia, E.; Armentano, I.; Luzi, F.; Bertoglio, F.; Restivo, E.; Visai, L.; Torre, L.; Puglia, D. Synergic effect of nanolignin and metal oxide nanoparticles into Poly(l-lactide) bionanocomposites: Material properties, antioxidant activity and antibacterial performance. *ACS Appl. Bio Mater.* **2020**, *3*, 5263–5274. [CrossRef]
50. Yang, W.; Qi, G.; Ding, H.; Xu, P.; Dong, W.; Zhu, X.; Zheng, T.; Ma, P.; Yang, W. Biodegradable poly (lactic acid)-poly (ε-caprolactone)-nanolignin composite films with excellent flexibility and UV barrier performance. *Compos. Commun.* **2020**, *22*, 100497. [CrossRef]
51. Cavallo, E.; He, X.; Luzi, F.; Dominici, F.; Cerrutti, P.; Bernal, C.; Foresti, M.L.; Torre, L.; Puglia, D. UV Protective, antioxidant, antibacterial and compostable polylactic acid composites containing pristine and chemically modified lignin nanoparticles. *Molecules* **2020**, *26*, 126. [CrossRef]
52. Pahlevanneshan, Z.; Deypour, M.; Kefayat, A.; Rafienia, M.; Sajkiewicz, P.; Neisiany, R.E.; Enayati, M.S. Polyurethane-Nanolignin Composite Foam Coated with Propolis as a Platform for Wound Dressing: Synthesis and Characterization. *Polymers* **2021**, *13*, 3191. [CrossRef] [PubMed]
53. Rahman, O.U.; Shi, S.; Ding, J.; Wang, D.; Ahmad, S.; Yu, H. Lignin nanoparticles: Synthesis, characterization and corrosion protection performance. *New J. Chem.* **2018**, *42*, 3415–3425. [CrossRef]
54. Wang, H.; Qiu, X.; Liu, W.; Fu, F.; Yang, D. A Novel Lignin/ZnO Hybrid Nanocomposite with Excellent UV-Absorption Ability and Its Application in Transparent Polyurethane Coating. *Ind. Eng. Chem. Res.* **2017**, *56*, 11133–11141. [CrossRef]
55. Garcia Gonzalez, M.N.; Levi, M.; Turri, S.; Griffini, G. Lignin nanoparticles by ultrasonication and their incorporation in waterborne polymer nanocomposites. *J. Appl. Polym. Sci.* **2017**, *134*, 45318. [CrossRef]
56. Wang, Z.; Gnanasekar, P.; Nair, S.S.; Farnood, R.; Yi, S.; Yan, N. Biobased Epoxy Synthesized from a Vanillin Derivative and Its Reinforcement Using Lignin-Containing Cellulose Nanofibrils. *ACS Sustain. Chem. Eng.* **2020**, *8*, 11215–11223. [CrossRef]
57. Chen, Y.; Gong, X.; Yang, G.; Li, Q.; Zhou, N. Preparation and characterization of a nanolignin phenol formaldehyde resin by replacing phenol partially with lignin nanoparticles. *RSC Adv.* **2019**, *9*, 29255–29262. [CrossRef]
58. Li, S.; Zhang, Y.; Ma, X.; Qiu, S.; Chen, J.; Lu, G.; Jia, Z.; Zhu, J.; Yang, Q.; Chen, J.; et al. Antimicrobial Lignin-Based Polyurethane/Ag Composite Foams for Improving Wound Healing. *Biomacromolecules* **2022**, *23*, 1622–1632. [CrossRef]
59. Henn, K.A.; Forsman, N.; Zou, T.; Österberg, M. Colloidal Lignin Particles and Epoxies for Bio-Based, Durable, and Multiresistant Nanostructured Coatings. *ACS Appl. Mater. Interfaces* **2021**, *13*, 34793–34806. [CrossRef]
60. Available online: https://www.aiche.org/sites/default/files/cep/20220410.pdf (accessed on 2 December 2022).

Review

Sustainable Materials Containing Biochar Particles: A Review

Giulia Infurna *, Gabriele Caruso and Nadka Tz. Dintcheva *

Dipartimento di Ingegneria, Università degli Studi di Palermo, Edificio 8, 90128 Palermo, Italy
* Correspondence: giulia.infurna@unipa.it (G.I.); nadka.dintcheva@unipa.it (N.T.D.)

Abstract: The conversion of polymer waste, food waste, and biomasses through thermochemical decomposition to fuels, syngas, and solid phase, named char/biochar particles, gives a second life to these waste materials, and this process has been widely investigated in the last two decades. The main thermochemical decomposition processes that have been explored are slow, fast, and flash pyrolysis, torrefaction, gasification, and hydrothermal liquefaction, which produce char/biochar particles that differ in their chemical and physical properties, i.e., their carbon-content, CHNOS compositions, porosity, and adsorption ability. Currently, the main proposed applications of the char/biochar particles are in the agricultural sector as fertilizers for soil retirement and water treatment, as well as use as high adsorption particles. Therefore, according to recently published papers, char/biochar particles could be successfully considered for the formulation of sustainable polymer and biopolymer-based composites. Additionally, in the last decade, these particles have also been proposed as suitable fillers for asphalts. Based on these findings, the current review gives a critical overview that highlights the advantages in using these novel particles as suitable additives and fillers, and at the same time, it shows some drawbacks in their use. Adding char/biochar particles in polymers and biopolymers significantly increases their elastic modulus, tensile strength, and flame and oxygen resistance, although composite ductility is significantly penalized. Unfortunately, due to the dark color of the char/biochar particles, all composites show brown-black coloration, and this issue limits the applications.

Keywords: biochar particles; sustainable materials; polymers; biopolymers; asphalts

1. Introduction

Nowadays, the conversion of polymer waste, food waste, and biomass aimed at reducing their impact on the environment gives them a second life, and changing from a linear economy to a circular economy is being widely investigated [1,2]. Different thermochemical decomposition processes leading to the recovery of fuels and residual solid phase have been taken into consideration, including those methods that could be profitable for some applications, such as for soil remediation, as additives, for use as synthetic carbonaceous particles, for the formulation of polymer and biopolymer-based composites, as additives for asphalts, etc.

Therefore, this review reports on the use of biochar particles, coming from different sources, for the formulation of composites and asphalts. See Figure 1 for more detail. The first part of the review deals with the considered methods to produce biochar particles and their main properties; the second and thirst parts are related to the formulation of polymer-based and biopolymer-based composites, respectively; and the fourth part is focused on the use of biochar particles as new additives for asphalts systems.

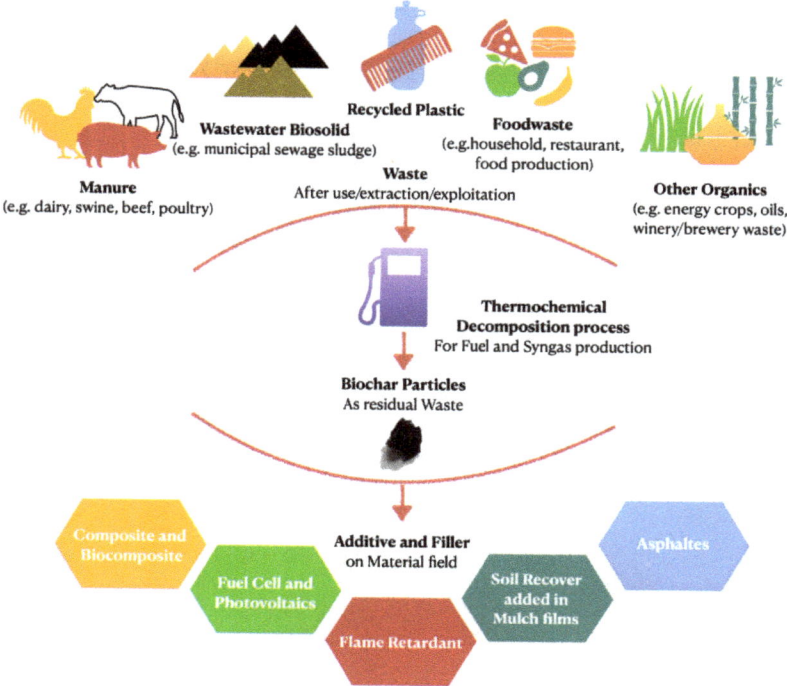

Figure 1. Different feedstocks used for the production of biochar particles and their adoption as suitable additives and fillers in polymers, biopolymers, and asphalts.

2. Biochar Particles: Production, Characteristics, and Properties

An opportunity to convert solid/food waste and biomass includes the thermochemical decomposition processes that are being used with increasing frequency. The main thermochemical decomposition processes explored are slow, fast, and flash pyrolysis, torrefaction, gasification, and hydrothermal liquefaction. All of these processes essentially generate: *i.* a solid phase, named char or biochar (in the case of biomass feedstock), *ii.* fuel, a mixed liquid phase of the heaviest hydrocarbon, *iii.* syngas, and a mixed gas phase of the lightest hydrocarbons are produced [3–6]. Of course, depending on the chemical composition of the treated materials (i.e., biomass, mixed waste, synthetic polymers), and depending on the operative condition (i.e., temperature process, heating rate, presence or absence of oxygen, residence time), the relative ratio between these three main products could change. Slow pyrolysis, conducted in the absence of oxygen, is characterized by slow heating rates and long residence times, as well as atmospheric pressure with an operating temperature that can vary from 350 to 800 °C; the necessary energy to pyrolyze the feedstock is usually provided internally by combusting a portion of the feedstock. The main product is a high-carbon solid char, and the coproducts are watery, low molecular weight liquid and a low energy combustible gas [7–10]. Fast pyrolysis, like slow pyrolysis, is conducted in the absence of oxygen with a temperature range between 400 and 600 °C. In contrast to slow pyrolysis, it uses a very high heating rate under a vacuum atmosphere, a short residence time, and the rapid quenching of vapor, since the main goal of this process is to produce bio-oil [7,11,12]. Flash pyrolysis is a batch process with an operative temperature range between 300 and 800 °C, and is similar to slow pyrolysis but with a high heating rate that uses moderate pressure (between 2 and 25 atm) to condense volatile elements and to promote secondary formation, since the aim of this process is to produce a biocarbon liquid fraction or biochar solid phase [13]. Torrefaction is a slow pyrolysis method with a lower temperature range, between 200 and 300 °C, that mainly removes water and some volatiles from the biomass to produce a "brown" char that is easy to ground and is a stabilized and friable biomass. Gasification is characterized by a high process temperature (between 750 and 1800 °C) with a limited and controlled oxygen concentration (normally calculated as the amount relative to stoichiometric combustion) [14] and/or steam [15]; as the name suggests, the primary products are a non-condensable gas mixture, called syngas, which is essentially composed by the presence of CO, H_2, with a smaller amount of carbon dioxide, methane, and other low molecular weight hydrocarbons [16]. Lastly, hydrothermal liquefaction is a process conducted in the presence of water, with a 250–450 °C temperature range under 100–300 bar; the main product of this process is called bio-crude, which is an energy-dense intermediate renewable source equivalent to oil that can be fractionated to a variety of liquid fuels [17]. Under this thermochemical process, the biomass is involved in depolymerization reactions (hydrolysis, dehydration, or decarboxylation, which produce insoluble products, such as bio-crude oil or bio-carbon, as well as volatile components (CO_2, CO, H_2 or CH_4) or soluble organic substances (mainly acids or phenols). All these processes and their differences are summarized in Table 1.

Table 1. Thermochemical processes and their main differences in terms of operative conditions, time of reactions, and primary products.

Thermochemical Process	Temperature Range [°C]	Heating Rate	Pressure	Residence Time	Primary Product
Slow Pyrolysis	350–800	Slow (<10 °C/min)	Atmospheric	Hours—Days	Char
Fast Pyrolysis	400–600	Very Fast (~1000 °C/s)	Vacuum-Atmospheric	Seconds	Oil
Flash Pyrolysis	300–800	Fast	Moderate (2–25 atm)	Minutes	Biocarbon/Char
Torrefaction	200–300	Slow (<10 °C/min)	Atmospheric	Minutes—Hours	Friable Biomass
Gasification	700–1800	Moderate-Vary Fast	Atmospheric Moderate	Seconds—Minutes	Syngas/Producer gas
Hydrothermal Liquefaction	250–450	Moderate	Elevated 100–300 atm	Minutes—Hours	Bio-crude (oil)

The focus of this work is biochar, which is essentially a carbon-made material that can potentially be produced through any thermochemical process, as a primary or auxiliary co-product, and from any feedstock. Feedstocks could include building materials, agricultural waste, forestry residues, municipal solid waste etc.

Biochar could be described as being divided into a "carbon" fraction, which includes carbon, hydrogen, and oxygen bonded together in different forms, and an ash inorganic fraction. For each thermochemical process employed, the temperature process, heating rate, and residence time affect the quality and the quantity of primary products and auxiliary co-products, and an operative parameter needs to be tailored to the feedstock, since the composition of potential biochar results may be affected by the feedstock characteristics. The primary analysis normally performed to characterize the feedstock is the operative temperature, and the relative char quality is the proximate analysis. This thermogravimetric analysis gives information about feedstock moisture content relative to the mass lost until 110 °C; volatile matter relative to the mass lost in an inert atmosphere at 950 °C; fixed carbon relative to mass lost in the air at 750 °C; and the remaining part relative to ash amount. Elemental analysis is normally employed to characterize the quality of char in terms of carbon content. This is a technique in which a sample is combusted at a very high temperature in a little chamber with an excess oxygen content, and the gasses relative to the combustion are trapped and, depending on the number of sensors available, it is possible to have, in terms of percentage in weight, information about carbon, hydrogen, nitrogen,

CHN element amount (relative CO_2, H_2O, NO), sulfur content, CHNS, oxygen/sulfur contents, and CHNOS.

As discussed above, biochar is a carbon-rich material which can be prepared from various waste feedstock. Municipal solid waste and agricultural waste are only two examples of the many organic wastes that may be utilized as feedstock to create biochar. Sludge is a solid waste that must be treated and disposed of, since it is produced during the wastewater treatment process. However, because it includes abundant carbon and nutrients such as ammonia, it is a viable feedstock for the synthesis of biochar [18]. The high carbon content, high cation exchange capacity, vast surface area, and stable structure of biochar are only a few of its benefits [19].

In general, organic or synthetic material can be used as feedstock with different processes depending on the physiochemical characteristics and the product composition.

The value of a particular type of biomass depends on the chemical and physical properties of the molecules from which it is made. Biomass is the main feedstock used in the literature for BC production because of different advantageous reasons. First, for environmental reasons, biomass is more readily available in a renewable way, either through natural processes or as a product of human activities. Furthermore, when produced by sustainable means, biomass produces approximately the same amount of carbon during conversion as is taken up during plant growth, which reduces the CO_2 amount in the atmosphere [20].

Biomass is mainly composed of three different organic compounds: cellulose, hemicellulose, and lignin, which give different mechanical and physiochemical properties to the woods. Cellulose makes up between 40% and 50% of the weight of dried wood and gives the biomass its strength [21]. Hetero polymers coexist with cellulose in plant cell walls to form hemicellulose. They contain several sugar monomers, including glucose, mannose, galactose, and xylose, and have lower molecular weights than glucose. Hemicellulose makes up anywhere from 20% to 35% of the bulk of dried wood [22]. The secondary cell wall of plants is made of lignin, which is a complex chemical compound. It is a kind of cross-linked resin that is amorphous, and it accounts for 15% to 30% of the mass of hardwoods. Depending on how much cellulose, hemicellulose, and lignin they contain, various biomass feedstocks have variable volatile matter concentrations and heating values, as well as different feedstock properties [23,24].

Biochar has received increasing attention due to its specific characteristics, such as high carbon content, cation exchange capacity, large specific surface area, and stable structure.

With different types of feedstocks, biochar has different physiochemical characteristics. The most typical processes for producing biochar are pyrolysis, gasification, and hydrothermal carbonization. Acid, alkali, oxidizing substances, metal ions, carbonaceous compounds, steam, and gas purging can all modify biochar. The environmental application fields determine the modification techniques to use.

The primary method used by biochar to remove organic and heavy metal contaminants is adsorption. The physiochemical characteristics of biochar, such as surface area, pore size distribution, functional groups, and cation exchange capacity, are strongly related to its adsorption ability, whereas physiochemical characteristics alter according to the production circumstances [25].

In general, biochar produced at high temperatures has a higher surface area and carbon content, mainly due to the increase in micro-pore volume caused by the removal of volatile organic compounds [26]. However, biochar yields decrease with temperature increases [27]. Therefore, an optimal strategy is required in terms of biochar yields and adsorption capacity. To sum up, the direct chemical composition of products and bioproducts is strictly connected to operative conditions (i.e., temperature, pressure and heating rate), which depend on the thermochemical process employed.

The physiochemical characteristics of biochar have been adjusted using metal ions, acids, alkalis, and oxidizing agents to make them better for various environmental processes [28].

Biochar has been widely employed in environmental applications, such as soil remediation, carbon sequestration, water treatment, and wastewater treatment because of its unique properties, which include high surface area, recalcitrant, and catalysis.

Common wastes, such as sludge and agricultural wastes, are produced in great quantities in the world. Sludge production alone reached 6.25 million tons in 2013 in China [29].

Converting common household wastes into biochar could be an option for environmental sustainability. Different feedstock has different proportions of element composition, and thus exhibits different properties, so the biochar derived from different feedstocks has various performances. The ways to deal with these wastes are directly linked to the impact they have on the environment.

Distinct feedstocks show varied qualities due to the different proportions of their elemental makeups, and, as a result, the biochar produced from those feedstocks performs differently. For instance, the pH (9.5) and potassium content (961 mg kg^{-1}) of straw-derived biochar were greater than those of wood biochar (349 mg kg^{-1}) [30]. Additionally, the biochar made from straw had more volatile material than non-volatile material, which is easier to remove during the pyrolysis process. Therefore, the high volatile component of the feedstock may contribute to poor biochar yields. Additionally, the content of pig and cow manures differed in terms of proportions [31]. Moreover, volatile content can be more easily removed than non-volatile content during pyrolysis. Therefore, the feedstock containing a high content of volatile content may result in low yields of biochar.

The type of feedstock has a significant effect on the physiochemical properties of biochar [32]. Therefore, the content of carbon in biochar is an important parameter, and different feedstocks can be converted into char using thermochemical decomposition processes, as was already described before (see Table 2).

Table 2. Proximate analysis of different biomass raw materials and relative elemental composition after pyrolysis process.

Feedstock	Proximate Analysis			Elemental Analysis			
	Volatile Matter [wt.%]	Fixed Carbon [wt.%]	Ash [wt.%]	C [wt.%]	H [wt.%]	N [wt.%]	O [wt.%]
Alfalfa [33] (Medicago sativa)	78.90	15.80	5.30	49.90	6.30	2.80	40.80
Almond Shell [34]	74.90	21.80	3.30	50.30	6.20	1.00	42.50
Bagasse [35]	71.00	13.70	2.10	51.71	5.32	0.33	42.64
Bamboo [36]	81.60	17.50	0.90	52.00	5.10	0.40	42.50
Carob Waste [8,37]	38.80	52.80	3.80	46.94	1.63	5.44	-
Coconut Fiber [38]	80.85	11.10	8.05	47.75	5.61	0.90	45.51
Corncob [39]	69.50	15.90	2.90	48.12	6.48	-	43.51
Cornstalk [39]	65.30	15.60	11.70	46.21	6.01	-	45.87
Cocopeat [35]	49.10	25.30	4.60	61.57	4.37	1.02	33.04
Dead Eucalyptus leaves [40]	77.60	16.90	0.80	52.90	8.10	0.30	47.90
Hamlin citrus [33]	77.90	17.80	9.40	50.70	6.60	1.60	42.90
Hornbeam Shell [41]	78.83	9.37	9.52	41.78	5.36	0.60	52.26

Table 2. Cont.

Feedstock	Proximate Analysis			Elemental Analysis			
	Volatile Matter [wt.%]	Fixed Carbon [wt.%]	Ash [wt.%]	C [wt.%]	H [wt.%]	N [wt.%]	O [wt.%]
Loblolly Pine [33] (*Pinus taeda*)	77.60	14.50	2.30	55.50	5.60	0.40	45.90
Maize straw [42]	-	-	5.31	42.20	7.21	1.28	49.20
Mesocarp Fiber [40] (*Oil palm*)	72.80	18.90	8.30	51.50	6.60	1.50	40.10
Oak sawdust [43]	69.24	16.51	0.81	52.28	5.74	0.06	41.92
Olive wood [44]	79.60	17.20	3.20	49.00	5.40	0.70	44.90
Paddy straw [35]	56.40	15.40	20.90	48.75	5.98	1.99	43.28
Pinewood [38]	85.45	13.15	1.40	48.15	6.70	1.35	43.60
Palm Kernel Shell [35]	66.80	17.90	3.40	55.82	5.62	0.84	37.73
Raw Pine sawdust [42]	83.10	16.80	3.76	50.60	6.18	0.05	43.10
Rice husk	62.80	19.20	18.00	49.30	6.10	0.80	43.70
Sawdust [39]	70.40	18.50	1.20	48.37	4.98	-	46.27
Tea waste [45]	70.29	18.57	3.88	48.60	5.43	3.80	42.17
Wood Stem [35]	80.10	10.70	0.40	50.52	5.81	0.23	43.44
Wood Bark [35]	68.90	16.30	4.90	53.42	6.12	1.40	39.06

Due to different compositions (carbon with the presence of alkali metals, e.g., Li, Na, and K or alkaline metals, e.g., Ca, Mg, and Ba metals) depending on the nature of the feedstock, biochar can have versatile properties leading to many applications, including bioenergy (co-gasification, co-firing, and combustion), chemical use (as a catalyst or catalyst support), agronomy (regarding water retention, plant nutrients, or soil conditioner), pharmacological use (regarding the adsorption of drugs and toxins), environment remediation (regarding carbon sequestration and the sorption of pollutants), and as biomaterials for the production of bio-composites, fuel cells, and photovoltaic plants [46].

3. Polymer-Based Composites Containing BCp

As was already explained above, the final properties of BC particles depend on several factors, such as the nature of feedstock used to produce BC particles, the type of process employed, and the relative operative condition. The final content of fixed carbon and ash (which involves the milling and sieving process used to control the dimension of the final particles), their surface area, and pore volume consequently determine the final properties of the composites, and this also depends on the polymer matrix employed in terms of interfacial adhesion, dispersion, thermal and mechanical stability, and ageing protection efficiency. In order to assess how the pyrolysis temperature and type of feedstock could determine a difference in the final properties of composites, Das et al. [47] added biochar particles as a co-filler for producing wood plastic polypropylene-based composites. The authors also pyrolyzed different feedstocks (landfill pine sawdust, sewage sludge, and poultry litter) at different temperatures, with the aim of identifying a routing process for waste employing. In fact, when keeping the landfill pine wood weight percentage constant at 30 wt.%, 24 wt.% of biochar particles were obtained through the following procedures: *i.* pine wood was pyrolyzed through a two-step parallel reactor chamber with a retention time of 60 min at 900 °C for a high-temperature activation regime (TCP 900); *ii.* using the same reactor, a torrefaction regime reached pyrolyzing at 350 °C (TCP 350); *iii.* the same feedstock was pyrolyzed by means of an auger reactor with a retention time of 10 min at 470 °C (PSD470); iv. the same pilot plant was used for the same feedstock with the same retention time at 420 °C (PSD420); v. sewage sludge biochar was produced with a pyrolysis temperature of 680 °C and a retention time of 10 min; vi. biochar from chicken litter was produced at 450 °C and 20 min of residence time. The fixed biochar concentration of 24 wt.% was determined thanks to a previous work of the same research group [48], in which BC produced from pine wood was added to the PP matrix at different loadings that ranged from 6 to 30 wt.%. In that study, a BC content of 24 wt.% showed a general improvement in tensile and flexural strengths, as well as in the young modulus of the final composite. Taking into account these results, it was found that an increase in tensile strength and the moduli was strongly related to the increase in surface area. Moreover, the presence of residual minerals (i.e., $CACO_3$ found in BC from chicken litter and the relative ash content) increased the impact strength of the composite and exhibited a lower heat release rate under the combustion regime compared to other composites. These results can be explained by considering that inorganic particles generally hinder the diffusion of oxygen through the matter, which creates a physical barrier between the combustible and the oxidizing agent, thus allowing the BC composites a possibility of being used in a flame-retardant field. Additionally, thanks to maleic anhydride grafted polypropylene/maleated anhydride polypropylene (MAPP) being used as a coupling agent, a general good dispersion of BC particles and an infiltration of polypropylene into biochar pores was observed for all composites. Without wood presence for the PP-based composites, a flame-retardant ability in pine wood biochar was established through a study by Das et al. [49], in which various BC loads, from 0 to 35 wt.% of the composites, exhibited increasingly stable compact char structures during controlled combustion tests that hid the O_2 diffusion in a polypropylene matrix. Moreover, the addition of char significantly reduced peak heat release and smoke production. The increase in flame retardant ability conferred by the presence of biochar particles in wood polypropylene composites was also explored in the presence of conventional inorganic flame retardants, such as magnesium hydroxide and ammonium polyphosphates [50,51]. On the other hand, in the presence of biochar, the two flame retardants particles were trapped into BC pores instead of in polypropylene with a final reduction in PP flow during processing and consequent reduction in interfacial adhesion and relative mechanical stability. Furthermore, the addition of biochar particles in wood polypropylene composites bestowed the composite with resistance toward water. This result remained valid without exceeding a threshold concentration, up to which the composites became more susceptible to water. In addition, it has been found that high pyrolysis temperatures generate more hydrophilic particles, due to absorption through the capillary action of pores [52]. Moreover, the reason for adding biochar particles to a Wood Polypropylene Composite (WPC) is that WPC usually suffers of thermal instability and thickness swelling, due to the high hydrophilic behavior of wood dust. Ayrilmis et al. [53] progressively reduced wood dust concentrations from 60 to 0 wt.% while respectively increasing commercial *Quercus* char flour concentrations from 0 to 60 wt.%, wherein thickness swelling was reduced by the 50% after 30 days when wood dust was completely substituted by BC dust, which increased the global dimensional stability. The same stabilization behavior was established for the water absorption after 30 days of WPC substation with BC dust, which decreased from ca. 21% for the PP/wood composite to ca. 15% for the PP/char composites. The same result of dimensional stabilization and global improvement of resistance to thermal degradation by adding BC particles to WPC was also confirmed through a study conducted by DeVallance et al. [54,55]. A variation in final composite properties related to pyrolysis temperature was also highlighted in a polypropylene/poly (octene-ethylene) copolymer (POE) (70/30 wt.%) blend [56], in which 10 and 20 wt.% of high-temperature pyrolyzed biocarbon (HTBioC) and low temperature pyrolyzed biocarbon (LTBioC) from *Miscanthus* were added. The HTBioC showed a lower presence of functional groups on the char surface, as well as a higher porosity with a relative increase in surface area that promoted better compatibility to the polymer blend, while also having a significantly better stiffness–toughness balance in the composite compared to the LTBioC.

Giorcelli et al. [57] also studied the relationship between the pyrolysis temperature and the electrical conductivity of biochar particles with the intention to use biochar particles as a filler in epoxy resin-based conductive composites. The residues of *Miscanthus* were pyrolyzed at 650, 700, and 750 °C and activated by CO_2 to increase surface area. The residues were characterized, and, then, 20 wt.% of the particles were added into epoxy resins for electrical characterization. As was already shown by other studies, it was found that an increase in pyrolysis temperature corresponded to and increase in carbon content with a corresponding reduction in other elements (i.e., O, Mg, Si, K, Ca) [46], and the ratio between the disordered and graphitized structure of the carbon structure increased with the increase in pyrolysis temperature. All of these properties led to an increase in the conductivity of biochar particles as a function of pyrolysis temperature, with a consequent increase in electrical performance for composites obtained with the addition of particles produced at higher pyrolysis temperatures. In the same lox viscosity epoxy resin LPL (Cores Ocean), two different biochars obtained by pyrolyzing Maple tree waste at low and high temperatures (600 and 1000 °C) were added at different weight percentages in order to improve the mechanical properties of the resin [58]. It was observed that the addition of a small amount of carbon fillers, lower than 2 wt.%, increased the load bearing capacity of the epoxy matrix, while also modifying the mechanical properties of the polymer matrix; on the other hand, a concentration equal to or higher to 2 wt.% transformed the pristine epoxy resin from brittle to a ductile composite, which was different from what was already seen for polypropylene-based composites. The optimum filler level depends on the type of polymer, the pyrolysis temperature, the type of feedstock used for biochar production, and the presence of other additives in composite production. When adding a curing agent (i.e., cycloaliphatic polyamine) and an embedding medium during epoxy-based composite processing, it is possible to increase the filler content above a critical level that normally lowers tensile strength. This configuration was found at a critical level of chars obtained from natural substances equal to 25 wt.%, and, in composites with plastic waste char, the critical level appeared to be reasonably low, equal to 15 wt.% [59]. Nevertheless, plastic waste char, or PWC (made from the pyrolysis of polyethylene terephthalate, PET), due to the terephthalic acid in the char structure, increased the global conductivity of a polymer composite [60].

Another way to activate biochar particles has been explored by Zhang et al. [61] by means of the impregnation of biomass feedstock before the carbonization process in an H_3PO_4 solution. In that work, biochar from rice husk, obtained pyrolyzing at 600 °C, was compared with activated biochar by varying the H_3PO_4 concentration in the activation solution. Generally, the activation of BC improved the thermal stability of the resulting composites, but a different concentration of activating agent affected the characteristic of the biochar in terms of chemical and morphological structure. In fact, a low concentration of H_3PO_4 improved the porous structure, which improved the resulting mechanical properties, thanks to better adhesion between particles and the HDPE polymer matrix, including flexural properties, rigidity elasticity, creep resistance, and anti-stress relaxation. On the other hand, a high concentration of H_3PO_4 in the activation solution generated fouling in the porous structure, which reduced all mechanical properties.

To further improve electrical properties, a carbonization process of charcoal from three different biomasses [62] has been performed with high fill ultra-high molecular weight polyethylene/linear low density poly ethylene UHMWPE/LLDPE [63]. For example, starting with charcoal coming from bamboo pyrolysis, which was further carbonized at 1100 °C in a muffle furnace in the absence of air, particles with irregular shapes have been obtained with a global transformation of their amorphous structure into a graphite-like structure with a higher crystallinity grade. This result was simultaneously confirmed by an increase in the three diffraction peaks at around 24.6°, 43.7°, and 50.1°, which were associated with the C (002), C (100), and C (004) diffractions, respectively, of the graphitic structure through XRD analysis and with an increase in the intensity ratio of the D-to-G peak obtained through Raman spectroscopy, which revealed a defective graphitic structure and turbostratic crystallites in the BC1100 particles. Moreover, the high temperature carbonization generated a high specific surface, due to the creation of a nanoporous structure. These particles were added into a UHMWPE blended with a LLDPE as a flow accelerator for reducing the melt viscosity of the UHMDPE and improving the processability of the composite and final particle dispersions. The highly filled composites (with a carbon load up to 80 wt.%) showed excellent electromagnetic interference shielding performance, and one of the highest values reported for conductive polymer composites was found; in fact, at maximum biochar concentration, a conductivity of 107.6 S/m was found. The same further-carbonized bamboo charcoal particles were used with high fill UHMWPE to produce scaffolds for cells proliferation. In that study [64], the raw bamboo charcoal and bamboo charcoal carbonized at 800 °C and 1100 °C were over pyrolyzed in a muffle furnace in the absence of air. Additionally, in that case, thanks to high temperature and the relative development of the surface into a nanoporous structure, crystallinity, hardness, and thermal stability were found to be higher for the biochar carbonized at the higher temperature. On the other hand, for better biocompatibility, achieving a high temperature of pyrolysis is not useful, because, at low temperature, biochar enables the composite to exhibit better hydrophilicity and higher specific surface energy, which promotes protein adhesion and cell proliferation. Moreover, globally, the composite obtained by adding to the UHMWPE showed good mechanical properties and friction performance that make it appropriate for use in orthopedic applications.

Arrigo et al. [65] performed an extra carbonization process of a torrefied coffee powder was into a tubular furnace, and the waste was pyrolyzed at 700 °C for 1 h under a nitrogen atmosphere, which was then added to high density polyethylene, HDPE, to understand the interaction between HDPE and BC from spent coffee grounds, as well as how the filler content influenced the rheological and thermal behavior of the resulting composites. The authors subjected BC/HDPE composites with different BC loads (up to 7.5 wt.%) to SEM analysis and rheological characterization that employed different flow fields, including linear and non-linear dynamic shear flow, which resulted in clear confinement of the polymer chains onto the surface of particles and into the porous structure of particles, as well as a pseudo solid-like behavior of the BC/HDPE composites due to the formation of a network.

To reduce waste for environmental purposes, Kane et al. [66] recently compared recycled high-density polyethylene, rHDPE, with and without biochar to look at both improving mechanical properties and environmental impact. From a mechanical point of view, the tensile behavior of the rHDPE was significantly altered by the addition of biochar particles coming from wood forestry residues, which increased the strength and stiffness through the global increase of crystallinity of the rHDPE through the reinforcing action of the polymer matrix, and, thanks to a good interface adhesion, that led to a polymer interlocking with the porous structure of the biochar. Moreover, by means of life cycle assessment, it has been noticed that the addition of biochar as a filler reduces the global amount of plastic spent to produce a product, which of course provides a benefit in terms of global warming potential when referring to the CO_2 emitted for plastic production. It has been calculated that rHDPE reached a 0 kg CO2 equivalent by adding less than 40 wt.% of biochar particles, which obtained a composite with a similar strength and stiffness obtained by adding 40–50 wt.% of biochar particles to virgin HDPE [67].

Another way to reduce the amount of polyolefin waste and reduce the amount of virgin polyolefin employed in the industrial field has been addressed in a study conducted by Idress et al. [68], in which a recycled poly-ethylene-terephthalate rPET-based composite was produced by adding biochar. The biochar employed in that work came from the high temperature pyrolysis (1100 °C) of PET waste under an autogenic pressure of ca. 150 bar. In that study, the researchers were able to extrudate recycled PET and PET/BC composites, which highlighted that the incorporation of biochar enhanced the mechanical properties and provided the PET with thermal properties, which suggests that BC could supply the necessity for commercial graphene materials in polymer composites. As was noticed with other polyolefins, the responsibility for the improvement in mechanical properties must be referred to the high surface porosity of BC particles and the high affinity between BC particles and the polymer matrix. The improvement in thermal properties must be referred to the known barrier effect of BC.

4. Bio-Polymer-Based Composites Containing BCp

Bioplastics should be intended as polymers that meet any of two criteria: the polymer is bio-based and/or biodegradable [69]. In the context of a sustainable and circular economy, the recovery of bio-waste and the addition of them in biopolymers, intended as bio-based and biodegradable, for sustainable bio-composites formulation is a challenging issue. Among biopolymer-based composites containing biochar particles, the literature reports a significant number of studies. One of the most biochar-added polyester matrixes is polylactic acid (PLA), which suffers from poor thermal stability and high brittleness that reduces its employment in many fields, i.e., textile, biomedicine, and food packaging [70]. Briefly, the use of BC in PLA can lead to growth in the PLA market, thanks to a global improvement in the mechanical stability of this polymer [33,71,72]. Kane et al. [73] investigated BC-added PLA composites and compared them to high density polyethylene HDPE composites. In contrast to HDPE composites, for PLA/BC composites, the work highlighted an impact of BC in thermal degradation behavior, which was shown through a decrease in onset degradation temperature and a global reduction in melt viscosity of the PLA, which was probably due to the presence of an inorganic element of the BC surface being responsible for catalyzing PLA thermal decomposition. The same behavior has been found by Arrigo et al. [74], in which BC particles derived from spent ground coffee were added in the PLA matrix by processing the composites through melt mixing and solvent casting methods. It was found that the PLA rheological behavior underwent significant alteration when the composites were obtained by melt processing. In fact, the authors reported (see Figure 2) a progressive increase in melt viscosity in composites obtained by solvent casting and a progressive decrease in melt viscosity in composites obtained by melt mixing, as the BC content increased, which, in the last case, suggested a severe reduction in polymer molar mass, due to thermal degradation [75], and PLA preservation when the processing was carried out at room temperature by means of solvent casting. In any case, a strong polymer-filler or filler-filler interaction has been found, which was demonstrated by the appearance of a yield stress behavior.

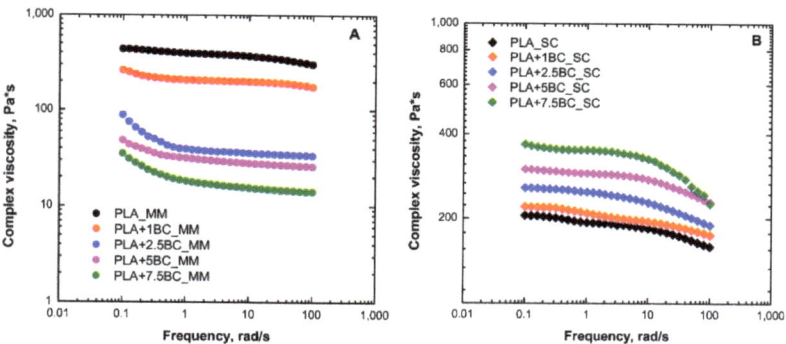

Figure 2. Complex viscosity as a function of frequency for neat poly(lactic acid) (PLA) and biochar (BC)-containing composites obtained through melt mixing (MM) (**A**) and solvent casting (SC) (**B**) [74].

A significant reduction in the molecular mass of poly(3-hydroxybutyrate) (PHB) processed at high temperatures with the addition of biochar particles has been demonstrated by Haeldermans et al. [76]. In their study, different PHB/char with varying BC loads (from 20 wt.% to 50 wt.%) were produced, and their formulation was compared with PHB/thermoplastic starch (TPS)/BC composites. Despite having the best biodegradability compared to other biopolymers, it is well known in the literature that PHB suffers from a significant reduction in molecular weight after processing, and this limits its application, due to a really small operational processing window [77], e.g., an unprocessed PHB-M_w of 611 Kg/mol and a melt-processed PHB-M_w of ca. 463 Kg/mol [76]. Regrettably, increasing the amount of biochar particles from 20 to 50 wt.% further reduced the PHB-M_w, which achieved a reduction to 218 Kg/mol for 50 wt.% of BC and reduced the global thermal stability of the BC bio-composites. Consistent with molecular weight reduction, a reduction in thermal properties was found, such as a decrease in melting point with a decrease in molecular weight [78]. Thanks to the presence of thermoplastic starch in bio-composites, the decrease in M_w is more gradual and controlled, and molecular weight analysis has shown that, at low BC loads, TPS can act as an intermediator between PHB and PHB by controlling the reduction in molecular weight.

In contrast, the addition of BC particles as a filler in an Ecovio commercial polymer blend containing poly(1,4-butylene adiphat-co-1,4-butylene terephthalate), PBAT, 47 mol% of an aromatic segment, and PLA, 25 mol%, significant increased the application field of BC particles in the biopolymers matrix [79]. In fact, a significant reduction in the resistivity of obtained bio-composites was found as the BC load increased by up to 30 wt.%, which suggests employment of the composites in equipment elements in laboratories for precise measurement, or as an antistatic agent in the packaging industry. Moreover, thermal stability has not been affected by the presence of BC compared to the Ecovio polymer matrix. Moreover, thanks to a global improvement of modulus shown in DMA analysis for all temperature ranges (−50 to 120 °C), mechanical properties were higher for composites with respect to the neat matrix, which suggested better mechanical stability.

Regarding the PBAT matrix containing biochar particles, several works have been published. Botta et al. [80] investigated the properties and the filmability of PBAT-based materials that were added to commercial biochar powder used in the food industry that was formed from birch and beech wood pyrolysis. The team performed a preliminary investigation of the prepared PBAT/BC composites by melt mixing with BC loads from 5 wt.% to 20 wt.% of commercial BC, which showed a uniform filler dispersion and a good adhesion within the selected biopolymer matrix, which led to an increase in global mechanical properties. Moreover, DSC analysis clarified how the BC did not influence the PBAT chain structure, which remained almost amorphous despite filler addition, even with the increase in T_m as the filler content increased and revealed compatibility between the filler and matrix. Instead, the rheological behavior of PBAT-based composite results were affected by the presence and the increase in carbonaceous filler, which resulted in a relative increase in melt viscosity, in all ranges of frequency, and suggested an influence of embedded filler on the long-range and short-range dynamics of polymer chains, especially when the BC load was equal to 20 wt.%. At that carbon load, the PBAT underwent a dramatic reduction in its intrinsic ductility and a significant decrease in the break–stretching ratio (BSR), which resulted in a composite with no filmability properties. The same rheological and mechanical behavior has been found by Infurna et al. [37], in which agricultural carob waste was pyrolyzed at three different temperatures (BC280, BC340, and BC400 respectively pyrolyzed at 280, 340, and 400 °C) and then added to PBAT at two different concentrations, i.e., 10 and 20 wt.%. In their work, an ageing protection assessment was performed on both the pristine particles and on the BC composites. First, the authors characterized the radical scavenging efficiency by means of 1,1-diphenyl-2-pycryl (DPPH) free radical analysis, in which the three different BC particles were added at constant loads to a methanol solution of DPPH, a stable free radical, and they monitored the disappearance of the free radical UV absorption peak at 517 nm. From their analysis, thanks to residual functional groups on the BC surface after 24 h, the particles obtained at a lower pyrolysis temperature achieved about 100% radical scavenging efficiency, despite the values obtained at 400 °C, with higher scavenging kinetics of the BC280. The monitoring of the DPPH UV absorption peak was also performed by increasing the amount of BC in the solution. In that case, it was demonstrated that, from a limited concentration onwards after 24 h, the radical scavenging efficiency results were comparable between the three different particles. This result is consistent with what has been found in the photooxidation assay of biopolymer composites, as partly shown in Figure 3, in which the variation in mechanical properties as a function of irradiation time was monitored while also extrapolating the half time as the time at which the elongation at break was half of the initial one.

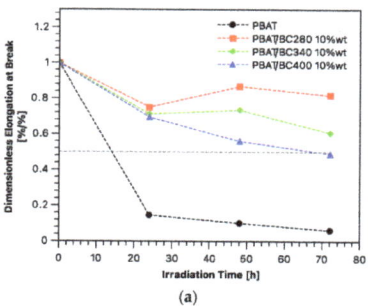 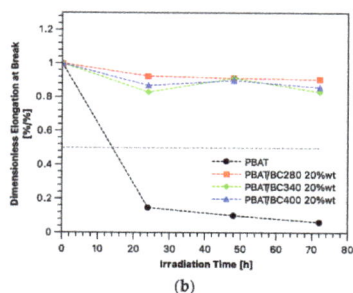

Figure 3. The trend of dimension elongation at break with (**a**) 10 wt.% and (**b**) 20 wt.% of filler content in the PBAT matrix [37].

A significant reduction in the ductility of pristine PBAT is shown, which achieved a half time of 14 h. In Figure 3a, the same trend of the DPPH assay is shown, in which the lower the pyrolysis temperature was of the obtained BC particles, the higher the concentration of functional groups on the surface were able to scavenge free radicals from the accelerating weathering test, which resulted in a higher shown resistance of the bio-composites. In conclusion, it was enough to increase the BC load from 10 wt.% to 20 wt.% to lead to a comparable ageing resistance for all bio-composites, and the same results have been found by ATR-FTIR analysis as a function of irradiation time.

Polyvinyl alcohol (PVA)/corn starch/BC bio-composites were successfully formulated by means of the solvent casing method in the presence of citric acid and glutaraldehyde, which was added for fixative effect before the casting period [81]. The interaction between the biopolymer blend and BC particles significantly affected the degradation path of the PVA/starch composites. In fact, a significant decrease in the narrowing of the peak relative to the hydroxy band with the increments of BC load was noticeable. The authors attributed this phenomenon to the good compatibility of the PVA, starch, and BC [82]. As had happened for PLA and PHB-based composites, the global thermal stability of the bio-composites was lower when BC was added to the blend.

A noticeable improvement in mechanical properties has been obtained by introducing a suitable concentration of biochar particles in eco-friendly bio-composites manufactured by a green epoxy matrix reinforced with short agave fibers for replacing synthetic materials in structural applications [83]. In that case, with the optimum amount of BC particles (in this case found to be equal to 2 wt.%) added by means of the synergic effect of short fiber and biochar particles, a better adhesion has been achieved between the epoxy matrix and fiber, which was demonstrated by the fiber pull-out test. This aspect involved an increase in the global Young's modulus and tensile strength, as well as an increase in fatigue performance with an increase in fatigue strength by about 67% and fatigue lifetime by at least three orders of magnitude. This remarkable enhancement of mechanical performance increases the possibility of employing a green epoxy matrix fiber reinforced for structural and semi-structural applications, especially in automotive and naval applications.

Green composites have been formulated by adding biochar particles in partially bio-based polymers or bio-based polymers that are non-biodegradable. Nagarajan et al. [6] performed a study of the varying particle size distributions of biochar particles produced from the low temperature slow pyrolysis process of *Miscanthus* fibers. After pyrolysis, size-fractionation of the BC was performed with sieves having different openings, e.g., 300, 212, 150, 125, 75, and 20 µm. The different size-fractioned BCs were added to a poly(trimethylene terephthalate) (PTT) 70/poly(lactic acid) (PLA) 30- ethylene-methyl acrylate-glycidyl methacrylate terpolymer (EMAGMA) polymer blend (85–15), using, in some formulation, an epoxy-functionalized chain extender (CE). A good range of particle size distribution, combined with the presence of the chain extender, helped to obtain a morphology with a better dispersion of the blend components, i.e., particle size distributions of 75–20 µm. Biochar particles under 20 µm in size diameter stabilized the blend morphologies, which showed a coalescence of PLA-EMAGMA particles that turned in smaller and finer morphologies in the presence of CE. This evaluation obtained from SEM image observations was consistent with rheological tests and mechanical analysis, which showed that, with an appropriate BC particle size and shape, morphologies and properties can be tailored to achieve desired properties with solid cost reduction. The partially bio-based PTT was combined with 20 wt.% high temperature bio-carbon from peanut hull pyrolysis, which resulted in superior mechanical performance that could be optimized for non-structural automotive components or electrical housing applications [84]. Also in this case, particle size distribution and particle size of the original biomass can play a crucial role in the resulting biochar, impact the concentration of volatiles and bio-oil from the pyrolysis process [85,86] and, as expected in bio-composite properties, call for a milling process before pyrolysis has been conducted [84]. The addition of peanut hull biochar, with its sheet-like surface morphology with high graphitic carbon content and its relatively low electrical conductivity, can contribute to the improved thermal stability of PTT-based green composites by increasing both flexural and tensile moduli, which suggest nonstructural and anti-static applications.

5. Asphalts Composites Containing BCP

Using biochar particles as an asphalt binder modifying filler is going to become a new and interesting application field. In fact, nowadays in the construction sector, it is already used as a substitute for cement in mortar or concrete, thanks to its help with accelerating cement hydrating [87] and due to global CO_2 mitigation [88]. The new approach regards the addition of asphalt binders as an ageing protector, which are an essential component of asphalt concrete in addition to being the heaviest coproduct of the petroleum refining system, after distillation, to obtain fuels and lubricants [89]. The oxidation of asphalt binders is an inevitable phenomenon that plays an enormous role in the deterioration of asphalt binders. In fact, the life expectancy of usual binders is susceptible to ultraviolet (UV) rays, which cause faster oxidation of asphalts, with a sensitive reduction in rheological characteristics and a loss of rutting properties, which can lead to pavement distress [90]. Walters et al. [91] investigated the impact of added biochar particles (coming from a thermochemical process used to convert swine manure in bio-oil) or nano-clay (Cloesite 30B) on the rheological properties and ageing susceptibility of asphalt binder, and compared the results with a control asphalt (PG 64-22) binder. The introduction of BC to the asphalt binder led to a reduction in asphalt temperature susceptibility, and, regarding the shear susceptibility, its sensitivity decreased by adding 10 wt.% of BC to PG 64-22, which achieved a lower value than control asphalt. Contrary to the addition of BC, the addition of nano-clay generated an impact on the layer spacing, which appeared to be responsible for enchaining the high temperature performance and ageing resistance of asphalt binders. In a second study by Walters et al. [90] a composite with both biochar (3 wt.%) particles and nano-clay (3 wt.%) was produced that resulted in a lower viscosity than the ones with only nano-clay, while the ageing susceptibility was improved significantly. This happens because biochar seems to have a role in the flow modifiers alleviating the stiffening effect of nano-clay, which help the nano-clay to disperse better in an asphalt binder.

Zhao et al. [92] evaluated the properties and performance of asphalt binders and mixtures by adding 5 wt.% and 10 wt.% of biochar from the fast pyrolysis of switchgrass for biofuel production. In their work, the authors found that biochar significantly increased the rutting resistances at high service temperatures of both asphalt binders and asphalt mixtures. Moreover, the addition of 5 wt.% of biochar may be the optimum in modifying binders in terms of cracking resistance, while 10 wt.% shows little effect in comparison with 5 wt.%. Another study was conducted by the same research group by adding different BC particles to a commonly used asphalt binder (PG 64–22), which resulted in a different type of pyrolysis of switchgrass, while taking into account that BC results as a by-product for biofuel production, and comparing these lab-made BC particles with a commercially activated carbon [93]. In particular, the BC particles were produced by the following techniques: *i.* a microwave reactor in which switchgrass was mixed with silicon carbide to absorb enough microwaves, then the mixture was heated up until 500 °C in less than 1 min and the temperature was maintained for 15 min, and, then, after cooling down, the silicon carbide particles were sieved out to finally obtain BC particles with size diameters between 75 and 150 µm; *ii.* a tube furnace method in which feedstock was heated up until 400 or 500 °C with a heating rate of 15 °C/min to result in BC particles with diameters smaller than 75 µm for particles obtained at 400 °C and 500 °C, and for the ones obtained at 500 °C a diameter range between 75 and 150 µm was obtained as well; *iii.* an activated commercially available carbon was selected for comparison. The composites were characterized in terms of viscosity modification, ageing and fatigue resistance, and rutting properties of un-aged and aged composites. The addition of all bio-modifiers increased the viscosity of the asphalt binder at a high service temperature and exploited a positive effect in ageing resistance at a long time of UV exposition. Globally, except for particles with smaller diameters (<75 µm) produced at 400 °C that showed a positive effect on the specific properties or performance of the asphalt binders, the pyrolysis method appeared to have a negligible effect on the degree of modification. Another study published by Zhang et al. [94] focused on varying biochar loads and biochar diameter distributions by comparing the biochar contribution on varying properties with the ones obtained by graphite adding. The biochar used in the work was obtained from waste wood resources, at a temperature ranging from 500 and 650 °C, through a pyrolysis plant able to heat at 10 °C/s. Then, BC particles were sieved to separate the different size ranges, and the size range of previous work was studied, i.e., between 75 and 150 µm and lower than 75 µm; the biochar contents in PG 58–28 control asphalt were 2 wt.%, 4 wt.%, and 8 wt.%, respectively. The flake graphite with a diameter lower than 75 µm and content of 4 wt.% was also added to PG 58–28 for comparison. As was expected, BC addition resulted in higher porosity and micro-structure compared with dense and smooth graphite. This aspect of course led to a larger and better adhesion interaction in the asphalt binders of BC than that of graphite, and, as a result, BC modified binders had better high-temperature rutting resistance and better anti-ageing properties, especially for the BC-modified binder with a lower BC diameter at a higher content.

In conclusion, it seems that the addition of BC to asphalt seems to increase the thermal resistance during asphalt preparation and their oxygen resistance in service. This is another important result of asphalt reducing viscosity during processing, which helps the dispersion of other asphalts constituents in order to obtain a high-performance pavement.

6. Conclusions

Char/biochar particles could be considered as a new kind of sustainable particle created from their "waste" feedstocks. Specifically, when also considering the circular principles, the conversion of polymer waste, food waste, and biomasses, through thermal treatment at high temperatures, gives an appropriate second life for these waste materials. Produced biochar particles differ in their physical and chemical properties, e.g., CHNOS compositions, porosity, and adsorption ability, because of the implementation of different thermal treatments, such as slow, fast, and flash pyrolysis; torrefaction, gasification; and hydrothermal liquefaction. Currently proposed applications of the new char/biochar particles are mainly as follows: *i.* as fertilizers for soil retirement, *ii-* as high adsorption particles for water remediation, and *iii.* as suitable fillers for the formulation of polymer/biopolymer-based composites and to produce asphalts. Therefore, this review critically reports on the current status of char/biochar particles in considering the main advantages and drawbacks in the use of these new particles as suitable fillers for polymers, biopolymers, and asphalts.

The char/biochar particles, being particles mainly composed of carbon atoms and having a large surface, are very useful to formulate composites with improved mechanical resistance, i.e., elastic modulus and tensile strength, as well as improved oxidative and photooxidative resistance, while also considering the particles' radicals scavenging abilities in comparison to the properties of neat matrices. Unfortunately, the main drawback is related to the particle color. Particularly, being black particles, their composites appear mainly with brawn-black coloration, and, obviously, this is a limitation for large scale esthetic applications.

Author Contributions: Conceptualization, G.I. and N.T.D.; methodology, G.I.; data curation, G.I. and G.C.; writing—original draft preparation, G.I. and G.C.; writing—review and editing, G.I. and N.T.D.; supervision, N.T.D.; funding acquisition, N.T.D. and G.I. All authors have read and agreed to the published version of the manuscript.

Funding: This research received no external funding.

Institutional Review Board Statement: Not applicable.

Data Availability Statement: Not applicable.

Acknowledgments: G.I. would like to thank MIUR—Italy (Ministry of Education, University and Research of Italy) for having joined with support by CLEAN—PRIN-20174FSRZS_002.

Conflicts of Interest: The authors declare no conflict of interest.

References

1. Ronzon, T.; M'Barek, R. Socioeconomic Indicators to Monitor the EU's Bioeconomy in Transition. *Sustain. Switz.* **2018**, *10*, 1745. [CrossRef]
2. European Commission. *Communication from the Commission to the European Parliament, the Council, the European Economic and Social Committee and the Committee of the Regions Closing the Loop—An EU Action Plan for the Circular Economy*; European Commission: Brussels, Belgium, 2018.
3. Tomczyk, A.; Sokołowska, Z.; Boguta, P. Biochar Physicochemical Properties: Pyrolysis Temperature and Feedstock Kind Effects. *Rev. Environ. Sci. Biotechnol.* **2020**, *19*, 191–215. [CrossRef]
4. Sizirici, B.; Fseha, Y.H.; Yildiz, I.; Delclos, T.; Khaleel, A. The Effect of Pyrolysis Temperature and Feedstock on Date Palm Waste Derived Biochar to Remove Single and Multi-Metals in Aqueous Solutions. *Sustain. Environ. Res.* **2021**, *31*, 9. [CrossRef]
5. Mohan, D.; Pittman, C.U.; Steele, P.H. Pyrolysis of Wood/Biomass for Bio-Oil: A Critical Review. *Energy Fuels* **2006**, *20*, 848–889. [CrossRef]
6. Nagarajan, V.; Mohanty, A.K.; Misra, M. Biocomposites with Size-Fractionated Biocarbon: Influence of the Microstructure on Macroscopic Properties. *ACS Omega* **2016**, *1*, 636–647. [CrossRef]
7. Al Arni, S. Comparison of Slow and Fast Pyrolysis for Converting Biomass into Fuel. *Renew. Energy* **2018**, *124*, 197–201. [CrossRef]
8. Maniscalco, M.; Infurra, G.; Caputo, G.; Botta, L.; Dintcheva, N.T. Slow Pyrolysis as a Method for Biochar Production from Carob Waste: Process Investigation and Products' Characterization. *Energies* **2021**, *14*, 8457. [CrossRef]
9. Volpe, M.; Panno, D.; Volpe, R.; Messineo, A. Upgrade of Citrus Waste as a Biofuel via Slow Pyrolysis. *J. Anal. Appl. Pyrolysis* **2015**, *115*, 66–76. [CrossRef]
10. Vardon, D.R.; Moser, B.R.; Zheng, W.; Witkin, K.; Evangelista, R.L.; Strathmann, T.J.; Rajagopalan, K.; Sharma, B.K. Complete Utilization of Spent Coffee Grounds to Produce Biodiesel, Bio-Oil, and Biochar. *ACS Sustain. Chem. Eng.* **2013**, *1*, 1286–1294. [CrossRef]
11. Palos, R.; Rodríguez, E.; Gutiérrez, A.; Bilbao, J.; Arandes, J.M. Cracking of Plastic Pyrolysis Oil over FCC Equilibrium Catalysts to Produce Fuels: Kinetic Modeling. *Fuel* **2022**, *316*, 123341. [CrossRef]
12. Sarkar, J.K.; Wang, Q. Different Pyrolysis Process Conditions of South Asian Waste Coconut Shell and Characterization of Gas, Bio-Char, and Bio-Oil. *Energies* **2020**, *13*, 1970. [CrossRef]
13. Antal, M.J.; Allen, S.G.; Dai, X.; Shimizu, B.; Tam, M.S.; Grønli, M. Attainment of the Theoretical Yield of Carbon from Biomass. *Ind. Eng. Chem. Res.* **2000**, *39*, 4024–4031. [CrossRef]
14. Guizani, C.; Javier, F.; Sanz, E.; Salvador, S.; Guizani, C.; Escudero Sanz, F.J.; Salvador, S. Influence of Temperature and Particle Size on the Single and Mixed Atmosphere Gasification of Biomass Char with H2O and CO2 Influence of Temperature and Particle Size on the Single and Mixed Atmosphere Gasification of Biomass Char with H_2O and CO_2. *Fuel Process. Technol.* **2015**, *134*, 175–188. [CrossRef]

15. Barisano, D.; Canneto, G.; Nanna, F.; Villone, A.; Fanelli, E.; Freda, C.; Grieco, M.; Lotierz, A.; Cornacchia, G.; Braccio, G.; et al. Investigation of an Intensified Thermo-Chemical Experimental Set-Up for Hydrogen Production from Biomass: Gasification Process Integrated to a Portable Purification System—Part II. *Energies* **2022**, *15*, 4580. [CrossRef]
16. Breault, R.W. Gasification Processes Old and New: A Basic Review of the Major Technologies. *Energies* **2010**, *3*, 216–240. [CrossRef]
17. Grande, L.; Pedroarena, I.; Korili, S.A.; Gil, A. Hydrothermal Liquefaction of Biomass as One of the Most Promising Alternatives for the Synthesis of Advanced Liquid Biofuels: A Review. *Materials* **2021**, *14*, 5286. [CrossRef]
18. Sepehri, A.; Sarrafzadeh, M.-H. Effect of Nitrifiers Community on Fouling Mitigation and Nitrification Efficiency in a Membrane Bioreactor. *Chem. Eng. Process. -Process Intensif.* **2018**, *128*, 10–18. [CrossRef]
19. Rizwan, M.; Ali, S.; Qayyum, M.F.; Ibrahim, M.; Zia-ur-Rehman, M.; Abbas, T.; Ok, Y.S. Mechanisms of Biochar-Mediated Alleviation of Toxicity of Trace Elements in Plants: A Critical Review. *Environ. Sci. Pollut. Res.* **2016**, *23*, 2230–2248. [CrossRef]
20. McKendry, P. Energy Production from Biomass (Part 1): Overview of Biomass. *Bioresour. Technol.* **2002**, *83*, 37–46. [CrossRef]
21. Ha, M.-A.; Apperley, D.C.; Evans, B.W.; Huxham, I.M.; Jardine, W.G.; Vietor, R.J.; Reis, D.; Vian, B.; Jarvis, M.C. Fine Structure in Cellulose Microfibrils: NMR Evidence from Onion and Quince. *Plant J.* **1998**, *16*, 183–190. [CrossRef]
22. Vassilev, S.V.; Baxter, D.; Andersen, L.K.; Vassileva, C.G.; Morgan, T.J. An Overview of the Organic and Inorganic Phase Composition of Biomass. *Fuel* **2012**, *94*, 1–33. [CrossRef]
23. Sami, M.; Annamalai, K.; Wooldridge, M. Co-FIring of Coal and Biomass Fuel Blends. *Prog. Energy Combust. Sci.* **2001**, *27*, 171–214. [CrossRef]
24. Stamatelatou, K.; Antonopoulou, G.; Ntaikou, I.; Lyberatos, G. The Effect of Physical, Chemical, and Biological Pretreatments of Biomass on Its Anaerobic Digestibility and Biogas Production. In *Biogas Production*; Mudhoo, A., Ed.; John Wiley & Sons, Inc.: Hoboken, NJ, USA, 2012; pp. 55–90. ISBN 978-1-118-40408-9.
25. Ahmad, M.; Lee, S.S.; Dou, X.; Mohan, D.; Sung, J.-K.; Yang, J.E.; Ok, Y.S. Effects of Pyrolysis Temperature on Soybean Stover- and Peanut Shell-Derived Biochar Properties and TCE Adsorption in Water. *Bioresour. Technol.* **2012**, *118*, 536–544. [CrossRef]
26. Chen, B.; Zhou, D.; Zhu, L. Transitional Adsorption and Partition of Nonpolar and Polar Aromatic Contaminants by Biochars of Pine Needles with Different Pyrolytic Temperatures. *Environ. Sci. Technol.* **2008**, *42*, 5137–5143. [CrossRef] [PubMed]
27. Xu, G.; Yang, X.; Spinosa, L. Development of Sludge-Based Adsorbents: Preparation, Characterization, Utilization and Its Feasibility Assessment. *J. Environ. Manag.* **2015**, *151*, 221–232. [CrossRef]
28. Ahmed, M.B.; Zhou, J.L.; Ngo, H.H.; Guo, W.; Chen, M. Progress in the Preparation and Application of Modified Biochar for Improved Contaminant Removal from Water and Wastewater. *Bioresour. Technol.* **2016**, *214*, 836–851. [CrossRef]
29. Yang, G.; Zhang, G.; Wang, H. Current State of Sludge Production, Management, Treatment and Disposal in China. *Water Res.* **2015**, *78*, 60–73. [CrossRef]
30. Vaughn, S.F.; Kenar, J.A.; Thompson, A.R.; Peterson, S.C. Comparison of Biochars Derived from Wood Pellets and Pelletized Wheat Straw as Replacements for Peat in Potting Substrates. *Ind. Crops Prod.* **2013**, *51*, 437–443. [CrossRef]
31. Kołodyńska, D.; Wnętrzak, R.; Leahy, J.J.; Hayes, M.H.B.; Kwapiński, W.; Hubicki, Z. Kinetic and Adsorptive Characterization of Biochar in Metal Ions Removal. *Chem. Eng. J.* **2012**, *197*, 295–305. [CrossRef]
32. Suliman, W.; Harsh, J.B.; Abu-Lail, N.I.; Fortuna, A.-M.; Dallmeyer, I.; Garcia-Perez, M. Influence of Feedstock Source and Pyrolysis Temperature on Biochar Bulk and Surface Properties. *Biomass Bioenergy* **2016**, *84*, 37–48. [CrossRef]
33. Wang, S.; Gao, B.; Zimmerman, A.R.; Li, Y.; Ma, L.; Harris, W.G.; Migliaccio, K.W. Physicochemical and Sorptive Properties of Biochars Derived from Woody and Herbaceous Biomass. *Chemosphere* **2015**, *134*, 257–262. [CrossRef] [PubMed]
34. Elleuch, A.; Boussetta, A.; Yu, J.; Halouani, K.; Li, Y. Experimental Investigation of Direct Carbon Fuel Cell Fueled by Almond Shell Biochar: Part I. Physico-Chemical Characterization of the Biochar Fuel and Cell Performance Examination. *Int. J. Hydrog. Energy* **2013**, *38*, 16590–16604. [CrossRef]
35. Lee, Y.; Park, J.; Ryu, C.; Gang, K.S.; Yang, W.; Park, Y.-K.; Jung, J.; Hyun, S. Comparison of Biochar Properties from Biomass Residues Produced by Slow Pyrolysis at 500 °C. *Bioresour. Technol.* **2013**, *148*, 196–201. [CrossRef] [PubMed]
36. Chen, D.; Liu, D.; Zhang, H.; Chen, Y.; Li, Q. Bamboo Pyrolysis Using TG–FTIR and a Lab-Scale Reactor: Analysis of Pyrolysis Behavior, Product Properties, and Carbon and Energy Yields. *Fuel* **2015**, *148*, 79–86. [CrossRef]
37. Infurna, G.; Botta, L.; Maniscalco, M.; Morici, E.; Caputo, G.; Marullo, S.; D'Anna, F.; Dintcheva, N.Tz. Biochar Particles Obtained from Agricultural Carob Waste as a Suitable Filler for Sustainable Biocomposite Formulations. *Polymers* **2022**, *14*, 3075. [CrossRef]
38. Liu, Z.; Han, G. Production of Solid Fuel Biochar from Waste Biomass by Low Temperature Pyrolysis. *Fuel* **2015**, *158*, 159–165. [CrossRef]
39. Liu, X.; Zhang, Y.; Li, Z.; Feng, R.; Zhang, Y. Characterization of Corncob-Derived Biochar and Pyrolysis Kinetics in Comparison with Corn Stalk and Sawdust. *Bioresour. Technol.* **2014**, *170*, 76–82. [CrossRef]
40. Jafri, N.; Wong, W.Y.; Doshi, V.; Yoon, L.W.; Cheah, K.H. A Review on Production and Characterization of Biochars for Application in Direct Carbon Fuel Cells. *Process Saf. Environ. Prot.* **2018**, *118*, 152–166. [CrossRef]
41. Moralı, U.; Şensöz, S. Pyrolysis of Hornbeam Shell (*Carpinus betulus* L.) in a Fixed Bed Reactor: Characterization of Bio-Oil and Bio-Char. *Fuel* **2015**, *150*, 672–678. [CrossRef]
42. Luo, L.; Xu, C.; Chen, Z.; Zhang, S. Properties of Biomass-Derived Biochars: Combined Effects of Operating Conditions and Biomass Types. *Bioresour. Technol.* **2015**, *192*, 83–89. [CrossRef] [PubMed]
43. Zhang, J.; Zhong, Z.; Zhao, J.; Yang, M.; Li, W.; Zhang, H. Study on the Preparation of Activated Carbon for Direct Carbon Fuel Cell with Oak Sawdust. *Can. J. Chem. Eng.* **2012**, *90*, 762–768. [CrossRef]

44. Hmid, A.; Mondelli, D.; Fiore, S.; Fanizzi, F.P.; Al Chami, Z.; Dumontet, S. Production and Characterization of Biochar from Three-Phase Olive Mill Waste through Slow Pyrolysis. *Biomass Bioenergy* **2014**, *71*, 330–339. [CrossRef]
45. Uzun, B.B.; Apaydin-Varol, E.; Ateş, F.; Özbay, N.; Pütün, A.E. Synthetic Fuel Production from Tea Waste: Characterisation of Bio-Oil and Bio-Char *Fuel* **2010**, *89*, 176–184. [CrossRef]
46. Nanda, S.; Dalai, A.K.; Berruti, F.; Kozinski, J.A. Biochar as an Exceptional Bioresource for Energy, Agronomy, Carbon Sequestration, Activated Carbon and Specialty Materials. *Waste Biomass Valorization* **2016**, *7*, 201–235. [CrossRef]
47. Das, O.; Sarmah, A.K.; Bhattacharyya, D. Biocomposites from Waste Derived Biochars: Mechanical, Thermal, Chemical, and Morphological Properties. *Waste Manag.* **2016**, *49*, 560–570. [CrossRef]
48. Das, O.; Sarmah, A.K.; Bhattacharyya, D. A Novel Approach in Organic Waste Utilization through Biochar Addition in Wood/Polypropylene Composites. *Waste Manag.* **2015**, *38*, 132–140. [CrossRef]
49. Das, O.; Bhattacharyya, D.; Hui, D.; Lau, K.T. Mechanical and Flammability Characterisations of Biochar/Polypropylene Biocomposites. *Compos. Part B Eng.* **2016**, *106*, 120–128. [CrossRef]
50. Das, O.; Kim, N.K.; Sarmah, A.K.; Bhattacharyya, D. Development of Waste Based Biochar/Wool Hybrid Biocomposites: Flammability Characteristics and Mechanical Properties. *J. Clean. Prod.* **2017**, *144*, 79–89. [CrossRef]
51. Das, O.; Kim, N.K.; Kalamkarov, A.L.; Sarmah, A.K.; Bhattacharyya, D. Biochar to the Rescue: Balancing the Fire Performance and Mechanical Properties of Polypropylene Composites. *Polym. Degrad. Stab.* **2017**, *144*, 485–496. [CrossRef]
52. Das, O.; Hedenqvist, M.S. Wettability Properties of Biochar Added Wood/Polypropylene Composites. *Acad. J. Polym. Sci.* **2018**, *1*, 66–69. [CrossRef]
53. Ayrilmis, N.; Kwon, J.H.; Han, T.H.; Durmus, A. Effect of Wood-Derived Charcoal Content on Properties of Wood Plastic Composites. *Mater. Res.* **2015**, *18*, 654–659. [CrossRef]
54. Devallance, D.B.; Oporto, G.S.; Quigley, P. Investigation of Hardwood Biochar as a Replacement for Wood Flour in Wood-Polypropylene Composites. *J. Elastomers Plast.* **2016**, *48*, 510–522. [CrossRef]
55. Zouari, M.; Devallance, D.B.; Marrot, L. Effect of Biochar Addition on Mechanical Properties, Thermal Stability, and Water Resistance of Hemp-Polylactic Acid (PLA) Composites. *Materials* **2022**, *15*, 2271. [CrossRef]
56. Behazin, E.; Misra, M.; Mohanty, A.K. Sustainable Biocarbon from Pyrolyzed Perennial Grasses and Their Effects on Impact Modified Polypropylene Biocomposites. *Compos. Part B Eng.* **2017**, *118*, 116–124. [CrossRef]
57. Giorcelli, M.; Savi, P.; Khan, A.; Tagliaferro, A. Analysis of Biochar with Different Pyrolysis Temperatures Used as Filler in Epoxy Resin Composites. *Biomass Bioenergy* **2019**, *122*, 466–471. [CrossRef]
58. Giorcelli, M.; Khan, A.; Pugno, N.M.; Rosso, C.; Tagliaferro, A. Biochar as a Cheap and Environmental Friendly Filler Able to Improve Polymer Mechanical Properties. *Biomass Bioenergy* **2019**, *120*, 219–223. [CrossRef]
59. Ahmetli, G.; Kocaman, S.; Ozaytekin, I.; Bozkurt, P. Epoxy Composites Based on Inexpensive Char Filler Obtained from Plastic Waste and Natural Resources. *Polym. Compos.* **2013**, *34*, 500–509. [CrossRef]
60. Özaytekin, İ.; Kar, Y. Synthesis and Properties of Composites of Oligoazomethine with Char. *J. Appl. Polym. Sci.* **2012**, *123*, 815–823. [CrossRef]
61. Zhang, Q.; Xu, H.; Lu, W.; Zhang, D.; Ren, X.; Yu, W.; Wu, J.; Zhou, L.; Han, X.; Yi, W.; et al. Properties Evaluation of Biochar/High-Density Polyethylene Composites: Emphasizing the Porous Structure of Biochar by Activation. *Sci. Total Environ.* **2020**, *737*, 139770. [CrossRef]
62. Li, S.; Li, X.; Deng, Q.; Li, D. Three Kinds of Charcoal Powder Reinforced Ultra-High Molecular Weight Polyethylene Composites with Excellent Mechanical and Electrical Properties. *Mater. Des.* **2015**, *85*, 54–59. [CrossRef]
63. Li, S.; Huang, A.; Chen, Y.J.; Li, D.; Turng, L.S. Highly Filled Biochar/Ultra-High Molecular Weight Polyethylene/Linear Low Density Polyethylene Composites for High-Performance Electromagnetic Interference Shielding. *Compos. Part B Eng.* **2018**, *153*, 277–284. [CrossRef]
64. Li, S.; Xu, Y.; Jing, X.; Yilmaz, G.; Li, D.; Turng, L.S. Effect of Carbonization Temperature on Mechanical Properties and Biocompatibility of Biochar/Ultra-High Molecular Weight Polyethylene Composites. *Compos. Part B Eng.* **2020**, *196*, 108120. [CrossRef]
65. Arrigo, R.; Jagdale, P.; Bartoli, M.; Tagliaferro, A.; Malucelli, G. Structure-Property Relationships in Polyethylene-Based Composites Filled with Biochar Derived from Waste Coffee Grounds. *Polymers* **2019**, *11*, 1336. [CrossRef]
66. Kane, S.; Van Roijen, E.; Ryan, C.; Miller, S. Reducing the Environmental Impacts of Plastics While Increasing Strength: Biochar Fillers in Biodegradable, Recycled, and Fossil-Fuel Derived Plastics. *Compos. Part C Open Access* **2022**, *8*, 100253. [CrossRef]
67. Zhang, Q.; Zhang, D.; Xu, H.; Lu, W.; Ren, X.; Cai, H.; Lei, H.; Huo, E.; Zhao, Y.; Qian, M.; et al. Biochar Filled High-Density Polyethylene Composites with Excellent Properties: Towards Maximizing the Utilization of Agricultural Wastes. *Ind. Crops Prod.* **2020**, *146*, 112185. [CrossRef]
68. Idrees, M.; Jeelani, S.; Rangari, V. Three-Dimensional-Printed Sustainable Biochar-Recycled PET Composites. *ACS Sustain. Chem. Eng.* **2018**, *6*, 13940–13948. [CrossRef]
69. Tokiwa, Y.; Calabia, B.; Ugwu, C.; Aiba, S. Biodegradability of Plastics. *Int. J. Mol. Sci.* **2009**, *10*, 3722–3742. [CrossRef]
70. Murariu, M.; Dubois, P. PLA Composites: From Production to Properties. *Adv. Drug Deliv. Rev.* **2016**, *107*, 17–46. [CrossRef]
71. Ho, M.; Lau, K.; Wang, H.; Hui, D. Improvement on the Properties of Polylactic Acid (PLA) Using Bamboo Charcoal Particles. *Compos. Part B Eng.* **2015**, *81*, 14–25. [CrossRef]

72. Qian, S.; Sheng, K.; Yao, W.; Yu, H. Poly(Lactic Acid) Biocomposites Reinforced with Ultrafine Bamboo-Char: Morphology, Mechanical, Thermal, and Water Absorption Properties. *J. Appl. Polym. Sci.* **2016**, *133*, 43425. [CrossRef]
73. Kane, S.; Ryan, C. Biochar from Food Waste as a Sustainable Replacement for Carbon Black in Upcycled or Compostable Composites. *Compos. Part C Open Access* **2022**, *8*, 100274. [CrossRef]
74. Arrigo, R.; Bartoli, M.; Malucelli, G. Poly(Lactic Acid)-Biochar Biocomposites: Effect of Processing and Filler Content on Rheological, Thermal, and Mechanical Properties. *Polymers* **2020**, *12*, 892. [CrossRef] [PubMed]
75. Valentina, I.; Haroutioun, A.; Fabrice, L.; Vincent, V.; Roberto, P. Poly(Lactic Acid)-Based Nanobiocomposites with Modulated Degradation Rates. *Materials* **2018**, *11*, 1943. [CrossRef] [PubMed]
76. Haeldermans, T.; Samyn, P.; Cardinaels, R.; Vandamme, D.; Vanreppelen, K.; Cuypers, A.; Schreurs, S. Bio-Based Poly(3-Hydroxybutyrate)/Thermoplastic Starch Composites as a Host Matrix for Biochar Fillers. *J. Polym. Environ.* **2021**, *29*, 2478–2491. [CrossRef]
77. Pachekoski, W.M.; Dalmolin, C.; Agnelli, J.A.M. The Influence of the Industrial Processing on the Degradation of Poly(Hidroxybutyrate)—PHB. *Mater. Res.* **2012**, *16*, 237–332. [CrossRef]
78. Luo, S.; Grubb, D.T.; Netravali, A.N. The Effect of Molecular Weight on the Lamellar Structure, Thermal and Mechanical Properties of Poly(Hydroxybutyrate-Co-Hydroxyvalerates). *Polymer* **2002**, *43*, 4159–4166. [CrossRef]
79. Musioł, M.; Rydz, J.; Janeczek, H.; Kordyka, A.; Andrzejewski, J.; Sterzyński, T.; Jurczyk, S.; Cristea, M.; Musioł, K.; Kampik, M.; et al. (Bio)Degradable Biochar Composites—Studies on Degradation and Electrostatic Properties. *Mater. Sci. Eng. B Solid-State Mater. Adv. Technol.* **2022**, *275*, 115515. [CrossRef]
80. Botta, L.; Teresi, R.; Titone, V.; Salvaggio, G.; La Mantia, F.P.; Lopresti, F. Use of Biochar as Filler for Biocomposite Blown Films: Structure-Processing-Properties Relationships. *Polymers* **2021**, *13*, 3953. [CrossRef]
81. Terzioğlu, P.; Parın, F.N. Biochar Reinforced Polyvinyl Alcohol /Corn Starch Biocomposites. *Süleyman Demirel Üniversitesi Fen Bilim. Enstitüsü Derg.* **2020**, *24*, 35–42. [CrossRef]
82. Abdullah, Z.W.; Dong, Y. Preparation and Characterisation of Poly(Vinyl Alcohol) (PVA)/Starch (ST)/Halloysite Nanotube (HNT) Nanocomposite Films as Renewable Materials. *J. Mater. Sci.* **2018**, *53*, 3455–3469. [CrossRef]
83. Zuccarello, B.; Bartoli, M.; Bongiorno, F.; Militello, C.; Tagliaferro, A.; Pantano, A. New Concept in Bioderived Composites: Biochar as Toughening Agent for Improving Performances and Durability of Agave-Based Epoxy Biocomposites. *Polymers* **2021**, *13*, 198. [CrossRef] [PubMed]
84. Picard, M.; Thakur, S.; Misra, M.; Mielewski, D.F.; Mohanty, A.K. Biocarbon from Peanut Hulls and Their Green Composites with Biobased Poly(Trimethylene Terephthalate) (PTT). *Sci. Rep.* **2020**, *10*, 3310. [CrossRef] [PubMed]
85. Suriapparao, D.V.; Vinu, R. Effects of Biomass Particle Size on Slow Pyrolysis Kinetics and Fast Pyrolysis Product Distribution. *Waste Biomass Valorization* **2018**, *9*, 465–477. [CrossRef]
86. Hu, X.; Gholizadeh, M. Biomass Pyrolysis: A Review of the Process Development and Challenges from Initial Researches up to the Commercialisation Stage. *J. Energy Chem.* **2019**, *39*, 109–143. [CrossRef]
87. Maljaee, H.; Madadi, R.; Paiva, H.; Tarelho, L.; Ferreira, V.M. Incorporation of Biochar in Cementitious Materials: A Roadmap of Biochar Selection. *Constr. Build. Mater.* **2021**, *283*, 122757. [CrossRef]
88. Tan, K.; Qin, Y.; Wang, J. Evaluation of the Properties and Carbon Sequestration Potential of Biochar-Modified Pervious Concrete. *Constr. Build. Mater.* **2022**, *314*, 125648. [CrossRef]
89. Speight, J.G. Asphalt Paving. In *Asphalt Materials Science and Technology*; Elsevier: Amsterdam, The Netherland, 2016; pp. 409–435. ISBN 978-0-12-800273-5.
90. Walters, R.; Begum, S.A.; Fini, E.H.; Abu-Lebdeh, T.M. Investigating Bio-Char as Flow Modifier and Water Treatment Agent for Sustainable Pavement Design. *Am. J. Eng. Appl. Sci.* **2015**, *8*, 138–146. [CrossRef]
91. Walters, R.; Fini, E.; Abu-Lebdeh, T. Enhancing Asphalt Rheological Behavior and Aging SuSceptibility Using Biochar and Nano-Clay. *Am. J. Eng. Appl. Sci.* **2014**, *7*, 66–76. [CrossRef]
92. Zhao, S.; Huang, B.; Ye, P. Laboratory Evaluation of Asphalt Cement and Mixture Modified by Bio-Char Produced through Fast Pyrolysis. In Proceedings of the Pavement Materials, Structures, and Performance, American Society of Civil Engineers, Shanghai, China, 5 May 2014; pp. 140–149.
93. Zhao, S.; Huang, B.; Ye, X.P.; Shu, X.; Jia, X. Utilizing Bio-Char as a Bio-Modifier for Asphalt Cement: A Sustainable Application of Bio-Fuel by-Product. *Fuel* **2014**, *133*, 52–62. [CrossRef]
94. Zhang, R.; Dai, Q.; You, Z.; Wang, H.; Peng, C. Rheological Performance of Bio-Char Modified Asphalt with Different Particle Sizes. *Appl. Sci.* **2018**, *8*, 1665. [CrossRef]

Disclaimer/Publisher's Note: The statements, opinions and data contained in all publications are solely those of the individual author(s) and contributor(s) and not of MDPI and/or the editor(s). MDPI and/or the editor(s) disclaim responsibility for any injury to people or property resulting from any ideas, methods, instructions or products referred to in the content.

Article

Fractional Calculus Approach to Reproduce Material Viscoelastic Behavior, including the Time–Temperature Superposition Phenomenon

Andrea Genovese, Flavio Farroni * and Aleksandr Sakhnevych

Department of Industrial Engineering, University of Naples Federico II, Via Claudio 21, 80125 Naples, NA, Italy
* Correspondence: flavio.farroni@unina.it

Abstract: The design of modern products and processes cannot prescind from the usage of viscoelastic materials that provide extreme design freedoms at relatively low cost. Correct and reliable modeling of these materials allows effective use that involves the design, maintenance, and monitoring phase and the possibility of reuse and recycling. Fractional models are becoming more and more popular in the reproduction of viscoelastic phenomena because of their capability to describe the behavior of such materials using a limited number of parameters with an acceptable accuracy over a vast range of excitation frequencies. A particularly reliable model parametrization procedure, using the poles–zeros formulation, allows researchers to considerably reduce the computational cost of the calibration process and avoid convergence issues typically occurring for rheological models. The aim of the presented work is to demonstrate that the poles–zeros identification methodology can be employed not only to identify the viscoelastic master curves but also the material parameters characterizing the time–temperature superposition phenomenon. The proposed technique, starting from the data concerning the isothermal experimental curves, makes use of the fractional derivative generalized model to reconstruct the master curves in the frequency domain and correctly identify the coefficients of the WLF function. To validate the methodology, three different viscoelastic materials have been employed, highlighting the potential of the material parameters' global identification. Furthermore, the paper points out a further possibility to employ only a limited number of the experimental curves to feed the identification methodology and predict the complete viscoelastic material behavior.

Keywords: viscoelasticity; material parametrization; WLF coefficients; pole–zero formulation; fractional model

1. Introduction

With the advancements in modern technology, the continuous evolution of materials, and more efficient manufacturing processes, the design of modern products and processes cannot prescind from the usage of viscoelastic materials. These kinds of materials provide extreme design freedoms at relatively low cost: high elasticity and impermeability, adequate chemical and heat resistance, insulation, and an ability to absorb shocks and dampen noise [1,2]. A holistic approach to knowledge-based material selection allows researchers to not only adequately select materials from the perspective of functional attributes but also consider their emotional and esthetic qualities, contemplating the added sustainable value, such as recyclability, energy efficiency, and solar-power capacity [3,4].

The knowledge of the material and the ability to properly model the material characteristics starting from the earliest design stages becomes mandatory in cases where the aim is to govern the material properties through the manufacturing processes and for the entire product lifecycle, taking into account its eventual changes during aging to overcome the typical limits and constraints in a design path where materials and transformation technologies are both variables of the creation process [5,6]. The crucial aspects in this context concern the choice of the proper mathematical formulation for the material modeling and the ability to calibrate the model with a limited amount of data explored in a particularly narrow frequency or temperature range, whose accuracy can be affected by the experimental technique employed. Indeed, the aim of both destructive and non-destructive testing techniques is to exploit the widest possible frequency–temperature working domain of the material to understand the material behavior within the exploited operating conditions and provide a sufficient amount of data for model-calibration purposes [7–10].

The constitutive law for linear viscoelasticity, based on Boltzmann's superposition principle [3], can be established by means of three different approaches: integral models, linear differential models, and fractional derivative models. The differential approach is used to describe the rheological properties by means of linear differential equations that link stress and strain [11]. A combination of mechanical elements, ideal springs, and dashpots are used to build suitable rheological models. The generalized models are often employed to describe the viscoelastic behavior of the materials in a wide range of frequencies and time scales. If, on the one hand, these models offer a good description of the viscoelastic materials, on the other hand, they involve a set of differential equations to describe the dynamic state of the system, which could considerably complicate the overall mathematical formulation, significantly increasing the computational load due to a larger set of motion equations to be solved. To overcome these issues, fractional models are becoming more and more popular because of their ability to reproduce the behavior of viscoelastic materials using a limited number of parameters with an acceptable accuracy level over a vast range of excitation frequencies, combining ideal spring and spring-pot elements [12].

Several applications take advantage of fractional models. A review regarding the application of fractional calculus in the models of linear viscoelasticity utilized in dynamic problems of mechanics of solids has been conducted by Shitikova [13]. Abouelregal [14] proposed a methodology to study thermoelastic vibrations in a homogeneous isotropic three-dimensional solid based on a fractional derivative Kelvin–Voigt model. In [15], Zhou et al. adopted a variable-order fractional derivative material model to numerically analyze the behavior of the frozen soil, including creep, stress relaxation, and strain rate effects. In [16], Wang et al. adopted the fractional derivative model to describe the hysteretic behavior of the magnetorheological elastomers, demonstrating a huge potential in the field of intelligent structures and devices.

Furthermore, in [2], the authors presented a particularly reliable procedure adopting the poles–zeros formulation, which considerably reduced the calculation effort of the identification/calibration process. Starting from these results, the aim of this work is to demonstrate that it is also possible to identify both the master curve and the WLF function by adopting the poles–zeros identification methodology. The technique involves the individual isothermal experimental curves and makes use of the fractional derivative generalized model poles–zeros identification to reconstruct the master curve and correctly identify the coefficients of the WLF function. Furthermore, this paper points out a further possibility to employ only a limited number of experimental isothermal curves to feed the proposed identification methodology (such as the one addressing the material glass transition zone). It has to be highlighted that at the current stage, the vertical shifts have been neglected (this hypothesis is fine if the experimental acquisitions do not cover particularly cold temperatures or, on the other side/dually, particularly high testing frequencies), but they could also be part of the parameters to be identified in further investigations.

The paper is organized as follows: in Section 2, the fractional derivative generalized model is defined in the frequency domain to obtain the relative poles–zeros formulation; in Section 3, the WLF time–temperature superposition principle is defined with a particular focus on the determination of the horizontal and vertical shifts and an analytic study to demonstrate the WLF validity in the poles–zeros domain; in Section 4, the possibility of adopting the fractional models and the poles–zeros formulation to determine not only the viscoelastic moduli but also the WLF shift factor starting from the experimental data is investigated; in Section 5, a specific study to understand how the minimum number of experiments affects the identification procedure is presented.

2. Fractional Derivative Model in Poles–Zeros Formulation

Viscoelastic materials, due to their intrinsic rheological stress–strain dependence on time, which is based on the fact that the deformation energy is not totally stored but partially dissipated through a hysteretic mechanism, exhibit both elastic and viscous characteristics concurrently [17]. To analyze the mechanical response of viscoelastic materials, three different testing methodologies are usually adopted: static, transient, and dynamic [18]. Whereas the static characterization regards the static or quasi-static application of load or deformation, transient (creep and stress relaxation experiments) and dynamic (DMA and DSC techniques) testing procedures concern the analysis of material response towards time once deformation or load functions (elongation or shear) are applied [10,19,20]. The dynamic behavior of this kind of material is characterized by the material dynamic stiffness E^*, which is a complex variable defined as (1):

$$\frac{\sigma(\omega)}{\varepsilon(\omega)} = E^* = E' + E'' \tag{1}$$

where E' is the storage modulus (Pa), E'' is the loss modulus (Pa), $\sigma(\omega)$ is the cyclic stress applied to the material (Pa) at given circular frequency ω, and $\varepsilon(\omega)$ is the corresponding strain response.

These quantities are linked to the way the material dissipates energy provided by means of a load/stress time function. Particularly, they are related to the phase angle δ, according to Equation (2):

$$\frac{E''(\omega)}{E'(\omega)} = tan\delta \tag{2}$$

It is worth noting that all the physical quantities, referred to as the properties of the viscoelastic material's behavior, are a function of the particular frequency levels at which sinusoidal load/deformation is applied during the test. More precisely, the modulus, the energy loss, and the hysteresis of a viscoelastic material change in relation to two parameters: the frequency the force is applied with and the temperature at which the phenomena are evaluated.

The correct modeling of the material viscoelastic properties is a key element in achieving reliable results from analytical models' outputs or finite-element-based analyses within the design of the desired dynamic behavior of mechanical systems. Several mathematical models can be found in the literature to help understand and describe material viscoelastic behavior [21]. The Maxwell and Kelvin models fail to represent the actual response of viscoelastic materials at low and high frequencies, respectively, while generalized models return more accurate results but imply a more complicated mathematical formulation and increase computational costs. As this paper is focused on viscoelastic solids, the Fractional Derivative Generalized model (FDGM) is adopted by the authors to model the viscoelastic behavior. This model, depicted in Figure 1, is obtained by connecting fractional Maxwell cells in series, where a fractional Maxwell cell is defined as a spring and a spring-pot element arranged in series.

From a mathematical point of view, a generic constitutive equation for viscoelastic materials, based on fractional derivative orders, is expressed in the following Equation (3):

$$\sum_{n=0}^{N} a_n \frac{d^{\alpha_n}\sigma(t)}{dt^{\alpha_n}} = \sum_{m=0}^{M} b_m \frac{d^{\beta_m}\varepsilon(t)}{dt^{\beta_m}} \tag{3}$$

where α_n and β_m are the fractional derivative orders included within the range $[0, 1]$, $N = M$ and $b_0 = 0$ for the considered Maxwell formulation. Turning to the frequency domain by applying the Fourier transform and assuming that $\alpha = \beta$, Equation (3) gives the following expression for the complex Moduli (4):

$$E^*(i\omega) = E_0 + \sum_{k=1}^{N} \frac{(i\omega)^{\alpha_k} E_k \eta_k}{E_k + (i\omega)^{\alpha_k} \eta_k} \tag{4}$$

where the parameters E_k and η_k represent the springs' stiffness and spring-pot coefficient, respectively, as represented in Figure 1, and ω is the angular frequency.

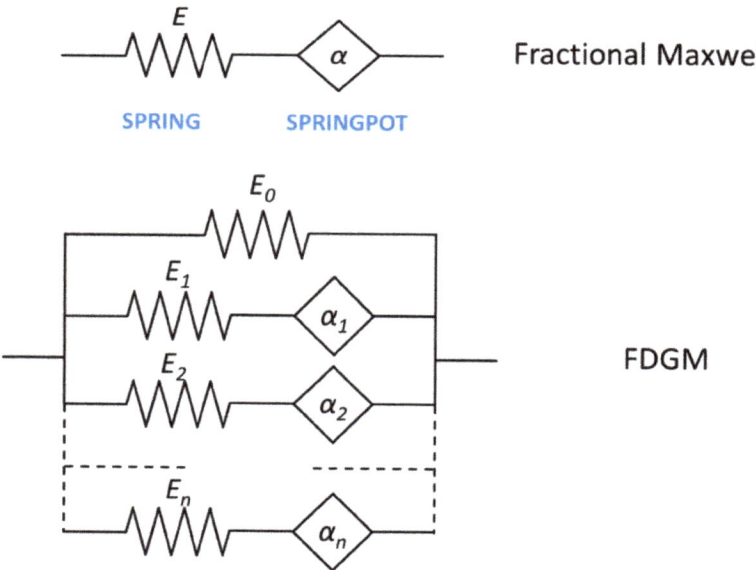

Figure 1. Scheme of the fractional derivative generalized model.

Renaud et al. [22] have shown that, in analogy with the generalized models, Expression (4) can be equivalently expressed in the poles–zeros formulation as (5):

$$E^*(i\omega) = E_0 \prod_{k=1}^{N} \frac{1 + (i\omega/\omega_{z,k})^{\alpha_k}}{1 + (i\omega/\omega_{p,k})^{\alpha_k}} \tag{5}$$

where ω_z and ω_p are the zeros and poles, respectively. The presented approach, consisting of a superposition of poles–zeros coupled behavior, overcomes the computational and convergence issues typically occurring in the parameterization of the rheological models in the time and frequency domain forms since it enables the determination of boundary conditions and initial starting guess for the optimization problem [23]. Figure 2 schematizes the main phases of the adopted constrained nonlinear optimization procedure aiming at identifying a robust set of poles–zeros coefficients, starting from a feasible set of initial conditions, detailed by the authors in [2].

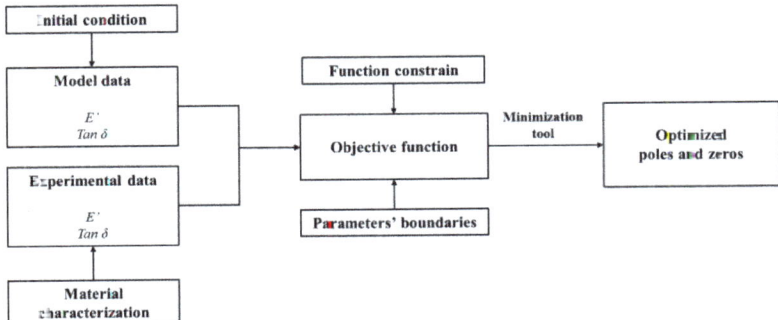

Figure 2. Constrained nonlinear optimization procedure to identify poles–zeros coefficients: functional scheme.

3. Time–Temperature Superposition

Due to the limitation of most commercial DMA equipment, the tests for the determination of the complex modulus are usually limited to the low-frequency range domain [24]. To extend the material properties to a broader frequency range, which is necessary for applications such as contact, friction, and wear modeling approaches, the time–temperature superposition (TTS) is applied [25]. Time–temperature superposition (also called frequency–temperature superposition or method of reduced variables) is a frequently applied procedure to determine the temperature dependence of the rheological behavior of a polymer and to expand the time or frequency regime at a given temperature at which the material behavior is studied.

As observed in [26,27], the frequency–temperature superposition is valid due to the fact that various relaxation times belonging to a given relaxation process have the same temperature dependence. This principle relates the material response at a given time t (or frequency ω), and at a given temperature T, to that at other conditions (denoted by subscript r):

$$\begin{aligned} \omega_r &= a_T(T, T_r)\, \omega \\ E'(\omega_r, T_r) &= b_T(T, T_r) E'(\omega, T) \\ E''(\omega_r, T_r) &= b_T(T, T_r) E''(\omega, T) \end{aligned} \qquad (6)$$

where $a_T(T, T_r)$ and $b_T(T, T_r)$ are coefficients that indicate the amount of horizontal and vertical shifting (respectively) to be applied to isotherms of storage and loss moduli measured at a temperature T in order to estimate the material properties at a reference temperature T_r, as qualitatively shown in Figure 3 for explanatory purposes.

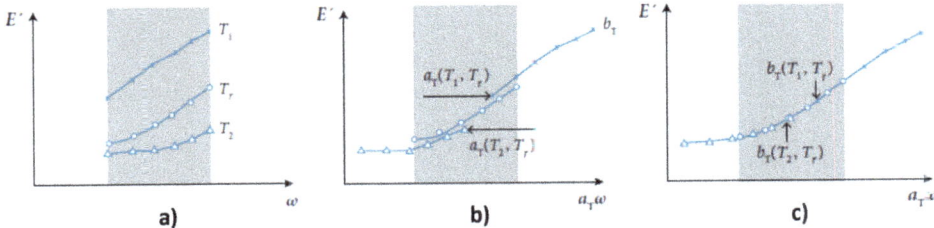

Figure 3. Determination of the frequency–temperature superposition shifting factor a_T and b_T: (a) Isotherms of storage modulus on the frequency range measurable by DMA, at temperatures T_1, T_2, and T_r, with $T_1 < T_r < T_2$; (b) Isotherms of storage modulus after application of the horizontal shift factors, taking T_r as the reference temperature; (c) Isotherms of storage modulus after application of both horizontal and vertical shift factors, taking T_r as the reference temperature [28].

The horizontal shift factors a_T (T, T_r) describe the temperature dependence of the relaxation time and usually follow the empirical Williams–Landel–Ferry (WLF) law (7):

$$\log_{10} a_T(T, T_r) = -\frac{C_1(T - T_r)}{C_2 + (T - T_r)} \tag{7}$$

where C_1 and C_2 are empirical constants whose order of magnitude is about 10 and 100 K, respectively.

The vertical shift factors b_T (T, T_r) are related to thermal expansion effects, which for most polymers can be neglected due to their small variation, and that, for this reason, will be neglected in the following. Here, it is worth noting that this hypothesis is highly acceptable in the viscoelastic regions where the frequency/time dependence of material functions is sharp. On the other hand, overlooking thermal vertical shifts in viscoelastic regions with weak frequency/time dependence may lead to different values of horizontal shifting whose accuracy depends on the material under investigation.

Figure 4 shows the typical qualitative trend of the vertical and horizontal shift factors as a function of the difference from a reference temperature T_R.

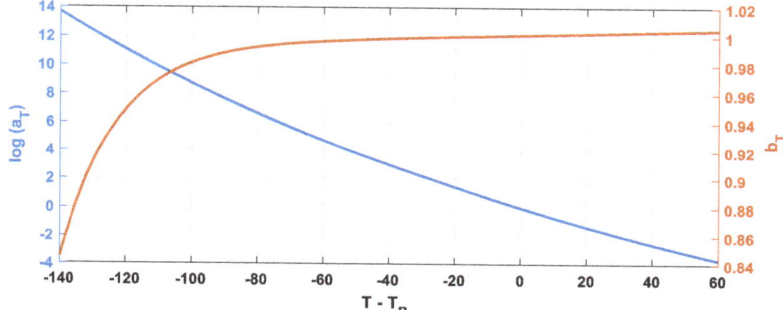

Figure 4. Typical trend of a_T and b_T.

The question arises regarding the eventual validity of the WLF principle through the material transformation to the poles–zeros formulation. To this end, the validity of the WLF principle has been investigated starting from a completely characterized polymeric material, whose characteristics in terms of the experimental curves (E' and $tan(\delta)$) and parameters in terms of WLF coefficients (C_1 and C_2) have been measured and calculated, respectively. To transform the FDGM in poles–zeros formulation for two different temperatures, T_r an T_1, the standard procedure consists of the following steps:

- Identification of poles and zeros at a reference temperature T_r, applying the scheme represented in Figure 2;
- Application of the WLF formulation to the experimental curves referring to a reference temperature T_r to obtain the experimental curves at the new temperature T_1, shifting all frequency vector ω_{exp};
- Identification of the poles and zeros starting from the experimental curves referring to the new temperature T_1 by means of the procedure summarized in Figure 2.

It should be highlighted that this procedure is computationally consuming since the identification procedure must be performed as many times as the temperatures of interest.

In this study, the authors also investigate an alternative procedure to perform the shift of the curves:

- Identification of the poles and zeros at reference temperature T_r, with the procedure described in Figure 2;
- Application of the WLF law directly on the identified poles and zeros, obtaining the master curves at the new temperature T_1.

The two procedures, validated by a significant amount of experimental data, provide the exact same results, which can be mathematically summarized as follows (8):

$$\begin{aligned} E'_{T1} &= f\left(\frac{\omega_{exp}(T_r)}{a_T}; \omega_{p,k}(T_1); \omega_{z,k}(T_1); \alpha_k\right) = f\left(\omega_{exp}, \frac{\omega_{p,k}(T_r)}{a_T}, \frac{\omega_{z,k}(T_r)}{a_T}, \alpha_k\right) \\ tan\delta_{T1} &= f\left(\frac{\omega_{exp}(T_r)}{a_T}; \omega_{p,k}(T_1); \omega_{z,k}(T_1), \alpha_k\right) = f\left(\omega_{exp}, \frac{\omega_{p,k}(T_r)}{a_T}, \frac{\omega_{z,k}(T_r)}{a_T}, \alpha_k\right) \end{aligned} \quad (8)$$

where ω_{exp} is the frequency of the experimental test.

Therefore, it is possible to generalize the poles–zeros Formulation (5) by including the WLF (9):

$$E^*(i\omega) = E_0 \prod_{k=1}^{N} \frac{1 + \left(\frac{i\omega}{\left(\frac{\omega_{z,k}}{a_T}\right)}\right)^{\alpha_k}}{1 + \left(\frac{i\omega}{\left(\frac{\omega_{p,k}}{a_T}\right)}\right)^{\alpha_k}} \quad (9)$$

where $a_T = 1$ when $T = T_{rif}$.

From Equations (8) and (9), indicating with $\omega_{p,k_{rif}}$ and $\omega_{z,k_{rif}}$, respectively, the poles and zeros identified at the reference temperature T_{rif}, it is possible to write the WLF function applied to the poles–zeros formulation as (10):

$$\frac{\omega_{p,k}(T)}{\omega_{p,k_{rif}}} = \frac{\omega_{z,k}(T)}{\omega_{z,k_{rif}}} = a_T\left(T - T_{rif}\right) = 10^{-\frac{C_1 \cdot (T - T_{rif})}{C_2 + (T - T_{rif})}} \quad (10)$$

The equivalence between the Equations (7) and (10) is highlighted in Figure 5 for a polymer of known parameters C_1 and C_2 and adopting FDGM with three elements (N = 3). It should be noted that for an FDGM with three elements, the subscript k in (10) is three. This means that, for each temperature, three poles and three zeros are identified, and each of them satisfies Equation (10).

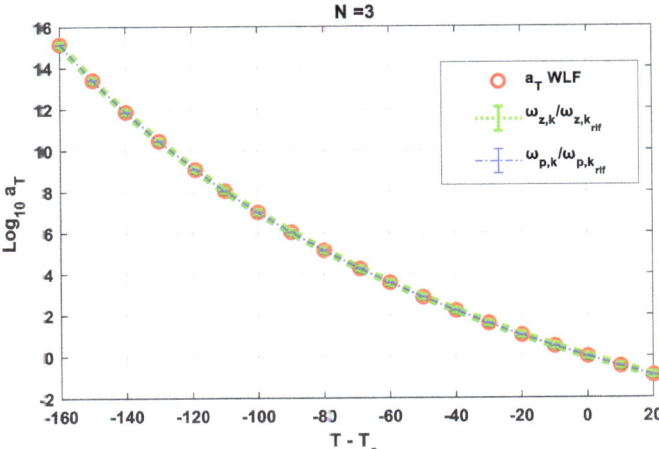

Figure 5. Relation between poles, zeros, and WLF law.

Therefore, the validity of the WLF principle through the material transformation to the poles–zeros formulation has been highlighted. Here, it is worth noting that all that has been demonstrated for the FDGM pole–zeros formulation is still valid for the generalized formulation (in Maxwell and Kelvin–Voight form).

4. Material Parameters' Global Identification

4.1. Identification Procedure

Since Equation (10) demonstrated the relation between the WLF function and the poles and zeros, the authors aimed to study the possibility of adopting the fractional models and the poles–zeros formulation to determine not only the viscoelastic moduli but also the WLF shift factor starting from the experimental data. With this purpose in mind, the authors, starting from the experimental data obtained by the DMA, make use of the FDGM to obtain the viscoelastic mater curves. Figure 6 shows, for a polymer compound named A, the curves obtained by the experimental DMA, consisting of a series of isotherms in a frequency range of 0–100 rad/s.

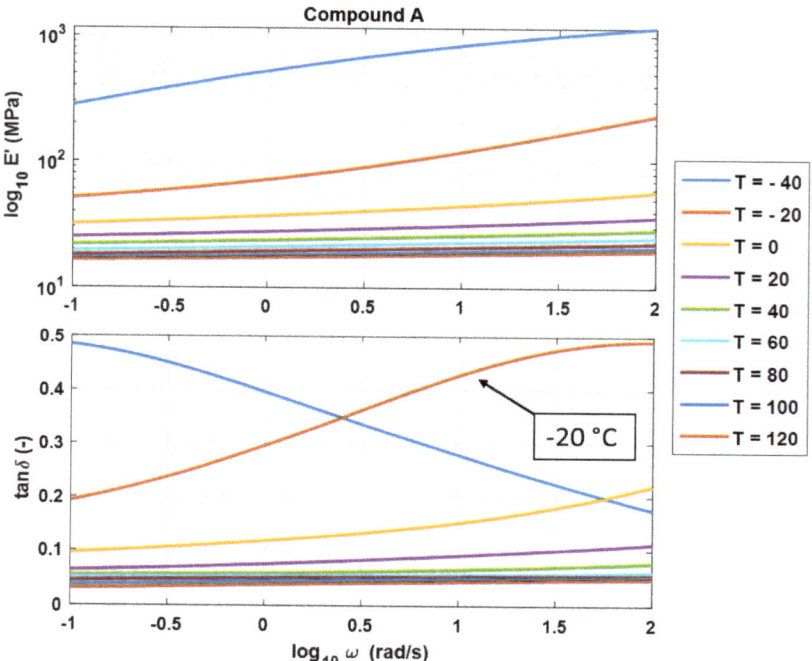

Figure 6. Experimental DMA isothermal curves.

To build the moduli master curves as function of the frequency at one temperature, an arbitrary reference temperature must be set. In order to minimize the randomness of the procedure, a unique way to establish the reference temperature has been defined. The following it has been assumed that T_{rif} is the temperature of the isotherm curve that presents the maximum value of the tan δ (i.e., for compound A, $T_{rif} = -20\ °C$).

The number of parameters to be identified is 3N + 3:

- 3N + 1 parameters for the pole, zero, and static modulus (9);
- Two parameters for the WLF (10).

Since in previous work, it has been shown that three is a sufficient number to obtain an excellent fitting [2], an FDGM with N = 3 fractional Maxwell cells has been considered in the following. According to the procedure defined in [2], the parameters' identification is performed by means of a constrained procedure, seeking the minimum of a nonlinear error function W (11): of the 3N + 3 real variables:

$$W = (1 - r) \cdot Err_E + r \cdot Err_{tan\delta} \qquad (11)$$

where $Err_{tan\delta}$ and Err_E are defined by (12) and (13), and r is the weight factor:

$$Err_E = \frac{\sqrt{\frac{\sum_{i=1}^{N}(E-E_{model})^2}{N}}}{mean(E)} \qquad (12)$$

$$Err_{tan\delta} = \frac{\sqrt{\frac{\sum_{i=1}^{N}(tan\delta-tan\delta_{model})^2}{N}}}{mean(tan\delta)} \qquad (13)$$

In other words, Error Function (11) is defined as the weighted sum of the normalized root mean square error (NRMSE) calculated for the storage modulus and the loss tangent as in (12) and (13). It is worth noting that an error function so defined allows fitting modulus, phase, and WLF coefficient at the same time. The weighting coefficient r has been assumed to be equal to 0.5 for all in the following to test the procedure in the same conditions.

Regarding the initial condition of the identification procedure, the initialization of the pole–zero parameters has been carried out by means of the method proposed by Renaud et al. [22]; while for initial C_1 and C_2 in (10), they have been assumed to be 17.44 and 51.60 respectively, which are generally accepted values when $T_{rif} = T_g$ [26].

4.2. Results

The identification procedure described in the previous sub-section has been employed by the authors for three different polymers' compounds, A, B, and C, to evaluate the capability of the FDGM models to describe three completely different viscoelastic materials in terms of time–temperature superposition.

The thermal properties of samples were investigated by using a TA DSCQ2000 differential scanning calorimeter equipped with a TA Instruments DSC cooling system. Dry nitrogen gas with a flow rate of 20 mL/min was purged through the cell during the measurements and the thermal treatments. Samples of approximately 8 mg were heated from −80 to 100 °C and kept at this temperature for 3 min, then cooled from 100 to −80 °C at 50 °C/min, kept at this temperature for 3 min, and re-heated from −80 to 100 °C at 20 °C/min. The heating rate was fixed to 20 °C/min, whereas the cooling was carried out at 50 °C/min.

In particular, the heat flows for different compounds are represented in Figure 7, and the identified glass transition temperatures (Tg), evaluated in the second heating run, are listed in Table 1.

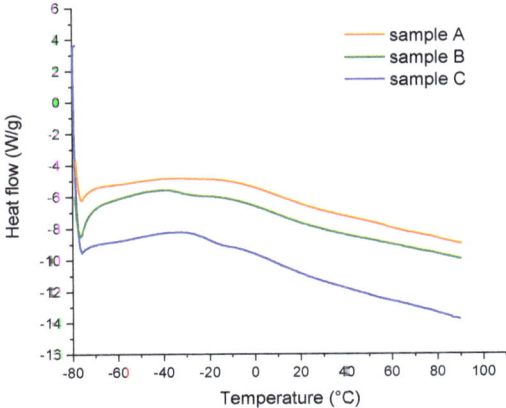

Figure 7. Differential scanning calorimeter testing routine.

Table 1. Experimentally identified glass transition temperatures for compounds A, B, and C.

	I T_g (°C)	II T_g (°C)
Compound A	-	9
Compound B	−33	8
Compound C	−20	9

All the samples exhibited a weak glass transition temperature around 9 °C, while only samples B and C showed an intense transition at −33 and −20 °C, respectively.

For compound A, Figure 8a compares the experimental master curves with the results obtained by the identification procedure, while in Figure 8b, the WLF law is shown.

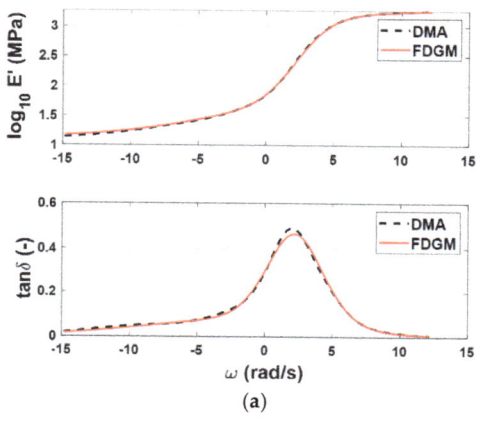
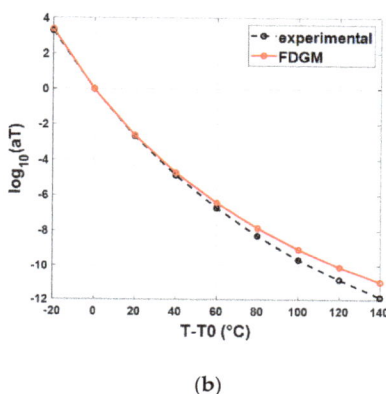

Figure 8. Compound A: Storage modulus and loss tangent (**a**), WLF law (**b**)—experimental data (dashed line) vs. FDGM model.

From a qualitative point of view, the FDGM with three fractional elements is able to provide an acceptable representation of the master curves' shapes and the WLF law. Particularly in Figure 8b, it is possible to note that for temperatures close to $T_{rif} = -20°$ ($T - T_0 = 0$), the procedure is quite accurate, while for temperatures far from this value, some differences appear from the experimental data. These small differences (<10%) are due to the vertical shift factors $b_T(T, T_r)$ that, in this work, were neglected. Figures 9a,b and 10a,b show the results for the compound B and C, respectively. For both compounds, $T_{rif} = -20°$.

To quantify the accuracy of the FDGM approach, the NRMSE is also evaluated, both for the storage modulus and for the loss tangent quantities defined in Equations (12) and (13). The purpose of these indicators is to quantify the goodness of the models' behavior toward the experimental data reproduction. Table 2, in addition to the NRMSE, reports the identified coefficients C_1 and C_2 of the WLF law (10).

Table 2. NRMSEs and identified WLF coefficients for compounds A, B, and C.

	NRMSE E' (MPa)	NRMSE tan δ (-)	NRMSE Total	T_{rif} (°C)	C_1	C_2
Compound A	0.0108	0.0637	0.0372	−20	23.36	158.21
Compound B	0.0486	0.0442	0.0464	−20	13.19	55.12
Compound C	0.0402	0.0406	0.0404	−20	17.31	56.24

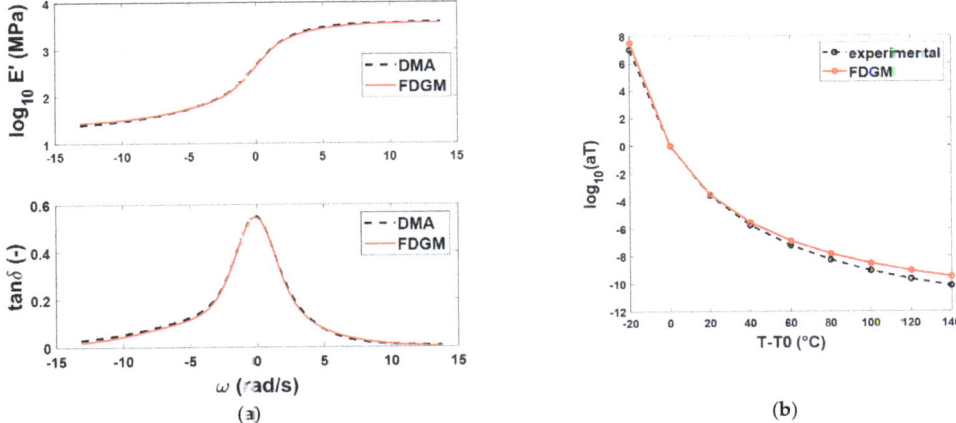

Figure 9. Compound B: Storage modulus and loss tangent (**a**), WLF law (**b**)—experimental data (dashed line) vs. FDGM model.

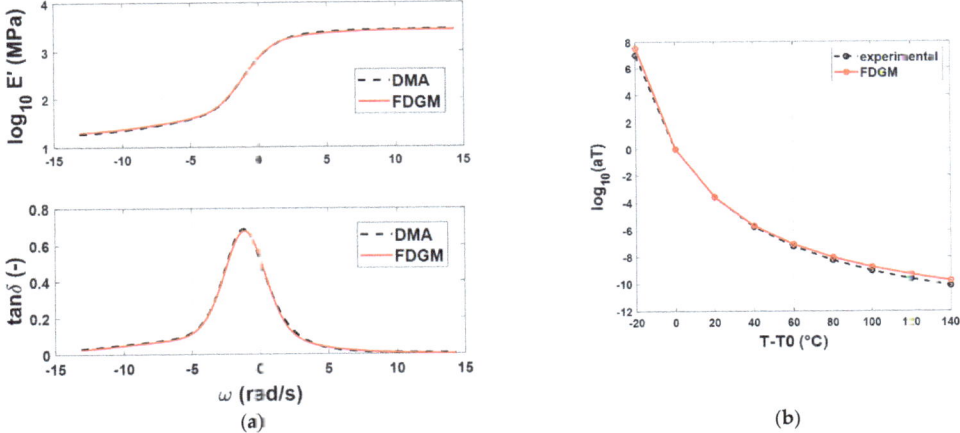

Figure 10. Compound C: Storage modulus and loss tangent (**a**), WLF law (**b**)—experimental data (dashed line) vs. FDGM model.

The identified parameters of these models are detailed in Table A1 in Appendix A.1. Analyzing the results, it is possible to note that the FDGM returns a very good fitting of experimental data with an NRMSE < 0.05 for all the compounds.

Regarding the WLF coefficients, the procedure is able to identify three different couples of parameters, one for each polymer compound. We can note that each identified couple is different from the starting set.

5. Master Curves and Time–Temperature Superposition Parameters' Estimation with Partial Experimental Data

In the previous sections, it has been demonstrated how the fractional calculus in poles–zeros formulation is a powerful analytical model able to reproduce the material viscoelastic behavior, including the time–temperature superposition phenomenon. The FDGM ensures not only a high correlation with the viscoelastic master curves but also is able to estimate the different behavior of each polymer compound in terms of time–temperature superposition, returning different WLF laws for each of them. Taking advantage of this methodology, the question arises about the possibility of reproducing the viscoelastic behavior with a minimum number of experimental data. The aim of the following analysis is to test the capability of the FDGM model in moduli and WLF coefficients estimation, starting from a reduced set of experimental isotherms from DMA data on which to make the fitting. It is worth noting that a reduction in the experimental test necessary to characterize the viscoelastic behavior would lead to savings in resources, money, and time.

In [2], it has been demonstrated that an excellent prediction of the master curves can be achieved by adopting data coming from the lower and upper frequencies plateau of the storage modulus, the peak of the loss tangent curve, and the curvature change of both curves. Moreover, satisfying results can also be obtained by considering only the data available of the upper and low-frequency plateaus plus those at the loss tangent peak. Borrowing these results, three isotherms have been chosen for the analysis: an isotherm at high temperature, an isotherm at low temperature, and one that presents the maximum value of tan δ. For explanatory purposes, Figure 10 reports the experimental starting set for the identification of compound C.

The parameters of the three-element FDGM and of the WLF law have been identified, and the models' results have been compared with the experimental results in the entire frequency range. Figure 11 depicts the curve comparisons for compound C: Figure 11a shows the master curves estimated (FDGM reduced) compared with the full experimental DMA. Furthermore, in the same figure, the partial experimental data used for the identification are highlighted in the frequency domain. Figure 11b shows the experimental WLF law in comparison with the identified value of the FDGM by adopting both all and partial experimental data. These results highlight how the proposed technique is able to give a good estimation of the master curves and the time–temperature superposition parameters, while also using a reduced dataset of experimental data. To quantify the approximation in the latter case, the NRMSE has been calculated in the entire frequency range and for the three compounds. The results are reported in Table 3, where it is possible to note that the NRMSE < 0.06 in all cases. Some differences can be found in the estimation of parameters C_1 and C_2; however, this difference implies a small difference in the calculation of a_T in the temperature range exanimated, as shown in Figure 12a,b.

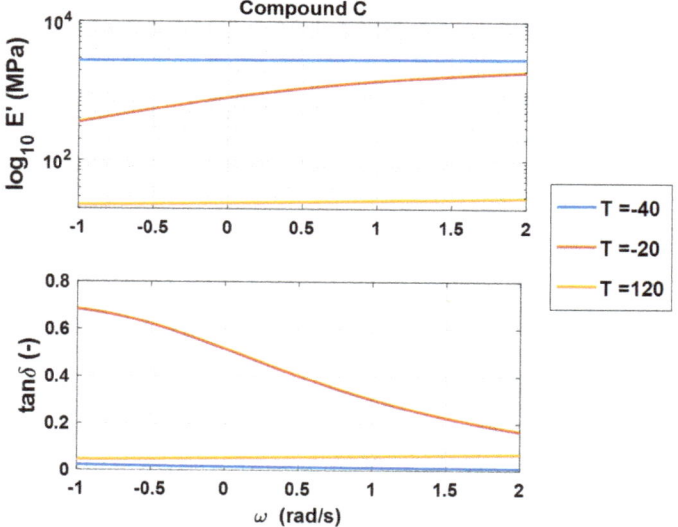

Figure 11. Example of the experimental starting set for the identification with partial DMA for compound C.

Table 3. NRMSEs and identified WLF coefficients for compounds A, B, and C in case of partial DMA data.

	NRMSE E′ (MPa)	NRMSE tanδ (-)	NRMSE Total	T_{rif} (°C)	C_1	C_2
Compound A	0.0339	0.0568	0.0454	−20	30.24	203.53
Compound B	0.012	0.057	0.0345	−20	16.13	65.04
Compound C	0.011	0.11	0.0617	−20	12.15	53.64

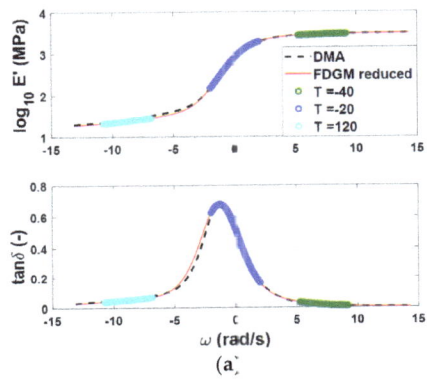
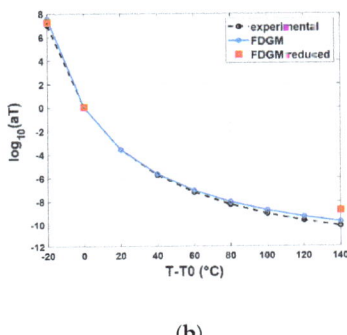

Figure 12. Compound C: Storage modulus and loss tangent (a), WLF law (b)—experimental data (dashed line) vs. FDGM model complete (light blue) and FDGM obtained with partial staring experimental dataset (red).

The identified parameters for the three compounds, A, B, and C, using partial experimental data, are detailed in Table A2 in Appendix A.

6. Conclusions

The search for rheological models to be used for the design and study of components made of viscoelastic material is a topic of great interest both from the industrial and the academic point of view, involving crucial aspects such as the reliability, lifecycle, and environmental sustainability of these materials. In this context, fractional models are becoming more and more popular because of their capability to describe the behavior of such materials using a limited number of parameters. These models can be expressed by adopting a pole–zero formulation. This formulation that in logarithmic scale becomes a superposition of pole–zero couple behavior allows us to overcome the computational and convergence issues that occur in parameter identification of a rheological model in the time domain and frequency domain form.

In this work, first, the validity of the WLF principle through the material transformation to the poles–zeros formulation was demonstrated. After that, a constrained pole–zero identification procedure was used not only to identify both the master curve but also the material WLF function. An extensive study was carried out on three different compounds to evaluate the capability of the FDGM in poles–zeros formulation to reconstruct the master curve in the frequency domain and correctly identify the coefficients of the WLF function, making use of the data concerning the individual isothermal experimental curves from a DMA test. The analysis shows that the fractional calculus in poles–zeros formulation is a powerful analytical model able to reproduce the material viscoelastic behavior, including the time–temperature superposition phenomenon. The FDGM ensures not only a high correlation with the viscoelastic master curves but is also able to estimate the differences in the WLF law of each polymer compound.

As an extension, the paper points out the further possibility of employing only a limited number of the experimental isothermal curves to feed the identification methodology, as the samples specifically concern the material glass transition zone.

These results highlight how the proposed methodology is able to give a good estimation of the master curves and the time–temperature superposition parameters while using a reduced dataset of experimental data. These results could open favorable scenarios both from an economic and environmental point of view: reducing the number of tests necessary for the preliminary characterization of materials, reducing development and design costs, and increasing predictive knowledge about the behavior of the material during its entire life for reuse and recycling.

Author Contributions: Conceptualization, A.G. and A.S.; methodology, A.G., F.F. and A.S.; software, A.G.; validation, A.S.; formal analysis, F.F.; writing—original draft preparation, A.G. and A.S.; writing-review and editing, F.F. and A.S.; funding acquisition, A.G., F.F. and A.S. All authors have read and agreed to the published version of the manuscript.

Funding: This work was partly supported by the project "FASTire (Foam Airless Spoked Tire): Smart Airless Tyres for Extremely-Low Rolling Resistance and Superior Passengers Comfort" funded by the Italian MIUR "Progetti di Ricerca di Rilevante Interesse Nazionale (PRIN)"–grant n. 2017948FEN.

Data Availability Statement: Not applicable.

Acknowledgments: Giovanna Gomez d'Ayala of the Institute for Polymers, Composites and Biomaterials of the National Council of Research (IPCB-CNR) is thanked for their help in performing the DSC analysis.

Conflicts of Interest: The authors declare no conflict of interest.

Appendix A. Parameter Values Obtained with Pole–Zero Identification Procedure

Appendix A.1. Fractal Derivative Generalized Maxwell Model Parameters

Table A1. Compounds A, B, and C: parameters of FDGM models composed of 3 elements.

Element	Compound A			Compound B			Compound C		
E0 (MPa)	13.24			25.38			17.3		
Element	1	2	3	1	2	3	1	2	3
Pole	0.30	4.13	13.00	−1.29	1.05	3.12	−3.94	0.30	1.83
Zero	−6.96	0.30	4.13	−6.50	−1.29	1.05	−8.15	−2.57	0.30
Gamma	0.10	0.37	0.01	0.15	0.44	0.18	0.13	0.49	0.17

Table A2. Compounds A, B, and C: parameters of FDGM models composed of 3 elements using a reduced number of DMA experimental data.

Element	Compound A			Compound B			Compound C		
E0 (MPa)	12.62			22.99			15.89		
Element	1	2	3	1	2	3	1	2	3
Pole	−1.50	3.59	5.27	−1.22	0.96	2.62	−2.92	0.35	1.32
Zero	−8.62	0.39	4.93	−8.30	−1.22	0.96	−8.02	−2.92	0.35
Gamma	0.11	0.40	0.41	0.13	0.45	0.23	0.11	0.46	0.21

References

1. Jankowski, R. Non-linear viscoelastic modelling of earthquake-induced structural pounding. *Earthq. Eng. Struct. Dyn.* **2005**, *34*, 595–611. [CrossRef]
2. Genovese, A.; Carputo, F.; Maiorano, A.; Timpone, F.; Farroni, F.; Sakhnevych, A. Study on the Generalized Formulations with the Aim to Reproduce the Viscoelastic Dynamic Behavior of Polymers. *Appl. Sci.* **2020**, *10*, 2321. [CrossRef]
3. Mark, J.E. *Physical Properties of Polymers Handbook*; Mark, J.E., Ed.; Springer: New York, NY, USA, 2007; ISBN 978-0-387-31235-4.
4. Sharma, V.; Agarwal, V. Polymer Composites Sustainability: Environmental Perspective, Future Trends and Minimization of Health Risk; IPCBEE Vol. 4; IACSIT Press: Singapore, 2011; Volume 4.
5. Lifshitz, J.M.; Leibowitz, M. Optimal sandwich beam design for maximum viscoelastic damping. *Int. J. Solids Struct.* **1987**, *23*, 1027–1034. [CrossRef]
6. Mottahed, B.D. Enhanced shielding effectiveness of polymer matrix composite enclosures utilizing constraint-based optimization. *Polym. Eng. Sci.* **2000**, *40*, 61–69. [CrossRef]
7. Farroni, F.; Genovese, A.; Maiorano, A.; Sakhnevych, A.; Timpone, F. *Development of an Innovative Instrument for Non-destructive Viscoelasticity Characterization: VESevo*; Springer: Cham, Switzerland, 2021; Volume 91.
8. Zhou, B.; Zhang, X. Comparison of five viscoelastic models for estimating viscoelastic parameters using ultrasound shear wave elastography. *J. Mech. Behav. Biomed. Mater.* **2018**, *85*, 109–116. [CrossRef]
9. Jrad, H.; Dion, J.L.; Renaud, F.; Tawfiq, I.; Haddar, M. Experimental characterization, modeling and parametric identification of the non linear dynamic behavior of viscoelastic components. *Eur. J. Mech. A/Solids* **2013**, *42*, 176–187. [CrossRef]
10. Alves, N.M.; Mano, J.F.; Gómez Ribelles, J.L. Molecular mobility in polymers studied with thermally stimulated recovery. II. Study of the glass transition of a semicrystalline PET and comparison with DSC and DMA results. *Polym. Guildf.* **2002**, *43*, 3627–3633. [CrossRef]
11. Palmeri, A.; Ricciardelli, F.; De Luca, A.; Muscolino, G. State Space Formulation for Linear Viscoelastic Dynamic Systems with Memory. *J. Eng. Mech.* **2003**, *129*, 715–724. [CrossRef]
12. Koeller, R.C. Applications of fractional calculus to the theory of viscoelasticity. *J. Appl. Mech. Trans. ASME* **1984**, *51*, 299–307. [CrossRef]
13. Shitikova, M.V. Fractional Operator Viscoelastic Models in Dynamic Problems of Mechanics of Solids: A Review. *Mech. Solids* **2022**, *57*, 1–33. [CrossRef]
14. Abouelregal, A.E. Thermo-viscoelastic properties in a non-simple three-dimensional material based on fractional derivative Kelvin–Voigt model. *Indian J. Phys.* **2022**, *96*, 399–441. [CrossRef]
15. Zhou, F.X.; Wang, L.; Liu, Z.; Zhao, W.C. A viscoelastic-viscoplastic mechanical model of time-dependent materials based on variable-order fractional derivative. *Mech. Time Depend. Mater.* **2021**, *26*, 699–717. [CrossRef]
16. Wang, P.; Yang, S.; Liu, Y.; Zhao, Y. Experimental Study and Fractional Derivative Model Prediction for Dynamic Viscoelasticity of Magnetorheological Elastomers. *J. Vib. Eng. Technol.* **2022**, *1*, 3. [CrossRef]
17. Friedrich, C. Understanding Viscoelasticity. *Appl. Rheol.* **2019**, *13*, 240–241. [CrossRef]
18. Flügge, W. *Viscoelasticity*; Springer: Berlin/Heidelberg, Germany, 1975; ISBN 978-3-662-02278-8.
19. Cucos, A.; Budrugeac, P.; Miu, L.; Mitrea, S.; Sbarcea, G. Dynamic mechanical analysis (DMA) of new and historical parchments and leathers: Correlations with DSC and XRD. *Thermochim. Acta* **2011**, *516*, 19–28. [CrossRef]
20. Ewoldt, R.H.; Hosoi, A.E.; McKinley, G.H. New measures for characterizing nonlinear viscoelasticity in large amplitude oscillatory shear. *J. Rheol.* **2008**, *52*, 1427–1458. [CrossRef]
21. Bonfanti, A.; Kaplan, J.L.; Charras, G.; Kabla, A. Fractional viscoelastic models for power-law materials. *Soft Matter* **2020**, *16*, 6002–6020. [CrossRef]
22. Renaud, F.; Dion, J.-L.; Chevallier, G.; Tawfiq, I.; Lemaire, R. A new identification method of viscoelastic behavior: Application to the generalized Maxwell model. *Mech. Syst. Signal Process.* **2011**, *25*, 991–1010. [CrossRef]
23. Carputo, F.; Genovese, A.; Sakhnevych, A. *Application of Generalized Models for Identification of Viscoelastic Behavior*; Springer: Cham, Switzerland, 2021; Volume 91.
24. Eftekhari, M.; Fatemi, A. On the strengthening effect of increasing cycling frequency on fatigue behavior of some polymers and their composites: Experiments and modeling. *Int. J. Fatigue* **2016**, *87*, 153–166. [CrossRef]
25. Ding, Y.; Sokolov, A.P. Breakdown of time—Temperature superposition principle and universality of chain dynamics in polymers. *Macromolecules* **2006**, *39*, 3322–3326. [CrossRef]
26. Williams, M.L.; Landel, R.F.; Ferry, J.D. The Temperature Dependence of Relaxation Mechanisms in Amorphous Polymers and Other Glass-forming Liquids. *J. Am. Chem. Soc.* **1955**, *77*, 3701–3707. [CrossRef]
27. Gaylord, N.G.; Van Wazer, J.R. *Viscoelastic Properties of Polymers*; John, D.F., Ed.; Wiley: New York, NY, USA, 1961.
28. Rouleau, L.; Pirk, R.; Pluymers, B.; Desmet, W. Characterization and modeling of the viscoelastic behavior of a self-adhesive rubber using dynamic mechanical analysis tests. *J. Aerosp. Technol. Manag.* **2015**, *7*, 200–208. [CrossRef]

Article

Implications of the Circular Economy in the Context of Plastic Recycling: The Case Study of Opaque PET

Noel León Albiter [1,*], Orlando Santana Pérez [1,*], Magali Klotz [1], Kishore Ganesan [2,3], Félix Carrasco [4], Sylvie Dagréou [5], María Lluïsa Maspoch [1] and César Valderrama [2,3]

1. Centre Català del Plàstic (CCP)—Universitat Politècnica de Catalunya Barcelona Tech (EEBE-UPC), ePLASCOM, Avda, Eduard Maristany, 14, 08019 Barcelona, Spain
2. Chemical Engineering Department, UPC-BarcelonaTECH, C/Eduard Maristany, 10-14 Campus Diagonal-Besòs, 08019 Barcelona, Spain
3. Barcelona Research Center for Multiscale Science and Engineering, C/Eduard Maristany, 10-14 (Campus Diagonal-Besòs), 08019 Barcelona, Spain
4. Department of Chemical Engineering, Universitat de Girona (UdG), C/Maria Aurèlia Capmany 61, 17003 Girona, Spain
5. CNRS, Institut Des Sciences Analytiques et de Physico-Chimie Pour l'Environnement et les Matériaux (IPREM), Université de Pau et des Pays de l'Adour, E2S UPPA, UMR5254, 64053 Pau, France
* Correspondence: noel.leon@upc.edu (N.L.A.); orlando.santana@upc.edu (O.S.P.)

Abstract: The use of recycled opaque PET (r-O-PET, with TiO_2) as a reinforcement for the recycled polypropylene matrix (r-PP) was evaluated through the life cycle assessment according to different scenarios corresponding to two different recycled blends and considered two virgin raw plastic material as reference materials when comparing the environmental performance of the proposed treatments. The results indicate that the environmental performance was quite different for each blend, since the additional extrusion process required in scenario 2 (blend with TiO_2) causes all impact categories analysed to report higher values when compared with scenario 1 (blend without TiO_2). The stage that contributes the most corresponds to the different extrusion processes included in both recycling blends, representing at least 80% of the total for global warming. Compared with virgin raw materials, the blend with TiO_2 showed better performance in all the impact categories analysed in comparison with virgin PA66, while the blend without TiO_2 showed the opposite trend when compared to PP. Furthermore, the fact that the upcycling treatment was carried out on a pilot scale provides room for improvement when implemented on a full scale. It is worth noting the high energy consumption of the treatment processes and their associated cost, in addition to the market cost of virgin raw materials, however, when considering the environmental cost of raw materials, it is observed that when substituting virgin materials PP and PA66 for the blends evaluated in this study results in a reduction of the environmental price of up to 2.5 times.

Keywords: circular economy; upcycling; sustainability; polyblends; rPP; PA66; rPET-O

1. Introduction

In the past decade, global awareness regarding environmental issues (global warming, climate change, resource depletion...) has increased considerably. Plastic pollution is of particular concern, given that due to their numerous benefits, plastics have become ubiquitous throughout society and, consequently, the amount of post-consumer plastic waste is increasing, leading to significant environmental drawbacks. In the absence of action, the amount of plastic waste produced globally is predicted to triple between 2015 and 2060, to between 155 and 265 million tonnes per year [1].

Governments and policymakers have started to understand the need to transition to more sustainable systems. Hence, in 2011 European Union designated resource efficiency as one of the flagships of its Europe 2020 Strategy (turning wastes into a resource) [2]

Citation: León Albiter, N.; Santana Pérez, O.; Klotz, M.; Ganesan, K.; Carrasco, F.; Dagréou, S.; Maspoch, M.L.; Valderrama, C. Implications of the Circular Economy in the Context of Plastic Recycling: The Case Study of Opaque PET. *Polymers* **2022**, *14*, 4639. https://doi.org/10.3390/polym14214639

Academic Editors: Jesús-María García-Martínez and Emilia P. Collar

Received: 7 July 2022
Accepted: 27 October 2022
Published: 31 October 2022

Publisher's Note: MDPI stays neutral with regard to jurisdictional claims in published maps and institutional affiliations.

Copyright: © 2022 by the authors. Licensee MDPI, Basel, Switzerland. This article is an open access article distributed under the terms and conditions of the Creative Commons Attribution (CC BY) license (https://creativecommons.org/licenses/by/4.0/).

and the EU settled in July 2014 the target of recycling at least 80% of plastic packaging waste by 2030 and banning burying recyclable waste in landfill as of 2025 [3]. In January 2018, Plastics Europe released its '2030 Plastic Voluntary Commitments', announcing its willingness to achieve the 100% re-use, recycling and/or recovery target of all plastics packaging in the EU-28, Norway and Switzerland by 2040 [4].

In this context, the circular economy (CE) approach promotes more effective use of materials by creating more value, for instance, promoting the cycle of high-value material instead of recycling only for law value raw materials as in traditional recycling [5,6]. Through various principled actions, it is possible to operate a closed-loop ecosystem and extend the life-cycle of products, equipment and infrastructure. Thus, improving the resource utilization, reducing waste, and energy consumption.

Regarding plastic packaging, the Ellen MacArthur Foundation's report [7] concreted a set of three priority actions to trigger the transition towards a new economy: (i) the fundamental redesign of 30% of plastic packaging that otherwise will never be reused or recycled; (ii) the reuse of at least 20% of plastic packaging; and (iii) the recycling of the remaining 50% of plastic packaging, radically improving their quality and economic attractiveness.

Despite the numerous solutions included in the CE approach, particular emphasis is still placed on the recycling of plastic packaging wastes in response to the EU targets, as well as the communities' aspirations towards "zero waste" cities. Nevertheless, the main technical difficulties of recycling are the variability of composition and the level of contamination of the waste stream.

Recycling of plastics involves many processes including collecting the wastes from the point of production or disposal, sorting, compressing, crushing and pelletizing them into raw materials. These procedures are followed by their thermal, chemical or mechanical processing to the final product. For this reason, the recycling of plastic waste is intricate and less preferred compared to other materials such as aluminium, glass, ceramics and paper [8]. Through shredding and grinding, plastic wastes are degraded during mechanical recycling [9]. The method is however not preferable if the mixture of wastes is complex and instead, incineration is preferred [10].

Recycled plastic can be reused in a closed loop material flow, reused in less critical applications (down-cycling), or upgraded for another use (up-cycling). The upgrading strategy generally comprises several elements of the polymer blend technology, viz alloying (i.e., compatibilization and/or impact modification), blending to the desired morphology, and compounding with other additives (e.g., stabilizers and fillers) [11].

Plastic milk bottles were traditionally made of high-density polyethylene (HDPE), but in order to achieve cost savings, the use of PET started to develop in the 2010s. In addition to the fact that it is cheaper than HDPE, the use of PET reduces the weight of the container by 25% and eliminates the aluminium seal on the bottle [12,13]. It also reduces water consumption by 20% and the energy consumption of the manufacturing process by 13%. Since PET bottles are normally transparent, between 10 to 20% w/w of TiO_2 is used as opacifying agent, conferring a screening effect from UV radiation of the content, minimising gas permeation, and a glossy white aspect when no other pigments are used.

The counterpart of opaque PET is its difficulty in being recycled by conventional processes used for transparent PET (r-T-PET), especially the bottle-to-fibre recycling process, which represents around 44% of its market share [14]. The French packaging compliance organisation Eco-Emballages stated that above a threshold of 15% by weight, the presence of recycled opaque PET (r-O-PET) in the PET bottle-to-fibre recycling stream leads to a series of difficulties during the filament manufacturing and significant deterioration of the mechanical properties of the fibres [13].

The presence of at least 0.35% w/w of TiO_2 is capable of retard strain hardening and accompanying stress-induced crystallization, when stretched in the rubbery state (temperatures between T_g and T_m), generating low levels of crystallinity and orientation. Apparently, the submicron TiO_2 particles interfere in the formation of the physical network necessary to promote the necessary hardening in the stabilization of the stretching stage during the manufacture of the fibre. Concentrations higher than 4% w/w generate a structure with poor mechanical properties [15]. That is why, as an alternative solution, recycling companies have chosen to market a grade of r-O-PET "diluted" with r-T-PET such that the TiO_2 content is around 2% w/w, which allows the manufacturing fibres with "acceptable" mechanical properties.

Taking into account that in 2016 opaque PET already represented an average of 12% by weight of the materials managed in the bottle-to-fibre recycling streams in France and that the market continues to grow. Therefore, if no action is taken recyclers will have no choice but to send opaque PET bottles to landfill or incineration, as the treatment systems cannot tolerate such quantities. Consequently, this waste stream will have a significant environmental impact and therefore will not follow the European strategies for plastics within the circular economy policy [16].

To address this growing issue, Eco-Emballages launched an action plan in 2017 [13], divided into three actions. One of these actions aims to search for an added-value market for r-O-PET. The case study presented in this work is framed within this context. The rationale behind the idea was to recover the opaque PET waste stream and improve its mechanical/physical properties and consequently its economic value while reducing its environmental impacts, which is in line with the CE approach.

In a previous study, it was observed that the presence of almost 2% w/w of TiO_2 on r-T-PET promotes an increase of 21% in the energy required for crack propagation. This effect is caused by the TiO_2 sub-micron particle cavitation, which allows a significant release of local triaxiality, promoting plastic the premature plastic tearing of the PET matrix [17]. Thus, one of the upgrading pathways proposed considered in this case study was to explore the use of r-O-PET as a reinforcement for recycled polypropylene (r-PP) matrix, inducing a microfibrillation of the PET phase during processing that acts as reinforcing fibres, under the manufacturing philosophy of in situ microfibrilated composites (MFC) [18]. According to his study, the blend composition that offers the best balance of mechanical properties and fracture behaviour of the MCM produced is the one with 20% w/w of r-O-PET. However, due to the intrinsic immiscibility of both polymeric phases, and considering the relatively low amount of TiO_2 compared to the actual content in the bottles, the inclusion of superficially modified TiO_2 in the blend was tested in order to evaluate its effectiveness to enhance the compatibility between the two polymeric phases. According to the results, the use of an hydrophobic treated particle until 12% w/w induces an emulsifying effect of the r-O-PET phase and good compatibilization in the elastic regime of mechanical performance [19].

The objective of this study is to evaluate two different solutions for the opaque PET waste stream, two different blends of r-O-PET have been proposed and produced on a pilot scale and then tested on an industrial scale to produce components for the electrical industry. In order to evaluate the environmental performance of the two blends, the life cycle assessment (LCA) methodology is used and, in addition, it is compared with virgin plastic materials of fossil origin that are commonly used to manufacture the same components. The aim is to evaluate the environmental cost over the life cycle of recycling the opaque PET stream and that of competing for virgin materials for the same application. The objective is also to illustrate the challenges that users/supporters of the CE approach must overcome when it comes to recycling plastic packaging, especially in terms of promoting high-added-value applications and creating new markets.

2. Materials and Methods

2.1. Materials and Scenarios

It is worth mentioning that this study is part of a project carried out in conjunction with the Institut Des Sciences Analytiques et de Physico-Chimie Pour l'Environnement et les Matériaux (IPREM) (Pau, France). The recycled raw material for the manufacture of the proposed blends corresponds to grades marketed by Suez RV Plastiques Atlantique

(Bayonne, France): flakes of a mixture of recycled O-PET mixed with recycled transparent PET to reach a nominal global content of 2% w/w TiO_2 (r-O-PET); and a recycled PP (r-PP) from the automotive industry. Both were sent from IPREM (Pau, France) to our facilities (Spain). However, for the life cycle inventory, the original origin of both (France and The Netherlands, respectively) was considered.

The composition of the blend was selected after a preliminary study carried out on a laboratory scale. In this study, it was established that the addition of 20% by weight of r-O-PET to r-PP offered the best balance of mechanical properties. The generation of a microfibrillated morphology of the PET phase during processing promotes a reinforcement in stiffness to the system. For this composition, a global amount of TiO_2 determined was 0.2% w/w. Additionally, and considering the possibility of increasing this TiO_2 content, a mixture with a global content of 4% w/w of TiO_2 was evaluated to compare its performance. Details of this study could be found in [18,19].

An industrial partner, specialising in the manufacture of injection moulded parts for different markets, carried out a study on possible parts that could be considered based on processability and mechanical performance. Two components used in the building construction industry were selected for a comparative study between the traditional petroleum-based plastic and the proposed recycled blends. The traditional virgin raw material employed in their manufacture is a Polypropylene heterophasic copolymer, ISPLEN 140 G2M (Repsol, Madrid, Spain) and a Polyamide 66, Zytel 101L BKB080 (Dupont, Wilmington, DE, USA).

Table 1 collects the values of the mechanical, physical and flow properties that were used for the selection of the comparison scenarios of this study, after the blends preparation in the pilot plant scale. It is important to note that the values reported for virgin materials (PP and PA66) correspond to the technical specifications provided by the raw material producers, while those of the blends were determined under the same conditions and following the same standards used for virgin PP.

Table 1. Physical, mechanical and flow properties of virgin raw materials and blends selected for the study.

Parameter	Material			
	PP	PA66	20w% r-O-PET/80% r-PP	20w% r-O-PET/76% r-PP/4% TiO_2
Density, ρ (@ 23 °C) (kg·m^{-3})	902	1120	940	1006
MFI [a] (dg·min^{-1})	5	–	3.8	8
Elastic modulus, E (MPa)	1200	1400	1220	1260
Yielding stress, σ_Y (MPa)	28	55	24	26
Elongation at yield (e_y) (%)	13	25	10	8
Specific modulus, E_{sp} (MPa·m^3·kg^{-1})	1.33	1.25	1.30	1.24

When selecting the scenarios for the present LCA analysis, an attempt was made to prioritize similarity in terms of:

(a) *Similarity of Specific elastic modulus (Esp) values:* Ratio between elastic modulus (E) and material density (ρ) which is a mechanical parameter used in structural design.

(b) *Similarity in the fluidity of the melt under normal processing conditions.* In the case of the parts made from virgin PP, the proximity between MFI was used as a criterion. In the case of PA66, as this parameter is not available, this comparison is meaningless. However, the conditions for the processing of both blends (similar to a PP) offer fewer drawbacks than those used for processing a typical PA66: there is no need for drying, there is no need to use screws with anti-return and anti-drip valves, and the temperature profile is more venébolo (230 °C, for PP vs. 290 °C for PA66).

In this way, the following study scenarios were selected:
Scenario 1: 20w% r-O-PET/80w% rPP vs. PP (Isplen)
Scenario 2: 20w% r-O-PET/76w% rPP/4% TiO$_2$ vs. PA66 (Zytel)

2.2. Blends Preparation at Pilot Plant Scale

The heterogeneity in the composition and geometry (flakes) of the received r-O-PET hinders any continuous forming process, which is why as the first step in blend preparation a homogenization of raw material by extrusion was performed. It used a KNETER 25 × 24D co-rotating twin screw extruder from COLLIN (Ebersberg, Germany) with a length-to-diameter ratio (L/D) of 36, a screw diameter (D) of 25 mm, and 7 heating zones (1 for die). A filament-type die of 3 mm nominal diameter was used to produce pellets after cooling in a water bathtub (1500 mm in length) at 20 °C. The processing conditions used were:

Temperature profile (°C): 180/220/240/245/245/250/250 °C (die)
Screws rotation speed: 55 rpm.

Under these processing conditions, the decrease in the Intrinsic Viscosity (IV) of the product was determined to be less than 10% [20].

The selected blend composition (r-PP/r-O-PET: 80/20) was prepared using the same twin screw extruder described above with the same extrusion conditions used in the homogenization step. Additionally, the same blend composition was prepared with an additional 4% by weight of TiO$_2$. For this, an r-PP/TiO$_2$ masterbatch was prepared and subsequently diluted with the required amounts of r-PP and r-O-PET until reach the desired proportion of additional TiO$_2$ in the resultant r-PP/r-O-PET-O. This methodology assures a better dispersion and distribution of the inorganic component.

The preparation of the masterbatch was carried out in an E-30/25 single screw extruder from IQAP-LAP (Barcelona, Spain) with an L/D ratio of 25, a screw diameter of 30 mm, and 4 heating zones (1 for die). A filament die diameter of 3 mm (for pelletizing) die was used. Prior to pellets cutting the filament was cooled in a water bathtube (1500 mm in length) at 20 °C. The following processing conditions:

Temperature profile (°C): 140/160/185/210 (die)
Screw rotation speed: 50 rpm.

It is important to quote several aspects on PET pre-conditioning before processing:

(a) Before each processing stage, PET must be dried to minimize its high tendency to hydrolytic thermos-degradation. In this case, drying was carried out in dried during 4 h at 120 °C in a PIOVAN hopper-dryer (DSN506HE, Venice, Italy) with a dew point of −40 °C, a common industrial device in PET processing. This procedure was carried out both for the homogenization of the r-O-PET flakes and for the preparation of the proposed blends.

(b) Due to the cooling conditions used during the flake homogenization stage, the r-O-PET obtained is in an amorphous state. In order to carry out its drying under the usual conditions (see point (a) above), a recrystallization process is required in order to avoid agglomeration of the pellets. Recrystallization was performed by heating the pellets in a Dry Big 2,003,740 convection oven (JP Selecta, Barcelona, Spain) at 90 °C for 4 h, taking care every 30 min to remove the pellets to avoid agglomeration in this step.

(c) To further minimize the inevitable hydrolytic thermodegradation of the PET phase during flake homogenization, an N$_2$ blanket was introduced into the feeding zone.

2.3. LCA Methodology

2.3.1. Functional Unit and System Boundaries

The functional unit is 1 kg of plastic granules produced at the gate of the treatment facility for both scenarios. It means that all the flows involved in the system were calculated or estimated in order to obtain 1 kg of the blends according to the composition listed in Table 1. The system boundaries include the processes from raw materials to plastic pellets, and the compounding with additives as well. The production of raw materials in both scenarios corresponds to the recycling stages of opaque PET and PP waste. The scope of the analysis is restricted from cradle to gate, as recycled blends and virgin materials are assumed to be equivalent for the market regarding their mechanical properties as well as their processing window.

The schematic flow charts of the processes under consideration in the LCI system boundaries for both scenarios are shown in Figure 1. As is conventionally conducted in literature for open-loop recycling processes, a "cut-off" approach is applied, therefore it is not considered the first life of plastic waste [21]. The main difference between the two scenarios is the presence of an additional extrusion process to prepare the r-PP/TiO$_2$ master batch in Scenario 2.

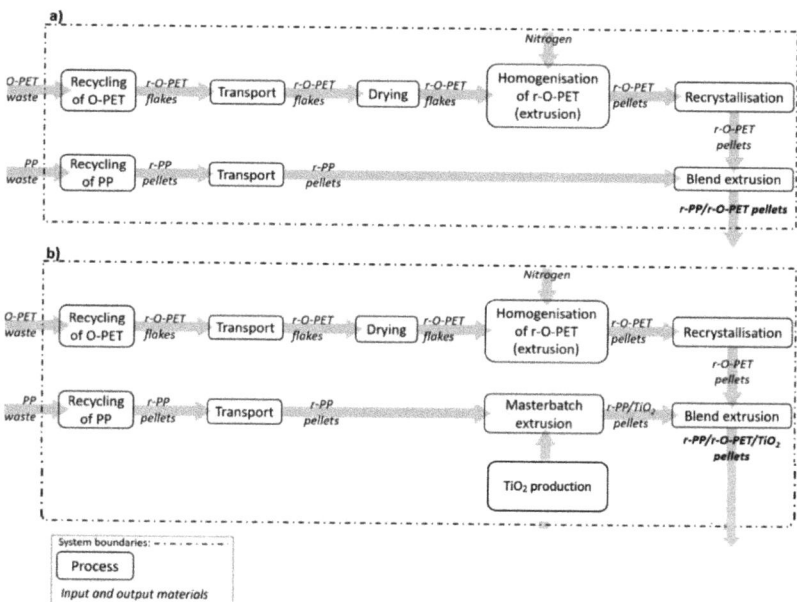

Figure 1. Schematic flow charts cradle-to-gate system boundaries for the recycling opaque PET waste for (**a**) Scenario 1 and (**b**) Scenario 2.

2.3.2. Life Cycle Inventory

The recycling stages of opaque PET (O-PET) and PP include the first phase of plastic waste collection, transport, initial sorting and balling, as well as the second phase of plastic waste reprocessing. First, the PET waste is unbaled, pre-washed in order to remove labels, sorted by colour and chopped into flakes. The obtained flakes are then washed, rinsed and dried. On the other hand, the PP waste is washed, sorted, shredded into flakes and pelletised. Therefore, the r-O-PET flakes are produced in France and r-PP pellets are produced in the Netherlands. It is assumed that they both are transported by lorry to Spain, where the blend preparations take place in a pilot-scale extrusion plant. The transport stages included in the system boundaries are listed in Table 2.

Table 2. Transport stages included in the system boundaries from plastic recyclers to pilot scale extrusion plants.

Material	Distance (km)	Transport Type
r-O-PET flakes	600	Transport, freight, lorry>32 metric tons, EURO5
r-PP pellets	1350	Transport, freight, lorry>32 metric tons, EURO5

The energy consumption during the pre-conditioning stage of r-O-PET (flakes and homogenized pellets) was estimated form the calibration curve Temperature vs. electrical power provided by the manufacturer of the Dryer hopper and convection oven.

The material loss for the extrusion processes is estimated between 7 and 10w%, due to the characteristics of the tests carried out at the pilot scale, the operation is not continuous and there are material losses every time the plant is started and during the stabilization process. The construction of plants and equipment, as well as the maintenance of plants and machinery and the end-of-life treatments of the waste generated by the blending processes, are out of the scope of the present study.

2.3.3. Input Data and Assumptions

The data for r-O-PET flakes and r-PP pellets production is derived from eco-profiles made by SRP (French syndicate of manufacturers of recycled plastics) in 2017 [22].

Although the data for recycled PET is estimated from the data for transparent PET, it is assumed that the results would be similar for the production of r-O-PET flakes. This assumption comes from the fact that uncoloured transparent PET and coloured transparent PET are already recycled separately, but no distinction is made between the two in the LCI data provided by SRP.

The dataset of r-PP pellets provided by SRP is estimated for France. It is assumed in this study that data would be similar for polypropylene recycled in the Netherlands.

On-site primary data collection was performed for the processes of flake homogenization, re-crystallization, masterbatch preparation, and r-O-PET blend preparation. In these processes, energy, water and resource consumption, waste generation and the quantities of intermediate products obtained in each process were determined. These values were used to prepare the inventory expressed in terms of the functional unit (1 kg of pellets) for the two scenarios.

The Ecoinvent database was used to model transports, electricity and heat production and the production of additional substances that are required for processes. The datasets provided by Ecoinvent used in the LCIA calculations are listed in Table 3. The data for petroleum-based polymers (virgin polymers) is derived from eco-profiles performed by Plastics Europe in 2014 [23,24].

Table 3. Database data used in the LCIA calculations.

Module Name	Source	Name of Dataset
Electricity, Spain	Ecoinvent 3.6	market for electricity, medium voltage | electricity, medium voltage | Cutoff, U-ES
Water	Ecoinvent 3.6	market for tap water | tap water | Cutoff, U
Non-hazardous waste (disposed)- O-PET	Ecoinvent 3.6	treatment of waste polyethylene terephthalate, sanitary landfill | waste polyethylene terephthalate | Cutoff, U
Non-hazardous waste (recycled) -O-PET	Ecoinvent 3.6	waste polyethylene terephthalate, for recycling, sorted
Electricity, France	Ecoinvent 3.6	market for electricity, medium voltage | electricity, medium voltage | Cutoff, U-FR
Non-hazardous waste (disposed) - PP waste	Ecoinvent 3.6	treatment of waste polypropylene, sanitary landfill | waste polypropylene | Cutoff, U

Table 3. Cont.

Module Name	Source	Name of Dataset
Electricity, Netherlands	Ecoinvent 3.6	market for electricity, medium voltage I electricity, medium voltage I Cutoff, U-NL
Transport, Spain	Ecoinvent 3.6	transport, freight, lorry >32 metric ton, EURO5 - RER
TiO_2	Ecoinvent 3.6	market for titanium dioxide I titanium dioxide I Cutoff, U
Material loss	Ecoinvent 3.6	treatment of waste polyethylene/polypropylene product, collection for final disposal I waste polyethylene/polypropylene product I Cutoff, U
Nitrogen	Ecoinvent 3.6	market for nitrogen, liquid I nitrogen, liquid I Cutoff, U

2.3.4. Environmental Impact Assessment

A conventional set of impact categories is defined in the CML IA baseline 2001< (Version 4.7) (Updated in August 2016) Method and is used to assess impacts on the mid-point levels: abiotic resource depletion potentials (ADPs) for fossil fuels and elements (ADP fossil, ADP elements), climate change (GWP100), eutrophication (EP), acidification (AP) and ozone layer depletion (ODP, steady state). This baseline characterisation method as well as the aforementioned impact categories were selected to allow a direct comparison with the environmental performances of virgin polymers, which were obtained through a secondary research database (Ecoinvent v3.8).

3. Results

3.1. Life Cycle Inventory and Environmental Performance of Recycling Scenarios

The flows of materials and products involved in the production of 1 kg of plastic blend pellets for both scenarios are summarized in Table 4. the results for ADP fossil, ADP elements, GWP100, EP, AP and ODP categories obtained for both scenarios are collected in Table 5.

Table 4. Inventory of inputs and outputs included in the system boundaries for the production of 1 kg of pellets (functional unit) for both recycling scenarios.

Process			Scenario 1	Scenario 2
O-PET waste recycling	Inputs	O-PET waste Auxiliary energy Auxiliary materials Water	0.35 kg 0.35 MJ 18.06 kg	0.35 kg 0.35 MJ 18.06 kg
	Outputs	Products r-O-PET flakes Waste Non-hazardous (disposed) Non-hazardous for recycling	 0.25 kg 0.19 kg 0.07 kg	 0.25 kg 0.19 kg 0.07 kg
PP waste recycling	Inputs	PP waste Auxiliary energy Auxiliary materials Water	0.90 kg 0.96 MJ 30.40 kg	0.96 kg 1.05 MJ 32.08 kg
	Outputs	Products r-PP pellets Waste Non-hazardous (disposed)	 0.89 kg 0.19 kg	 0.94 kg 0.20 kg
r-O-PET flakes transport	Inputs	r-O-PET flakes Transport	0.25 kg 0.15 t*km	0.25 kg 0.15 t*km
	Outputs	r-O-PET flakes (transported)	0.25 kg	0.25 kg

Table 4. Cont.

Process			Scenario 1	Scenario 2
r-PP pellets transport	Inputs	r-PP pellets Transport	0.89 kg 1.20 t*km	0.94 kg 1.27 t*km
	Outputs	r-PP pellets (transported)	0.89 kg	0.94 kg
Drying of r-O-PET flakes	Inputs	r-O-PET flakes (transported) Auxiliary energy Electric energy	0.25 kg 1.11 MJ	0.25 kg 1.11 MJ
	Outputs	r-O-PET flakes (dried)	0.25 kg	0.25 kg
Homogenisation of r-O-PET flakes	Inputs	r-O-PET flakes (dried) Auxiliary energy Electric energy Auxiliary materials Cooling water Nitrogen	0.25 kg 2.93 MJ 27.78 kg 0.07 kg	0.25 kg 2.93 MJ 27.78 kg 0.07 kg
	Outputs	Products r-O-PET pellets Residues Material loss	 0.22 kg 0.03 kg	 0.22 kg 0.03 kg
Re-crystallisation of r-O-PET pellets	Inputs	r-O-PET pellets Auxiliary energy Electric energy	0.22 kg 0.51 MJ	0.22 kg 0.51 MJ
	Outputs	r-O-PET pellets (crystallised)	0.22 kg	0.22 kg
Extrusion of r-PP/TiO2 masterbatch	Inputs	r-PP pellets (transported) TiO2 Auxiliary energy Electric energy Auxiliary materials Cooling water	- - - -	0.94 kg 0.05 kg 8.21 MJ 111.1 kg
	Outputs	Products r-PP/TiO2 masterbatch Residues Material loss	 - -	 0.89 kg 0.05 kg
Extrusion of final blend	Inputs	r-O-PET pellets (crystallised) r-PP pellets (transported) r-PP/TiO2 masterbatch Auxiliary energy Electric energy Auxiliary materials Cooling water	0.22 kg 0.89 kg - 12.6 MJ 125.0 kg	0.22 kg - 0.89 kg 12.6 MJ 125.0 kg
	Outputs	Products r-PP/r-O-PET pellets r-PP/r-O-PET/TiO2 pellets Residues Material loss	 1 kg - 0.11 kg	 - 1 kg 0.11 kg

The environmental performance of the blends is quite different for all the environmental categories summarized in Table 5, although the LCI for both is quite similar (Table 4). The additional extrusion process required in scenario 2 causes all impact categories assessed to report higher values when compared with scenario 1. This increase is significant across all impact categories. The contribution of each process involved in the treatment of the recycled blends is shown in Figure 2

Table 5. Life Cycle Impact Assessment Results for both recycling scenarios using the CML method.

Name	Scenario 1	Scenario 2	Units	Comparison (% of Increase of Impacts from Scenario 1 to Scenario 2)
Abiotic depletion	5.33×10^{-6}	9.38×10^{-6}	kg Sb eq	76%
Eutrophication	0.006	0.008	kg PO_4^{3-} eq	40%
Global warming (GWP100a)	2.39	3.57	kg CO_2 eq	50%
Acidification	0.01	0.03	kg SO_2 eq	80%
Ozone layer depletion (ODP)	1.29×10^{-7}	2.01×10^{-7}	kg CFC-11 eq	56%
Abiotic depletion (fossil fuels)	25.73	39.34	MJ	53%

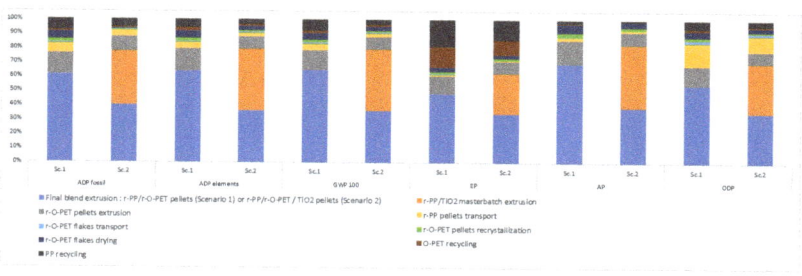

Figure 2. Contribution of the life cycle stages of the treatment processes to the environmental impact categories according to the CML method for both recycling scenarios.

The stage that contributes the most corresponds to the different extrusion processes included in both recycling blends, representing at least 80% of the total to the global warming, the abiotic elements, abiotic fossil, acidification potential, eutrophication and ozone depletion potential; it is due to the electricity consumption in these processes. It is worth mentioning that other stages such as recrystallization, drying and transportation of the pellets reported low contributions in all the impact categories evaluated.

In the case of Scenario 1, for all impact categories, the extrusion of the final blend represents approximately 60% of the total contribution of extrusion processes, meanwhile, the extrusion of r-O-PET pellets corresponds to 15%. The high contribution of extrusion processes is due to the electricity consumption, since it corresponds to approximately 84% of the total electric energy used in the system boundaries (Table 4).

In the case of Scenario 2, which includes an additional extrusion step, the total percentage of the extrusion processes contribution is higher than the one obtained for Scenario 1. Moreover, for all impact categories, an increase between 40% and 80% is observed (Table 5). The share of contribution between the different extrusion processes is also redistributed for all impact categories. The final blend extrusion step and the r-PP/TiO2 masterbatch extrusion are the major contributors (approximately 37% of the total contribution across categories), followed by the r-O-PET homogenization step (around 9%). It clearly illustrates the effect of the added extrusion step in Scenario 2, leading to an increase in extrusion processes contribution (Figure 2) and in consequence an increase in environmental impacts (Table 5). As in the case of Scenario 1, the aforementioned observations can be linked to the electrical energy consumption, the addition of the masterbatch extrusion process leads to an increase of 44% of the total electricity consumption (Table 4).

Regarding the contributions of the O-PET and PP recycling stages, it was observed they contribute mainly to the EP impact category, with each recycling stage contributing around 16% and 12% for scenarios 1 and 2, respectively. It is worth mentioning that these stages are carried out in France and that the French electricity mix is mainly dominated by nuclear energy (>70% of the electricity grid) [20]. Nuclear power plants use cooling water that is then discharged back into the environment at a temperature typically around 30–40 °C and this thermal water pollution can lead to eutrophication [25].

3.2. Comparison of Blends with Raw Virgin Materials

It is worth noting that the two blends (scenario 1 and scenario 2) have been assumed to be secondary raw materials that can replace virgin PP and virgin PA66, respectively, which means assuming that both products have the same quality and the necessary properties that allow their use in certain applications. It is also important to highlight the fact that the results presented in this study for recycled products were obtained experimentally in a pilot plant located in a specific location, while the results of the virgin polymers correspond to the ecoinvent database. Polypropylene granulate production for Europe has been used to represent virgin PP production and nylon 6 production for Europe (derived from the Eco-profiles of the European plastics industry (PlasticsEurope)) was used to represent virgin PA 66 production. Therefore, the results presented here mainly aim to provide elements for discussion and analysis on the conceptual development of the circular economy in the context of the recycling of plastics. The comparison of both blended products and the virgin raw materials is collected in Table 6 and depicted in Figure 3.

Table 6. Comparison between recycled blends and virgin polymers for Scenario 1 and Scenario 2.

Impact Category	Scenario 1		Scenario 2	
	r-PP/r-O-PET Blend	PP	r-PP/r-O-PET/TiO$_2$ Blend	PA66
ADP fossil (MJ)	25.73	68.35	39.34	104.31
ADP elements (kg Sb eq)	5.33×10^{-6}	1.33×10^{-5}	9.38×10^{-6}	6.53×10^{-5}
GWP 100 (kg CO$_2$ eq)	2.39	1.90	3.57	9.22
EP (kg PO$_4$ eq.)	5.94×10^{-3}	1.21×10^{-3}	8.33×10^{-3}	5.80×10^{-3}
AP (kg SO$_2$ eq.)	1.41×10^{-2}	5.43×10^{-3}	2.54×10^{-2}	2.96×10^{-2}
ODP (kg CFC11 eq.)	1.29×10^{-7}	2.24×10^{-8}	2.01×10^{-7}	7.19×10^{-9}

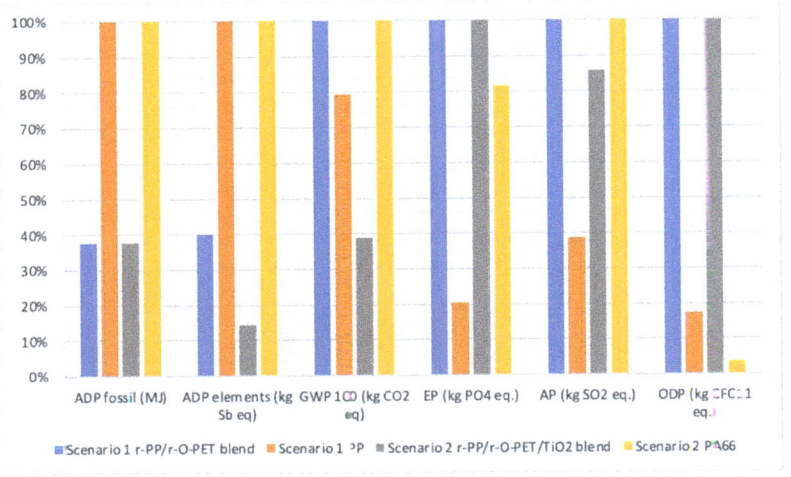

Figure 3. Relative indicator results of the impact categories for both scenarios. For each indicator, the maximum result is set to 100%.

Regarding Scenario 1, the blend without TiO$_2$ reduces fossil energy depletion by more than half compared to virgin PP. The recycled blend r-PP/r-O-PET shows a higher score than virgin PP for global warming (2.38 and 1.89 kg CO_2 eq, respectively). It can be explained, in the case of the recycled blend, by the high electricity consumption needed to produce 1 kg of plastic pellets and the low performance of the pilot plant. It is also important to stress the fact that data from the Spanish electricity grid were used for the recycled blend, meanwhile, a European mix was used for the virgin PP, as was reported by ecoinvent 3.8. The same trend is observed in the case of AP, EP, and ODP since the extrusion processes are the ones that contribute to these environmental impacts (Figure 2). Regarding the virgin PP, for all impact categories, the monomer production (i.e., crude oil, natural gas extraction and transport, the refinery, and the steam cracking and fluid catalytic cracking processes) contributes more than 80% of the total impact (ecoinvent v3.8). Despite the fact that the impact scores of the recycled blend of scenario 1 are higher than PP for most of the impact categories analysed, results are in the same order of magnitude, and the impact in these categories is expected to decrease when using full-scale industrial extruders with yield rates greater than 100 kg/h.

In the case of Scenario 2, the blend with TiO_2 shows better performance in almost all the impact categories analysed in comparison with virgin PA66, especially low in terms of ADP fossil and ADP elements, and nearly half reduction in GWP100. Here the same reflection can be made on the use of industrial extruders, since extrusion processes are the ones that contribute most to almost all the impact categories analysed. In view of these results and from the environmental perspective, the blend with TiO_2 (scenario 2) could be considered as a good alternative to replace virgin PA66, assuming that the characteristics and properties are adjusted to the requirements of the final application. As can be seen in Table 6, the environmental impact of PA66 is higher for all impact categories analysed in comparison with PP, scenario 1 and scenario 2. It is because PA66 is obtained from two monomers, adipic acid and hexametilendiamine (HMDA). The production stages of adipic acid and HMDA represent 44 and 49%, respectively, of the energy demand to produce PA66, and they contribute more than 60% of the total impact in all environmental impacts analysed [4].

4. Discussion: The Circular Economy Barriers and Challenges

Recycling is not the only element of the CE, but especially in the case of plastics, it will play an important role in the search for sustainable development, since it allows addressing multiple challenges that equally affect the economic, environmental and social dimensions. Recently, several authors have raised the barriers associated with the circular economy implementation and specifically for the plastic sector [26–29], due to the different types of plastics used and the quality and purity of the recycled secondary raw materials, some industrial branches demand high standards of raw material quality; hence, the acceptance of recycled plastic is rather low. However, other barriers must be taken into account: (i) economic, due to non-competitive prices of recycled polymers, oil price fluctuation and cost of waste management; (ii) legislative, the current regulatory framework does not yet adequately support the use of secondary plastics; (iii) social, manufactures are also sceptical about using resources coming from waste [30].

This study reports on the effort made at the pilot scale demonstration to propose two treatment routes for a specific type of plastic waste stream in order to upgrade the recycled material and generate secondary raw material that can be competitive, in terms of quality, with virgin materials available in the market. The environmental performance analysis indicates that it is necessary to use more resources (e.g., energy) in the material upcycling which represents a significant environmental impact of the recycling processes, despite that, the overall performance is better than the references virgin materials, especially for the scenario 2 (blend with TiO_2). However, the analysis must take into account other elements of the aforementioned barriers. In economics terms, the general outlook is not promising since the market cost of the virgin materials is 0.8 and 2 euros/kg for PP and PA66, respectively; while the cost of the energy consumption for both recycling scenarios is approximately 0.6 and 0.9 euros/kg (estimate made only for stages carried out in Spain, the recycling stages are not taken into account). It denotes the necessary consumption of resources that must be allocated to improve the quality of recycled plastics and highlights the difficulty of competing economically with the virgin products derived from the oil industry.

In addition to the market cost, environmental impacts can be externalized using the values from Bruyn et al. (2018) [31,32]. The methodology proposes a representative factor for each midpoint impact category that transforms the impact as such into an economic cost [33]. The comparison of the weighting of the environmental prices for the midpoint indicators (Recipe 2013 Hierarchist LCIA method) for the two scenarios analyzed and the virgin materials are collected in Tables 7 and 8. As can be seen, substituting the virgin materials PP and PA66 with the recycled materials from scenario 1 and scenario 2 results in an environmental price reduction of up to 2.5 times.

This also emphasizes the need for well-designed regulations to overcome barriers to the CE, the market cost of virgin plastics does not reflect the real cost of the product, internalizing the cost of all environmental externalities as seen in Table 7 throughout the plastic supply chain would allow plastic recycling processes to be more competitive with more attractive business models that would allow for a rapid implementation of CE strategies.

Making recycled plastic suitable for technical applications needs to go beyond treatment by developing high value-added, long-lasting applications that justify its high treatment cost and would better fit CE principles than the down-cycling approach. The project currently focuses its efforts on finding a method to enhance the r-O-PET quality through a reactive extrusion process. This improved r-O-PET could then be used to formulate high-value-added recycled plastics. Applications such as 3D printing will be explored, the use of polyolefins in 3D printing remains limited and the enhanced r-O-PET could be used as reinforcement for recycled polyolefins in order to improve their performance. Other applications such as thermal insulation will be looked into, by testing the suitability of the improved r-O-PET for physical foaming processes.

Table 7. Comparison between recycled blend (rPP/r-O-PET) and virgin PP for **Scenario 1** on environmental pricing [31,32].

Impact Category	Blend for Scenario 1	PP (Virgin)	Unit	Environmental Price per Environmental Impact Indicator	Unit	Blend for Scenario 1	PP (Virgin)
Climate change	2.35	1.83	kg CO_2-eq.	€0.06	€/kg CO_2-eq.	€0.13	€0.10
Ozone depletion	1.28×10^{-7}	2.25×10^{-8}	kg CFC-eq.	€30.40	€/kg CFC-eq.	€0.00	€0.00
Acidification	0.012	0.004	kg SO_2-eq.	€7.48	€/kg SO_2-eq.	€0.10	€0.04
Freshwater eutrophication	8.43×10^{-4}	2.47×10^{-4}	kg P-eq.	€1.86	€/kg P-eq.	€0.00	€0.00
Marine eutrophication	0.004	0.001	kg N	€3.11	€/kg N	€0.01	€0.00
Human toxicity	0.94	0.22	kg 1.4 DB-eq.	€0.10	€/kg 1.4 DB-eq.	€0.09	€0.02
Photochemical oxidant formation	7.89×10^{-3}	6.05×10^{-3}	kg NMVOC-eq.	€1.15	€/kg NMVOC-eq.	€0.01	€0.01
Particulate matter formation	4.70×10^{-3}	1.81×10^{-3}	kg PM10-eq.	€39.20	€/kg PM10-eq.	€0.18	€0.07
Terrestrial ecotoxicity	2.19×10^{-4}	2.06×10^{-5}	kg 1.4 DB-eq.	€8.69	€/kg 1.4 DB-eq.	€0.00	€0.00
Freshwater ecotoxicity	7.09×10^{-2}	2.63×10^{-2}	kg 1.4 DB-eq.	€0.04	€/kg 1.4 DB-eq.	€0.00	€0.00
Marine ecotoxicity	6.69×10^{-2}	2.32×10^{-2}	kg 1.4 DB-eq.	€0.01	€/kg 1.4 DB-eq.	€0.00	€0.00
Ionizing radiation	1.23	0.01	kg kBq U235-eq.	€0.05	€/kg kBq U235-eq.	€0.06	€0.00
Land use	3.42×10^{-5}	1.54×10^{-5}	m^2	€0.08	€/m^2	€0.00	€0.00
Abiotic depletion (fossil fuels)	25.73	63.35	MJ	€0.16	€/MJ	€4.12	€10.94
Total weigh LCA score using environmental pricing						€4.71	€11.19

Table 8. Comparison between recycled blend (rPP/r-O-PET/TiO$_2$) and virgin PA66 for **Scenario 2** on environmental pricing [31,32].

Impact Category	Blend for Scenario 2	PA 66 (Virgin)	Unit	Environmental Price per Environmental Impact Indicator	Unit	Blend for Scenario 2	PA 66 (Virgin)
Climate change	3.52	9.27	kg CO$_2$-eq.	€0.06	€/kg CO$_2$-eq.	€0.20	€0.52
Ozone depletion	2.00×10^{-7}	7.26×10^{-9}	kg CFC-eq.	€30.40	€/kg CFC-eq.	€0.00	€0.00
Acidification	0.022	0.027	kg SO$_2$-eq.	€7.48	€/kg SO$_2$-eq.	€0.17	€0.21
Freshwater eutrophication	1.32×10^{-3}	2.07×10^{-4}	kg P-eq.	€1.86	€/kg P-eq.	€0.00	€0.00
Marine eutrophication	0.006	0.009	kg N	€3.11	€/kg N	€0.02	€0.03
Human toxicity	1.39	0.08	kg 1.4 DB-eq.	€0.10	€/kg 1.4 DB-eq.	€0.14	€0.01
Photochemical oxidant formation	1.25×10^{-2}	2.83×10^{-2}	kg NMVOC-eq.	€1.15	€/kg NMVOC-eq.	€0.01	€0.03
Particulate matter formation	7.88×10^{-3}	9.49×10^{-3}	kg PM10-eq.	€39.20	€/kg PM10-eq.	€0.31	€0.37
Terrestrial ecotoxicity	2.82×10^{-4}	4.40×10^{-5}	kg 1.4 DB-eq.	€8.69	€/kg 1.4 DB-eq.	€0.00	€0.00
Freshwater ecotoxicity	0.10	1.94×10^{-2}	kg 1.4 DB-eq.	€0.04	€/kg 1.4 DB-eq.	€0.00	€0.00
Marine ecotoxicity	9.42×10^{-2}	7.53×10^{-3}	kg 1.4 DB-eq.	€0.01	€/kg 1.4 DB-eq.	€0.00	€0.00
Ionizing radiation	1.81	2.59×10^{-3}	kg kBq U235-eq.	€0.05	€/kg kBq U235-eq.	€0.08	€0.00
Land use	6.73×10^{-6}	0	m^2	€0.08	€/m^2	€0.00	€0.00
Abiotic depletion (fossil fuels)	39.34	104.31	MJ	€0.16	€/MJ	€6.29	€16.69
Total weigh LCA score using environmental pricing						€7.24	€17.87

It is worth noting that this is an analysis carried out within a European project on a pilot plant scale. If it is scaled to an industrial level, the environmental impacts would be even lower than those obtained in this study, it can be noted that:

- The r-O-PET homogenization stage: this first extrusion can be eliminated since the industrial devices allow the feeding of flakes without clogging problems, which in this study was necessary given the dimensions. This would make subsequent recrystallization (prior to the preparation of the mixtures) not required.
- On an industrial scale, the energy efficiency of the production per kg of material and consequently the electrical consumption would be much lower.
- The recycled material would come from nearby areas, which would reduce the impact on transportation.

5. Conclusions

This study evaluates two routes for treating and upcycling the O-PET waste stream, which creates problems when recycled together with transparent PET in order to produce secondary raw materials that can be used for value-added applications following the circular economy approach. Two scenarios were proposed based on the composition of the recycled blends. The treatments included the processes from raw materials (recycling stages of opaque PET) to plastic pellets and were conducted experimentally at a pilot scale. Both recycled blends were evaluated using the LCA methodology with a scope from cradle to gate and considered two virgin raw materials (plastic) as reference materials when comparing the environmental performance of the proposed treatments.

The results indicate that the environmental performance of the two mixtures was slightly different, although the treatments were similar with an additional extrusion process in Scenario 2. Higher scores were reported for most of the impact categories for the blend with TiO$_2$ mainly due to the energy consumption in the extrusion process. In terms of the comparison with virgin materials, the results indicated that PP reported better performance than the plastic mixture without TiO$_2$, while PA66 reported high scores for all impact categories compared to the mixture with TiO$_2$. However, it is worth mentioning that both treatments were carried out on a pilot scale, which provides a significant margin for improvement when using high-performance large-scale industrial extruders In CE terms, it is confirmed that upcycling deserves a significant number of resources to improve the quality of secondary raw materials. In this sense and from the economic point of view, the market cost of raw materials is significantly low, compared to the cost of treating plastic blends and raises the need to address the challenges associated with the application of the CE.

The externality cost associated with the impact of raw materials appears as one of the elements to be considered through a well-designed regulation, since internalizing the environmental externality costs of the products derived from the oil industry would allow the production of secondary raw materials to be more competitive, and would promote the interest of developing new business models for recyclers and end users of these materials.

Author Contributions: The general conceptualization of the work described here was performed by O.S.P., N.L.A., M.L.M. and C.V. Processing and blending preparations were carried out by N.L.A. and F.C. under the supervision of O.S.P. All Inventory inputs were performed by N.L.A., M.K., F.C. and S.D. their analysis was done by C.V. and K.G. The paper was written by C.V., N.L.A., O.S.P. and M.L.M. All authors have read and agreed to the published version of the manuscript.

Funding: This research was funded by Spanish Ministerio de Ciencia e Innovación through the Project: EcoBlends'UP (Ref: PID2019-1006518RB-I00) and the European Regional Development Fund (FEDER) within the POCTEFA 2014-2020 program through the project: REVALPET'UP (Ref : EFA329/19).

Institutional Review Board Statement: Not applicable.

Informed Consent Statement: Not applicable.

Data Availability Statement: The data presented in this study are available on request from the corresponding author.

Acknowledgments: We thank to Suez (Bayonne, France) for r-PET-O commercial grade supplied and recycled PP.

Conflicts of Interest: The authors declare no conflict of interest.

References

1. Tayeh, B.A.; Almesha, I.; Magbool, H.M.; Alabduljabbar, H.; Alyousef, R. Performance of sustainable concrete containing different types of recycled plastic. *J. Clean. Prod.* **2021**, *328*, 129517. [CrossRef]
2. European Comission Document 52011DC0571- Roadmap to a Resource Efficient Europe. 2011. Available online: https://eurlex.europa.eu/legalcontent/EN/TXT/?uri=CELEX%3A52011DC0571 (accessed on 10 May 2022).
3. European Comission. Environment: Higher Recycling Targets to Drive Transition to a Circular Economy with New Jobs and Sustainable Growth. 2014. Available online: https://ec.europa.eu/commission/presscorner/detail/en/IP_14_763 (accessed on 10 May 2022).
4. Plastics Europe. Plastics 2030—PlasticsEurope's Voluntary Commitment to Increasing Circularity and Resource Efficiency. 2018. Available online: https://www.plasticseurope.org/application/files/7215/1715/2556/20180116121358-PlasticsEurope_Voluntary_Commitment_16012018_1.pdf (accessed on 10 May 2022).
5. Korhonen, J.; Nuur, C.; Feldmann, A.; Eshetu, S. Circular economy as an essentially contested concept. *J. Clean. Prod.* **2018**, *175*, 544–552. [CrossRef]
6. Laubscher, M.; Marinelli, T. Integration of Circular Economy in Business. In Proceedings of the Going Green—Care Innovation 2014, Vienna, Austria, 17–20 November 2014. [CrossRef]
7. Ellen MacArthur Foundation. The New Plastics Economy: Catalysing Action. 2017. Available online: https://ellenmacarthurfoundation.org/the-new-plastics-economy-catalysing-action (accessed on 10 May 2022).
8. Nyika, J.; Dinka, M. Recycling plastic waste materials for building and construction Materials: A minireview. *Mater. Today Proc.* **2022**, *62*, 3257–3262. [CrossRef]
9. Singh, N.; Hui, D.; Singh, R.; Ahuja, I.P.S.; Feo, L.; Fraternali, F. Recycling of plastic solid waste: A state of art review and future applications. *Compos. Part B Eng.* **2017**, *115*, 409–422. [CrossRef]

10. Khoo, H. LCA of plastic waste recovery into recycled materials, energy and fuels in Singapore. *Conserv. Recycl.* **2019**, *145*, 67–77. [CrossRef]
11. Utracki, L.A. Role of polymer blends' technology in polymer recycling. In *Polymer Blends Handbook*; Kluwer Academic Publishers: Amsterdam, The Netherlands, 2003; pp. 1117–1165.
12. Chauvot, M. Les Emballages Ménagers, une Illustration des Dérives du Système. Les Echos. 2016. Available online: https://www.lesechos.fr/2016/11/les-emballages-menagers-une-illustration-des-derives-du-systeme-228559 (accessed on 10 May 2022).
13. Eco-Emballages. PET Opaque: Le Programme D'actions d'Eco-Emballages pour 2017 [WWW Document]. 2017. Available online: http://www.ecoemballages.fr/sites/default/files/20170201-cp-eco-emballages-petopaque.pdf (accessed on 10 May 2022).
14. Kaynap, H.K.; Sarioglu, E. PET bottle recycling for sustainable textiles. In *Polyester—Production, Characterization and Innovative Applications*; Camblible, N.O., Ed.; IntechOpen: London, UK, 2017. [CrossRef]
15. Taniguchi, A.; Cakmak, M. The suppression of strain induced crystallization in PET through sub micron TiO_2 particle incorporation. *Polymer* **2004**, *45*, 6647–6654. [CrossRef]
16. European Commission. Document: 52018DC0028 A European Strategy for Plastics in a Circular Economy. COM(2018)28 Final. 2018. Available online: https://eur-lex.europa.eu/legal-content/EN/TXT/?uri=COM:2018:28:FIN (accessed on 10 May 2022).
17. Loaeza, D.; Cailloux, J.; Pérez, O.S.; Sánchez-Soto, M.; Maspoch, M. Impact of Titanium Dioxide in the Mechanical Recycling of Post-Consumer Polyethylene Terephthalate Bottle Waste: Tensile and Fracture Behavior. *Polymers* **2021**, *13*, 310. [CrossRef] [PubMed]
18. Loaeza, D.; Cailloux, J.; Pérez, O.S.; Sánchez-Soto, M.; Maspoch, M.L. Extruded-Calendered Sheets of Fully Recycled PP/Opaque PET Blends: Mechanical and Fracture Behaviour. *Polymers* **2021**, *13*, 2360. [CrossRef] [PubMed]
19. Matxinandiarena, E.; Múgica, A.; Zubitur, M.; Yus, C.; Sebastián, V.; Irusta, S.; Loaeza, A.D.; Santana, O.; Maspoch, M.L.; Puig, C.; et al. The Effect of Titanium Dioxide Surface Modification on the Dispersion, Morphology, and Mechanical Properties of Recycled PP/PET/TiO_2 PBNANOs. *Polymers* **2019**, *11*, 1692. [CrossRef] [PubMed]
20. IEA World Energy Balances, 2018. Total Primary Energy Supply (TPES) by Source—France. Available online: https://www.iea.org/statistics/?country=FRANCE&year=2016&category=Energysupply&indicator=TPESbySource&mode=chart&dataTable=BALANCES (accessed on 10 May 2022).
21. Ligthart, T.N.; Ansems, T.A.M.M. Modelling of Recycling in LCA. In *Post-Consumer Waste Recycling and Optimal Production*; IntechOpen: London, UK, 2012; pp. 186–210. [CrossRef]
22. ICV des MPR. 2017. Available online: http://www.srp-recyclage-plastiques.org/index.php/donnees-recyclage/icv-des-mpr.html (accessed on 15 January 2022).
23. Plastics Europe. Polyamide 6.6 (PA6.6). Eco-Profiles Environ. Prod. Declar. Eur. Plast. Manuf. 2014. Available online: https://plasticseurope.org/sustainability/circularity/life-cycle-thinking/eco-profiles-set/ (accessed on 26 October 2022).
24. Plastics Europe. Polypropylene (PP). Eco-Profiles Environ. Prod. Declar. Eur. Plast. Manuf. 2016. Available online: https://plasticseurope.org/sustainability/circularity/life-cycle-thinking/eco-profiles-set/ (accessed on 26 October 2022).
25. Kirillin, G.; Shatwell, T.; Kasprzak, P. Consequences of thermal pollution from a nuclear plant on lake temperature and mixing regime. *J. Hydrol.* **2013**, *496*, 47–56. [CrossRef]
26. Simon, B. What are the most significant aspects of supporting the circular economy in the plastic industry? *Resour. Conserv. Recycl.* **2019**, *141*, 299–300. [CrossRef]
27. Paletta, A.; Filho, W.L.; Balogun, A.-L.; Foschi, E.; Bonoli, A. Barriers and challenges to plastics valorisation in the context of a circular economy: Case studies form Italy. *J. Clean. Prod.* **2019**, *241*, 118149. [CrossRef]
28. Gall, M.; Wiener, M.; de Oliveira, C.C.; Lang, R.W.; Hansen, E.G. Building a circular plastics economy with informal waste pickers: Recyclate quality, business model, and societal impacts. *Resour. Conserv. Recycl.* **2020**, *156*, 108645. [CrossRef]
29. Robaina, M.; Murillo, K.; Rocha, E.; Villar, J. Circular economy in plastic waste–Efficiency analysis of European countries. *Sci. Total Environ.* **2020**, *730*, 139038. [CrossRef] [PubMed]
30. Polymer Comply Europe. The Usage of Recycled Plastics Materials by Plastics Converters in Europe: A Qualitative European Industry Survey. 2017. Available online: https://www.ahpi.gr/wp-content/uploads/2019/01/PCE-Report-2nd-EuPC-Survey-on-the-Use-of-rPM-by-European-Plastics-Converters-v.1_compressed.pdf (accessed on 10 May 2022).
31. Bruyn, S.; de Bijleveld, M.; Graaff, L.; de Schep, E.; Schroten, A.; Vergeer, R.; Ahdour, S. *Environmental Prices Handbook*; CE Delft Publication: Delft, The Netherlands, 2018.
32. Environmental Cost Indicator (ECI)—Overview. Ecochain 2022. Available online: https://ecochain.com/knowledge/environmental-cost-indicator-eci/ (accessed on 5 August 2022).
33. Carranza, G.; Do Nascimento, M.; Fanals, J.; Febrer, J.; Valderrama, C. Life cycle assessment and economic analysis of the electric motorcycle in the city of Barcelona and the impact on air pollution. *Sci. Total Environ.* **2022**, *821*, 153419. [CrossRef] [PubMed]

Article

Influence of Stabilization Additive on Rheological, Thermal and Mechanical Properties of Recycled Polypropylene

Mohor Mihelčič [1], Alen Oseli [1], Miroslav Huskić [2] and Lidija Slemenik Perše [1,*]

1 Faculty of Mechanical Engineering, University of Ljubljana, Aškerčeva Ulica 6, 1000 Ljubljana, Slovenia
2 Faculty of Polymer Technology, Ozare 19, 2380 Slovenj Gradec, Slovenia
* Correspondence: lidija.slemenik.perse@fs.uni-lj.si

Abstract: To decrease the amount of plastic waste, the use of recycling techniques become a necessity. However, numerous recycling cycles result in the mechanical, thermal, and chemical degradation of the polymer, which leads to an inefficient use of recycled polymers for the production of plastic products. In this study, the effects of recycling and the improvement of polymer performance with the incorporation of an additive into recycled polypropylene was studied by spectroscopic, rheological, optical, and mechanical characterization techniques. The results showed that after 20 recycling steps of mechanical processing of polypropylene, the main degradation processes of polypropylene are chain scission of polymer chains and oxidation, which can be improved by the addition of a stabilizing additive. It was shown that a small amount of an additive significantly improves the properties of the recycled polypropylene up to the 20th reprocessing cycle. The use of an additive improves the rheological properties of the recycled melt, surface properties, and time-dependent mechanical properties of solid polypropylene since it was shown that the additive acts as a hardener and additionally crosslinks the recycled polymer chains.

Keywords: recycling; polypropylene; stabilization additive; rheological properties; thermal properties; mechanical properties

1. Introduction

In today's world, with the growing global prosperity, plastics have become a major part of our everyday life. The raw materials that are necessary for the production of such large quantities of plastic are, however, limited. Moreover, an annual increase in plastic waste made from nonrenewable sources has a negative impact on the environment [1]. The easiest way to decrease the environmental impact is mechanical recycling, which enables the transformation of plastic waste back to useful products. However, studies show that less than 15% of plastics in the world have been recycled. One of the polymers, which is easy to recycle, is polypropylene. It is the world's second-most produced thermoplastic polymer. Due to its good mechanical properties, it is widely used in the automotive, aerospace, and packaging industries; furniture; textiles, etc. [2–5] Moreover, it is highly adaptive to a wide range of different applications and types of production; therefore, PP has maintained a significant market share compared to other polymers. The problem with PP and most of the thermoplastic polymers is a deterioration of their mechanical properties caused by polymer chain degradation induced by mechanical reprocessing.

The degradation of polyolefins could be, in general, divided into three types: mechanical degradation, thermal degradation, and thermal oxidative degradation [6,7]. The changes due to the recycling process are visible through continued deterioration of physical properties, such as molecular structure, viscosity, degree of crystallinity, or Young's modulus [8–11]. It is known that thermo-mechanical reprocessing causes chain scission of the PP, leading to a decrease in thermal stability, melting temperature, viscosity, and viscoelastic properties with increasing reprocessing cycles [12–14]. One of the solutions

for maintaining or improving the mechanical properties during recycling is the addition of various additives to recycled plastics. Similarly, the additives that can improve several properties could be included already to the virgin polymers.

The research on the improvement of polypropylene properties is vast; however, many studies have been performed on plastic additives focusing mostly on mineral-based (reinforcing fillers), i.e., $CaCO_3$, talc, kaolin, mica [15] [16–18], as well as antioxidant stabilizers [19,20], and others. One of the useful additives is also CaO, which is a water-absorbing additive and neutralizes acidity [21]. Among various polymers, the polypropylene is slightly more susceptible to attack by strong oxidizing agents. It is known that the addition of antioxidants protects polymers at their first processing cycle, while further research of antioxidant stability during further repetitive extrusion cycles has rarely been studied. It has been reported that some hindered amines (HALS), and phenolic and phosphorous compounds, such as Tinuvin, Irganox, or Irgafos, could be used to improve the antioxidant properties of PP stability [19]. Among many characterization methods, ATR-IR spectrometry has already become a routine analysis for the quantitative characterization of antioxidative PP stability [22]. The studies reported in the literature are mainly focused on the rheological and mechanical properties of neat or recycled polymeric materials; however, according to our knowledge, there is a lack of investigations on the degradative effects in stabilized recycled polypropylene.

For this reason, our study focuses on the effect of recycling on rheological properties (indicating structural changes in polymer melt), thermal properties (related to phase transitions and crystallinity), and mechanical properties (associated with the practical use of solid polymer products). Moreover, the improvement of polymer performance with the incorporation of an additive into recycled polypropylene was studied with polypropylene, which was recycled by extensive mechanical processing.

2. Materials and Methods

2.1. Materials and Sample Preparation

In this study, the polypropylene (PP) homopolymer, Buplene 6331, purchased by Lukoil (LUKOIL Bulgaria Ltd., Sofia, Bulgaria; MFI of 8–16 g/10 min at 230 °C and 2.16 kg) was used. Commercially available Recyclobyk 4371 additive (provided by BYK-Chemie GmbH, Wesel, Germany) was used for the stabilization of recycled PP. The additive is usually used for the stabilization of polyolefin blends, i.e., recyclates from the automotive industry and the packaging industry; fiber-reinforced plastics with recyclates; and for the reduction of VOC and odors. The additive was purchased in a pellet form as a mixture of antioxidants and co-stabilizers. According to the producer, the additive helps to improve the mechanical properties of recycled polymer and helps to neutralize acids that may be present in the polymer from previous processing and use.

Mechanical recycling of PP samples was performed by using a co-rotating twin-screw extruder, Polylab PTW 16/40 OS (Thermo Scientific, Karlsruhe, Germany), with the screw speed set at 90 rpm. The temperatures on the ten heaters of the extruder were set ascending with the temperature of the first heater at 150 °C, while the last heater was set to a process temperature (205 °C).

After drying at 90 °C for 4 h, the virgin (as received) PP was reprocessed 10 times into 2-mm-long granules. After the 10th cycle, the granulated rPP batch was halved into two batches. One-half of the batch was characterized as the batch without additive, while 0.75% wt. of Recyclobyk 4371 additive was added to the second half in order to improve the mechanical properties of the recycled material. The recycled batches without (denoted as rPP) and with additive (denoted as rPP_a) were extruded and regranulated for ten more cycles (Table 1). The additive was introduced to PP, which was recycled for 10 times since our intention was to test if a degraded PP could be re-stabilized.

Table 1. Sample classification and corresponding abbreviations.

Sample	Abbreviation
Virgin PP	rPP_0
5× reprocessed PP	rPP_5
10× reprocessed PP	rPP_10
15× reprocessed PP	rPP_15
20× reprocessed PP	rPP_20
11× reprocessed PP with additive	rPP_11a
15× reprocessed PP with additive	rPP_15a
20× reprocessed PP with additive	rPP_20a

Test bars (L = 60, w = 10, h = 1 mm) for the dynamic mechanical analysis (DMA) were injection-molded on a Haake MiniJet II (Thermo Scientific, Karlsruhe, Germany) injection-molding machine. Samples were prepared by using the processing parameters presented in Table 2.

Table 2. Process parameters for injection-molding of test bars for DMA.

Process	Parameter
Melt temperature (Tm)	190 °C
Waiting time	3 min
Mold temperature	50 °C
Holding pressure	500 bar
Injection time	10 s
Post-pressure	100 bar
Post-processing time	10 s
Melt temperature (Tm)	190 °C

DMA test bars were used for all measurements, except for rheological characterization, where virgin or extruded granules were directly placed in the sensor system. Prior to creep and DMA testing, the samples were thermally annealed in order to erase the residual stresses. The annealing was performed at 110 °C for 2 h followed by cooling to room temperature at 0.1 °C/min.

2.2. Methods
2.2.1. ATR-IR Characterization

ATR-IR measurements were performed on a Perkin Elmer, Spectrum 65 (Waltham, MA, USA) FTIR spectrometer, equipped with a single reflection diamond crystal at room temperature. ATR-IR spectra were recorded in a range from 4000 to 600 cm^{-1} at 64 scans per spectrum with a resolution of 4 cm^{-1}.

2.2.2. Polarized Optical Microscopy

The phase transformation of PP was investigated using a Carl Zeiss AxioScope polarizing optical microscope (Carl Zeiss, Braunschweig, Germany) equipped with a 200× objective and a Linkam TS600 Heating Stage. For optical characterization, the samples were prepared as 10-μm-thick films using a microtome. The films were heated between a glass slide and coverslip to 200 °C and kept at constant temperature for 5 min. Subsequently, the samples were rapidly cooled to 130 °C for isothermal crystallization.

2.2.3. Thermal Properties

Differential Scanning Calorimetry (DSC) measurements of the samples were carried out on TA Instruments DSC Q2500 (TA Instruments, New Castle, DE, USA). Samples weighing 6 ± 1 mg were cut out from injection-molded DMA bars and placed into aluminium cups. Measurements were performed from −50 °C to 250 °C at a rate of 10 K min^{-1} followed by a 5 min range of constant temperature at 250 °C and cooling to −50 °C at a rate of 10 K min^{-1}. At the lowest temperature, the samples were exposed to a constant temperature again for 5 min; in this case, at -50 °C. Second heating, which was also used for the analysis, was conducted to obtain results independent from the thermal and mechanical history of the samples. The inert atmosphere with a nitrogen gas flow of 50 mL·min^{-1} was provided during the measurements. All measurements were repeated 3 times and the average values were used for the analyses. The results of the DSC tests were used to determine the transition temperatures and the degree of crystallinity (X_C) as:

$$X_C = \frac{\Delta H_m}{\Delta H_m^0}, \qquad (1)$$

where ΔH_m is the experimentally-determined enthalpy of melting and ΔH_m^0 is the theoretical value of the melting enthalpy of 100% crystalline PP (207 J/g) [23].

2.2.4. Rheological Properties

The rheological characterization of PP and rPP polymer melts were conducted at a constant temperature of 190 °C using the Anton Paar MCR 302 rotational rheometer (Anton Paar, Graz, Austria) equipped with parallel-plate geometry (plate diameter of 25 mm with a 1 mm gap) in inert nitrogen atmosphere. For the analysis, the samples in granulated form were directly placed on the lower plate of the sensor system. The viscoelastic properties (storage modulus—G' and loss—G") of the studied polymer melts were determined using amplitude and frequency oscillatory tests. The frequency sweep experiments were carried out from 100 to 0.01 Hz using a constant strain of 5%, which was within the linear viscoelastic region (LVR) previously determined from amplitude tests. The average molecular weight (M_w) and polydispersity index (PDI = M_w/M_n) were determined from rheological frequency tests using the *Rheocompass* software (v.1.30).

2.2.5. Mechanical Characterization

Nanoindentation

The hardness and elastic modulus of the surface of injection-molded samples were determined using a Nanoindenter G200 XP instrument manufactured by Agilent (Santa Clara, CA, USA). For the measurements, a standard three-sided pyramidal Berkovich probe was used. The mechanical properties were determined by using the CSM (Continuous Stiffness Measurement) [24] method with a tip oscillation frequency of 45 Hz and a 2 nm harmonic amplitude. All measurements were conducted at room temperature. For virgin and recycled PP material, 36 indentations on each sample were performed with a 100 µm distance between the indentations, which enabled the exclusion of interaction effects. The measuring depth was 2000 nm and the average values between 800 and 1800 nm in depth were used in the presented results.

Time-Dependent Measurements

Creep tests were conducted on the Mars Haake II Rheometer (Thermo Scientific, Karlsruhe, Germany) equipped with a controlled-temperature test chamber and solids-clamping tool. Shear creep compliance *J(t)* tests were carried out in a linear viscoelastic region at a constant shear stress, τ, of 0.01 MPa. Measurements were performed in the segmental form at nine different temperatures between 20 and 100 °C with a step of 10 °C. A time-temperature superposition (tTS) principle was used for the characterization of the time-dependent behavior of rPP, with and without additive, for the improvement of mechanical properties (for a detailed analysis protocol, see Ref. [25]).

Dynamic Mechanical Thermal Analysis (DMTA)

For all the mechanical characterizations, at least three samples (L = 50, w = 10, h = 1 mm) for each recycling set were tested in the temperature range from -18 °C to 110 °C, with a heating rate of 3 °C/min and an oscillation frequency of 1 Hz. The shear strain during the measurements was logarithmically increased from 0.02 to 0.1%. DMTA tests were performed using the Anton Paar MCR 702 (Anton Paar, Graz, Austria). The results enabled the determination of the glass transition temperature (T_g), which was determined at the temperature where tan, δ, exhibited a maximum value.

3. Results and Discussion

3.1. ATR-IR Characterization

IR spectroscopy is one of the most common methods for studying the degradation and stabilization of PP under various conditions [26,27]. The degradation behavior of recycled PP with and without additive was studied using ATR-IR spectroscopy, and the results are shown in Figure 1. The virgin PP (rPP_0) shows characteristic groups in the polymer chain with CH_3 stretching bands at 2950 and 2867 cm^{-1}, and CH_2 stretching bands at 2917 and 2838 cm^{-1} [28]. The CH_3 banding vibration peak can be assigned to asymmetric $δ_{as}$(C-H) at 1456 cm^{-1} and the CH_2 banding vibration to symmetric deformation $δ_s$(C-H) at 1376 cm^{-1} (Figure 1). The bands at 841, 997, and 1168 cm^{-1} are ascribed to vibrations characteristic of isotactic polypropylene [29]. Moreover, the shoulder peak at 2960 cm^{-1} ($ν_{as}$(methyl group -CH_3) increases with the increasing number of recycles, and is much more pronounced at rPP_20 (without additive). The findings suggest that the bands, 2960 cm^{-1} and 2837 cm^{-1}, are the recycling-sensitive bands, which were already confirmed by other authors [30]. This could be an identification of the chains scission of PP during multiple processing cycles.

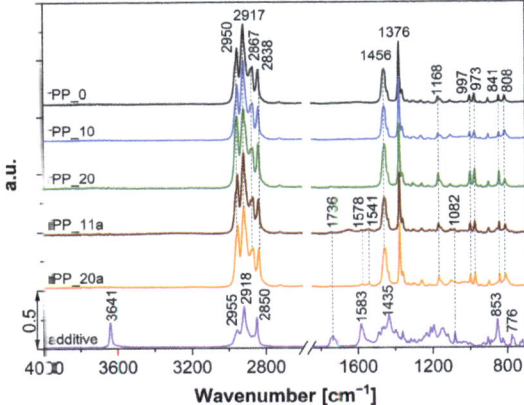

Figure 1. ATR-IR spectra of virgin PP, recycled with (rPP_a) and without (rPP) additive.

The ATR-IR spectra of the additive (Recyclobyk 4731), which is a mixture of antioxidants and co-stabilizers, showed a peak at 3641 cm^{-1}, which can be assigned to isolated OH groups present at the surface of the CaO solid and peak at 776 cm^{-1}, possibly due to the Ca–O bonds. The band at 1583 cm^{-1} corresponds to the carboxylate bond (–COO$^-$) and it disappeared after the first recycling using an additive (rPP_11a). New double bands at 1578 and 1541 cm^{-1} were observed for rPP_11a and rPP_20a, indicating the formation of carboxylate groups due to the addition of the additive.

3.2. Polarized Optical Microscopy

The polarized optical microscopy (POM) analysis is a well-established method to determine crystallization behavior. The effects of recycling PP with and without additive were studied isothermally at 130 °C. The results (Figure 2) show that the formation of crystals was the slowest for the virgin PP (rPP_0). The first individual spherulites appeared after 1 min (Figure 2A) while (Figure 2D) the crystallization process was finished after 10 min, resulting in a small amount of crystals of relatively large size. On the other hand, the crystallization rate with an increasing number of recycling cycles increases (Figure 2B) while the crystal size decreases. This is a consequence of the shorter polymeric chains resulting from the degradation during the recycling process [11]. A high number of recycles leads to densely-packed crystallites, with the boundaries hard to distinguish. High nuclei density hinders crystals to grow freely; moreover, the crystals are much smaller than those observed for the virgin polypropylene. The addition of an additive beneficially influences the growth of crystallites, also at higher processing cycles, as larger crystallites formed in rPP_20a compared to rPP_20 without additive (Figure 2C). During the observation of the crystallization process with optical microscopy, it was observed that the formation of crystals at rPP with additive occurs sooner compared to rPP without additive. In order to further make more detailed investigations, DSC measurements were performed.

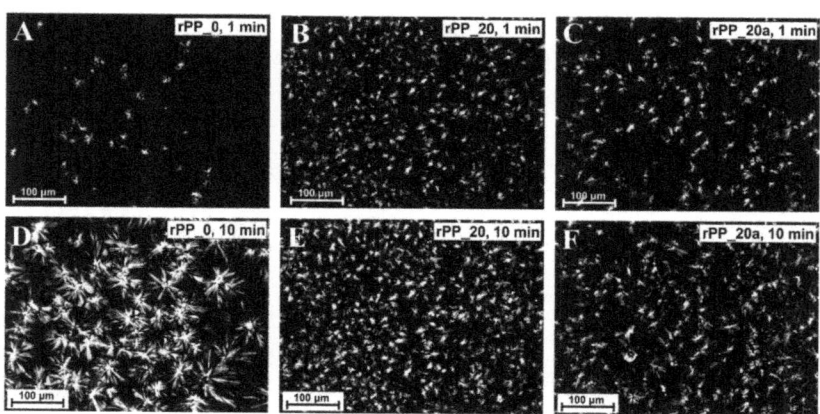

Figure 2. Polarizing optical microscopy of the nucleation and growth processes of virgin PP—rPP_0 (**A,D**), rPP_20 (**B,E**), and rPP_20a (**C,F**) after isothermal crystallization at 130 °C for 1 min (**A–C**) and 10 min (**D–F**).

3.3. Thermal Properties

The DSC is an important characterization method to determine the effects of recycling on the thermal behavior of semi-crystalline virgin and recycled PP. Additionally, the results enable the determination of the changes arising from the addition of the additive to recycled PP. The first heating run gives information on the material's thermal properties after the melt processing, including the mechanical and thermal history (influence of processing, crystallinity, ageing, heat treatment, etc.) For the determination of the crystallization temperature (T_c), the studied samples were investigated under controlled cooling. With second heating (third run), the obtained data provide information on the material's glass transition (T_g) and its melting temperature (T_m), independent of the history. The DSC analysis of virgin PP, recycled with additive and recycled without additive, is presented in Figure 3 and summarized in Table 3. The results of the 2nd DSC heating (Figure 3A) show that the melting temperature (T_m) of virgin PP (rPP_0) was 162.0 °C. As the number of recycles increased, the T_m increased up to the 10th cycle, while further increasing of the recycling process decreased the T_m. The addition of an additive to rPP_10 slightly decreased the T_m for rPP_11a, while further increasing of the recycles led to an increase of the T_m. However, the measured values of T_c and T_m were within the experimental error, calculated from three repeated measurements.

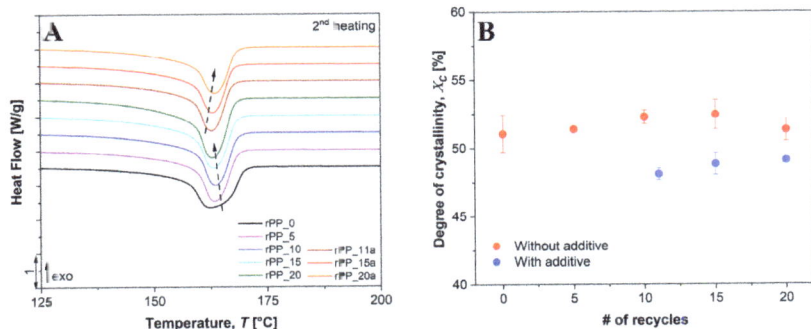

Figure 3. (**A**) DSC thermograms of melting peaks from the 2nd heating cycle of virgin PP, recycled PP without and with additive, and (**B**) the degree of crystallinity, determined by 2nd heating.

Table 3. Thermal properties, obtained from the DSC of virgin PP, recycled PP with and without additive.

# of Recycles	1st Heating		Cooling		2nd Heating		
	H_m (J/g)	T_{r1} (°C)	H_c (J/g)	T_c (°C)	H_m (J/g)	T_m (°C)	X_c (%)
rPP_0	99.0	166.0	100.9	112.9	105.7	162.0	51.1
rPP_5	99.5	166.1	101.8	120.2	106.4	163.3	51.4
rPP_10	99.7	165.7	103.1	120.4	108.2	163.5	52.3
rPP_15	98.8	166.1	103.6	121.2	108.5	163.0	52.4
rPP_20	94.0	165.5	102.3	121.5	106.7	162.5	51.3
rPP_11a	91.0	165.1	95.5	120.1	99.6	162.9	48.1
rPP_15a	91.2	165.2	96.8	120.6	101.1	163.3	48.8
rPP_20a	91.3	165.3	97.1	120.9	101.7	163.3	49.1

For the enthalpy of PP melting, the results (Table 3) show a constant upward trend with an increased number of recycles, while the addition of the additive to rPP_10 decreased the enthalpy sharply (from 108.2 J/g for rPP_10 to 99.6 J/g for rPP_11a). A similar result was observed for the determination of the degree of crystallinity (Figure 3B). With an increasing number of recycles, the crystallinity of additive-free samples increased from 51.1% to 51.3%. However, after the additive was added to rPP_10, the X_C first decreased by ~8%, while with a further nine recycles, the X_C increased only by ~0.2%. The increased crystallinity of the recycled PP could be attributed to the molecular weight reduction and smaller entanglements of chains due to several thermal-reprocessing cycles [31]. This can also be observed from the DSC measurements where the peaks with the same shape and size are only shifted, meaning that the changes are only due to the chain scission and not to changes in the molecular size of the formed structure. The lower molecular weight of PP contributes to a higher mobility of the chains, resulting in higher crystallinity; however, more imperfections in the resulting crystals could be expected. As mentioned before, such imperfections are mainly due to the formation of free radicals during the rupture of the molecular chains. These results are in agreement with the POM analysis, where a different kinetic rate between the virgin and recycled PP samples was observed.

The increase in the degree of crystallinity can also be explained by the fact that macromolecules with lower molecular weight act as nucleating agents, which enhance the crystallization of semicrystalline polymers by enabling the folding of the chains and building bigger crystal structures. Correspondingly, as a result of the decreased molecular weight, the crystallization temperature increases with the reprocessing cycles.

Figure 4 shows a main exothermic peak at 112.9 °C, corresponding to the crystallization of virgin PP. After recycling virgin PP 20 times (rPP_20), an increase in crystallization temperature (T_C) of 9.1 °C was observed for the sample without additive, while the increase for rPP with additive (rPP_20a) was 8 °C. These results show that thermal recycling promotes the formation of crystalline phases in the rPP samples.

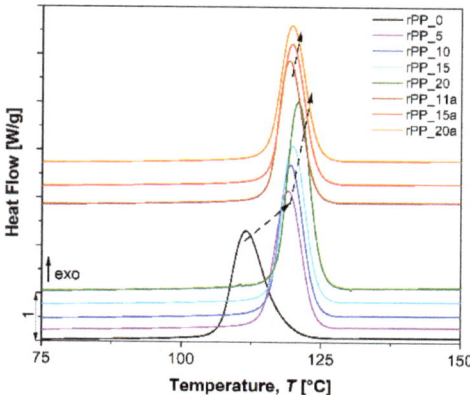

Figure 4. DSC thermograms of crystallization peaks of virgin PP, and recycled PP without and with additive.

3.4. Rheological Properties

The viscoelastic measurements are well-established and practical experimental methods for melt characterization. The rheological behavior, presented by storage modulus (G'), loss modulus (G''), and complex viscosity (η^*) of virgin and recycled PP with and without additive have been studied. The results of viscoelastic (G' and G'') properties in the linear viscoelastic region (LVR) at different frequencies (Figure 5) show that the values of both moduli (G' and G') increased with an increasing frequency of oscillation with G'' prevailing in the range of low frequencies. Moreover, the moduli of the recycled PP without additive decreased with increasing reprocessing cycles, indicating a degradation of the structure. After a higher number of recycling steps, PP exhibited a more liquid-like character due to chain-breaking and a reduction of molecular weight.

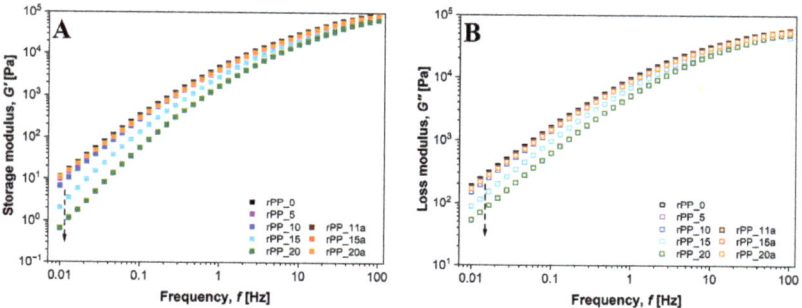

Figure 5. Frequency dependency of (**A**) storage modulus G' and (**B**) loss modulus G'' of virgin PP, recycled PP without and with additive depending on the processing cycle (at 190 °C).

On the other hand, a major improvement in viscoelastic properties occurred when the additive was added to PP, which had already been recycled 10 times. Compared to virgin PP, the G' and G'' of the 10-times-recycled PP with the addition of additive increased in the whole frequency range examined. This indicates that the additive propagated crosslinking between polymer chains with chain scissions, which improved viscoelastic behavior. The results suggest that the additive limits the mobility of the polymer chain and thus, reinforces the internal network structure of rPP.

If we compare the frequency dependency of G' and G'', respectively, we can investigate the point where the moduli have the same value. At this point, the behavior of the material changes; in our case, from liquid-like in the low frequency range to solid-like at higher frequencies. Figure 6 shows the intersection points between G' and G'' with respect to the frequency. This point is called the crossover modulus point ($G = G' = G''$) and the frequency at which the moduli are the same is called crossover frequency (ω_{co}). If all samples are compared, it becomes clear that for rPP samples without an additive, the crossover point of G' and G'' shift towards higher frequencies where the relaxation times of the molecules are lower. This is characteristic of shorter molecules, and it reflects the chain scission during the extensive recycling process. On the other hand, when the additive was added to the sample, recycled for 10 times, the crossover frequency remained the same regardless of the number of further processing cycles.

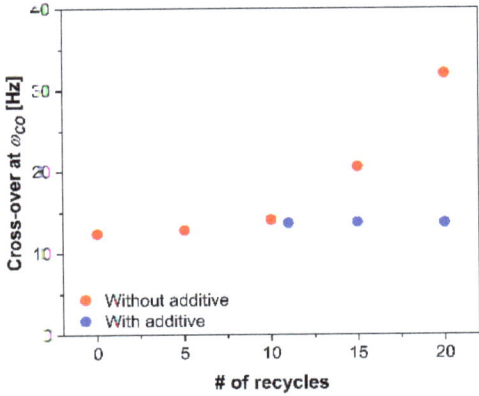

Figure 6. Crossover frequency, ω_{co}, of virgin PP, and recycled PP without and with additive, depending on the processing cycle.

In addition to the storage modulus (G') and the loss modulus (G''), the complex viscosity is also an important parameter that provides information about the viscous or elastic effects. As presented in Figure 7, the complex viscosity for all rPP samples decreased with increasing frequency, showing a typical shear-thinning behavior [10]. The zero-shear viscosity (η_0), i.e., the constant melt viscosity at low frequencies, decreased with the number of recycling cycles of rPP without additive. This can be correlated to the decreased molecular weight and polymer chain length where fewer entanglements between chains and intermolecular interactions occurred [13,32,33]. The shear-thinning behavior of rPP recycled for 20 times without additive was the least pronounced with the lowest zero-shear viscosity of 845 Pa·s, which is almost four times lower than the η_0 of virgin PP. The transition between the Newtonian and the shear-thinning region of the viscosity curve was smoother and occurred at lower frequencies. Furthermore, the addition of an additive increases the entanglements of the molecules due to the promotion of crosslinking of broken polymer chains, which were reflected in a higher η_0 for recycled samples compared to the virgin PP. Further recycling (up to 10 more times) did not affect the η_0.

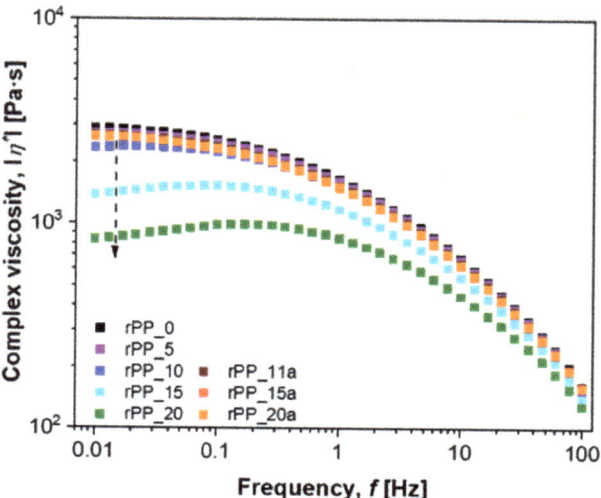

Figure 7. Dynamic rheological behavior of virgin PP, recycled PP without and with additive, complex viscosity, measured at 190 °C.

The changes in the average molecular weight (M_w) and polydispersity index (PDI) with an increasing number of recycling steps for PP without and with additive (Figure 8AB) were obtained from the rheological measurements of frequency sweep analysis [34], which has been reported in the literature to be in good agreement with the results of gel permeation chromatography (GPC), the most commonly used technique to determine M_w [35]. It has been observed that the peak of M_w shifts to lower values with increasing recycling steps, which can be attributed to degradation and chain scission [36]. A significant decrease in M_w was observed after the 10th recycling step without the addition of the additive. Furthermore, our results show that the M_w did not decrease further after the addition of the additive. The results are in very good agreement with the rheological characterization data, which also show a significant decrease in complex viscosity after the 10th recycling step for rPP without additive. The decrease in PDI values indicates a narrowing of the molecular weight distribution and the formation of shorter chains caused by chain scission (Figure 8B). In addition, lower M_w of the recycled polymer affects the mechanical performance of the material, which is characterized below.

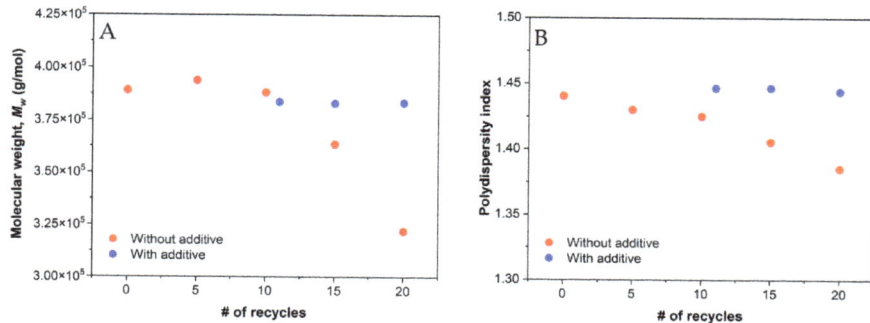

Figure 8. (**A**) Molecular weight distribution and (**B**) polydispersity of virgin PP, recycled PP without and with additive, respectively.

3.5. Mechanical Properties

3.5.1. Nanoindentation

Nanoindentation is a method used for the measurement of the local mechanical properties, such as the elastic modulus (E) and hardness (H) [37]. The range of microhardness values that each polymer exhibits is mainly determined by the nature of the molecular chains [38], which includes the structure, length, and polydispersity. As the recycling affected the structure, i.e., the length of the polymer chains, we used this technique to evaluate the effect of recycling on the micromechanical properties.

The values of the elastic modulus (E) for the recycled PP samples, with and without additive, are presented in Figure 9A, while the values for surface hardness (H) are shown in Figure 9B. The average values of the elastic modulus and hardness are presented at a penetration depth of 800–1800 nm. The results show that after five reprocessing cycles, the E and H values remained the same; however, with further recycling, the E and H values decreased, indicating that the surface of the extensively recycled PP became more rigid. The results were found to be in close agreement with the work carried out by Zciri et al. [39] and Bourmaud et al. [40], showing a slight reduction of elastic modulus after recycling, which could be due to physical ageing and thermo-oxidative degradation, i.e., decreasing of molecular weight [31]. On the other hand, the addition of an additive restores the elastic properties and hardness to higher values as detected for the virgin PP.

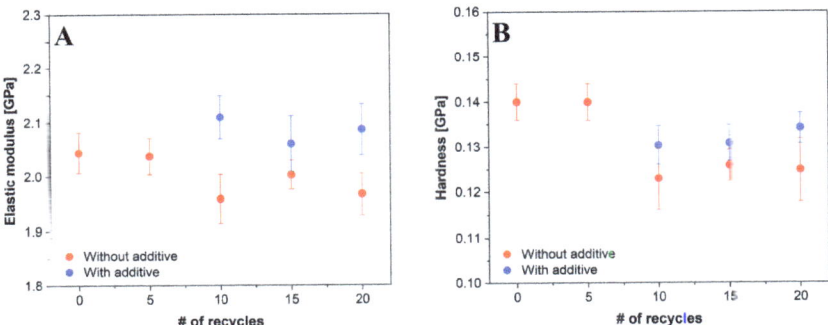

Figure 9. Average values of (**A**) elastic modulus and (**B**) hardness of virgin PP, recycled PP without and with additive. The average analysis depth was from 800 to 1800 nm.

3.5.2. Time-Dependent Behavior

The polymers suffer structural changes when used for a long time. However, exposure to natural operating conditions can last for a very long time before some observable changes occur. Hence, the accelerated procedures should be used for the prediction of long-term behavior of such materials. One of the possibilities to determine time-dependent mechanical properties of polymers is the use of a predictive method based on an increase in the test temperature, the so-called time-temperature superposition (tTS) principle [41]. The master curve of the long-term creep response can be composed of the short-term isothermal creep tests at different temperatures $J(t, T)$ with an individual shifting of segments. In Figure 10, the average master curve of creep behavior (the dependence of the creep modulus, J, on time, t) of virgin and recycled PP without and with additive are presented. The results obtained from the creep modulus do not show any changes up to 10th recycle (rPP_10), but with further reprocessing cycles (rPP_20) the values of the creep modulus decreased compared to virgin PP. Furthermore, the addition of an additive increased the creep modulus to higher values, but the shape of the creep compliance curve did not change significantly. Creep tests showed that the addition of the additive improved the creep resistance of recycled PP in both time and temperature, with this benefit becoming more significant with increasing temperatures and longer times, respectively.

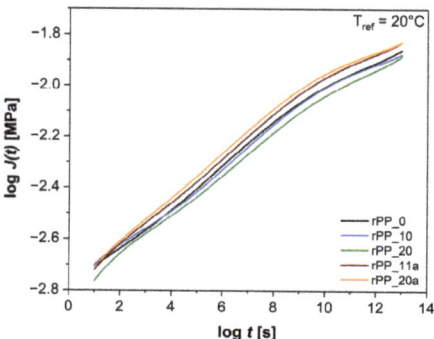

Figure 10. Creep master curve at the reference temperature (T_{ref} = 20 °C) for virgin PP, recycled PP without and with the additive.

3.5.3. Dynamic Mechanical Thermal Analysis (DMTA)

Dynamic mechanical thermal analysis (DMTA) is a widely used technique enabling the characterization of the mechanical behavior of materials as a function of temperature and/or frequency of oscillation. Figure 11A shows the temperature dependence of the storage modulus (G') and loss modulus (G'') of virgin and recycled PP without and with the additive. It can be seen that for all the samples, G' was approximately one decade higher than G'' in the whole temperature range examined. The increasing temperature caused the softening of the PP matrix and increased molecular mobility [42]. At low temperatures, the PP recycled 20 times (rPP_20) showed the same values of G' and G'' as the virgin PP, but at temperatures higher than 60 °C, the values of the moduli for rPP_20 slightly prevailed over the values of virgin PP. The increase in G'' with extensive recycling of PP (rPP_20) leads to lower stiffness of the polymer. On the other hand, the recycled PP samples with the additive exhibited the lowest G' and G'', respectively, meaning that these samples became more flexible compared to virgin and recycled PP without the additive.

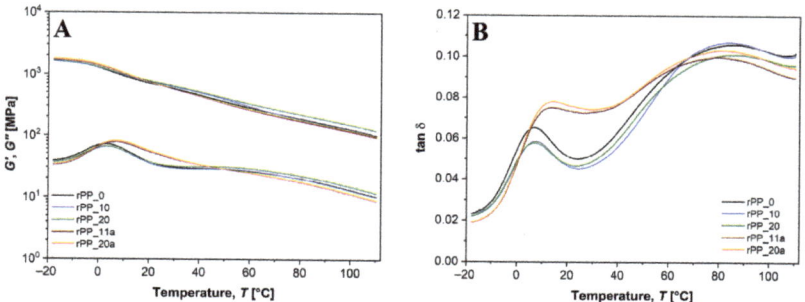

Figure 11. DMA curves: (**A**) storage modulus (G') and loss modulus (G''); (**B**) tan δ of virgin PP and recycled PP without and with the additive.

Tan δ is defined as the ratio between the loss and storage modulus (Tan δ = G''/G') and is related to the damping properties of the polymer. The Tan δ curve of PP exhibits three relaxations peaks at about −80 °C (γ), 10 °C (β), and 100 °C (α). With our instruments the measurements could only be performed from the temperatures of −18 to 110 °C; therefore, the γ relaxation peak, which originates from the motion of the short segments in the amorphous phase, was too low to determine. In Figure 11B, the first peak within a temperature range from 0 °C to 20 °C corresponds to the glass transition temperature (T_g) or β relaxation of PP [43]. T_g of virgin PP was thus detected at 6.3 °C and this value remained constant up to the 20th reprocessing cycle. Furthermore, with the addition of the additive,

T_g increased to 13.5 °C. This indicates that the PP polymer chains in the amorphous region became more rigid. The T_g values were obtained also from loss modulus curves, where the obtained values for rPP without additive were close to ~3 °C and ~7 °C for rPP_a samples, respectively. The second peak in the elastic-rubbery region, characteristic for α relaxation (long-range coordinated molecular motions), was located between 60 and 100 °C [44,45]. The α relaxation process for rPP_a samples occurred at lower temperatures (~81–82 °C) compared to rPP samples without the additive (85–88 °C). The T_g values for the β and α relaxation process, determined from tan δ and loss modulus curves, respectively, are summarized in Table 4.

Table 4. Dynamic mechanical analysis (DMTA) results of virgin PP and recycled PP without and with the additive.

Sample	T_g (°C) [1]	T_g (°C) [2]	α Relaxation Peak (°C)
rPP_0	6.3	2.5	86.7
rPP_10	7.5	3.2	84.5
rPP_20	6.7	3.0	86.7
rPP_11a	13.5	6.2	81.0
rPP_20a	13.5	6.7	82.0

[1] T_g obtained from tan δ; [2] T_g obtained from storage modulus (G').

For studying the viscoelastic behavior, the Cole–Cole plots can provide additional information about the relaxation processes occurring by reprocessing polymers. In our study, the Cole–Cole plot showed the relationship between the loss modulus (G'') and storage modulus (G'), and helped to understand the relaxation processes of a semi-crystalline PP polymer (Figure 12). For the virgin and recycled PP without additive, two peaks, observed at lower and higher temperatures, can be attributed to the β and α relaxation processes [46]. The first peak is related to the β relaxation process and could be ascribed to the segmental relaxation mechanisms in the amorphous region. This peak can also be related to glass transition (T_g). The second peak, attributed to the $α_c$ relaxation process, can be associated with relaxations appearing from the crystalline phases of the PP polymer matrix [45,47]. The addition of the additive to rPP (rPP_a), on the other hand, mimics the first peak and changed the shape of the Cole–Cole curves. The more pronounced second peak of the α relaxation process could be considered as an increasing exchange of stereo defects between the amorphous and crystal phases.

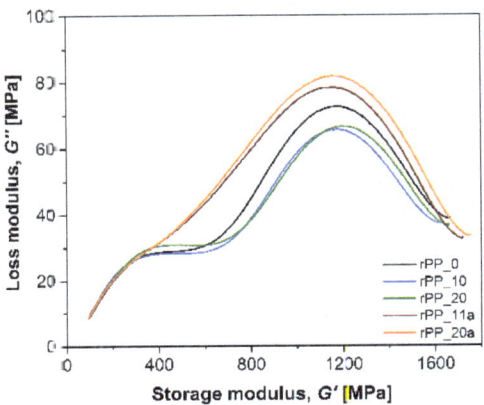

Figure 12. Cole–Cole plot of virgin PP and recycled rPP and rPP_a samples.

4. Conclusions

The extensive recycling of polymers leads to changes in the rheological, morphological, mechanical, thermal, and other properties of the processed polymer. In our study, it was shown that after 20 recycling steps of mechanical processing of polypropylene, thermo-oxidative degradation occurred. This was observed with spectroscopic measurements, which showed the appearance of an absorption band at 1735 cm^{-1}. The addition of a stabilization additive led to the formation of new bands at 1575 and 1541 cm^{-1}, suggesting interactions between the additive and the PP polymer.

The viscoelastic analysis showed that an increase in the frequency at which the polymer structure changes ($G' = G''$) from liquid- to solid-like was observed for the extensive recycling of the additive-free samples, while the frequency of the crossover point, ω_{co}, for the samples with the additive were lower compared to the virgin PP and only slightly changed with the extensive recycling. The increasing ω_{co} suggests that the molecular weight of the recycled polymer decreased. Lower molecular weight is associated with shorter molecules that disentangle faster and therefore, exhibit a higher crossover frequency, while longer molecules need more time to disentangle and exhibit a liquid-like behavior at lower frequencies. When the additive was added to the recycled PP polymer, the values of ω_{co} decreased, indicating that the molecular weight was higher compared to the virgin PP. The increasing slopes of the G' and G'' curves at low frequencies for the samples with the addition of the additive can be associated with a narrowing of the average molecular weight distribution. Furthermore, the results showed that when the additive was added to the recycled PP, the rheological properties were improved already after the first recycle (rPP_11a) and remained unchanged for the further ten recycling cycles.

The results of the thermal analysis revealed that the main mechanism of degradation during recycling of PP is the chain scission, as the thermograms did not show changes in the microstructure. With increasing recycling of the additive-free samples, the increase in crystallinity for the samples from the 10th to the 20th recycle was much higher (~5%) compared to the samples with the additive (~0.2%). We can conclude that shorter chains acted as nucleating agents causing easier formation of crystalline regions. These results were also confirmed by the POM technique.

The narrower distribution of polymer molecules after the addition of the additive can also be associated with the improved mechanical properties of the surface, determined by nanoindentation measurements. The results showed that the additive-free rPP samples exhibited a slightly lower elastic modulus and hardness than rPP_a samples, suggesting a hardening of the surface during extensive recycling without the additive.

The DMA analysis was used to determine the T_g of the PP and rPP samples. The results showed two tan δ peaks, the first which was for the virgin and recycled rPP without additive close to ~7 °C and was attributed to α relaxation (long-range coordinated molecular motions). For recycled rPP with additive, this peak was observed at higher values, i.e., ~13.5 °C. However, the β relaxation process for rPP_a samples occurred at lower temperatures (~81–82 °C) compared to the samples without the additive (85–88 °C).

The results showed that the use of an additive improves the time-dependent mechanical properties of solid polypropylene since it was shown that the additive acts as a hardener and additionally crosslinks the recycled polymer chains. On the other hand, the decrease in creep compliance modulus indicates a shorter lifespan with an increasing number of recycles due to chain scission of polymer chains. In summary, the main degradation processes of polypropylene that occur after extensive recycling are chain scission of polymer chains and oxidation, which, however, can be improved by the addition of a stabilization additive.

Author Contributions: Conceptualization, M.M. and L.S.P.; methodology, M.M., A.O. and L.S.P.; formal analysis, M.M., A.O. and M.H.; investigation, M.M.; writing—original draft preparation, M.M. and L.S.P.; writing—review and editing, M.M., A.O., M.H. and L.S.P.; visualization, M.M. and L.S.P.; supervision, L.S.P.; project administration, M.M. All authors have read and agreed to the published version of the manuscript.

Funding: The authors acknowledge the financial support from the Slovenian Research Agency (research core funding No. P2-0264).

Institutional Review Board Statement: Not applicable.

Informed Consent Statement: Not applicable.

Data Availability Statement: Not applicable.

Acknowledgments: The authors acknowledge the financial support from the Slovenian Research Agency (research core funding No. P2-0264). The authors would like to thank to Marko Bek for his helpful advice on the various technical issues examined in this paper, and the company, BYK-Chemie GmbH, for providing the additive.

Conflicts of Interest: The authors declare no conflict of interest. The funders had no role in the design of the study; in the collection, analyses, or interpretation of data; in the writing of the manuscript; or in the decision to publish the results.

References

1. Roland, G.; Jambeck, J.R.; Lavender, L.K. Production, use, and fate of all plastics ever made. *Sci. Adv.* **2017**, *3*, e1700782. [CrossRef]
2. Ajorloo, M.; Ghodrat, M.; Kang, W.-H. Incorporation of Recycled Polypropylene and Fly Ash in Polypropylene-Based Composites for Automotive Applications. *J. Polym. Environ.* **2021**, *29*, 1298–1309. [CrossRef]
3. Khalaj, M.-J.; Ahmadi, H.; Lesankhosh, R.; Khalaj, G. Study of physical and mechanical properties of polypropylene nanocomposites for food packaging application: Nano-clay modified with iron nanoparticles. *Trends Food Sci. Technol.* **2016**, *51*, 41–48. [CrossRef]
4. Aizenshtein, E. Polypropylene fibres and yarns in the current state of development. *Fibre Chem.* **2008**, *40*, 399–405. [CrossRef]
5. Karian, H. *Handbook of Polypropylene and Polypropylene Composites*; Marcel Dekker: New York, NY, USA, 2003; ISBN 0824740645/9780824740641.
6. Ragaert, K.; Delva, L.; Van Geem, K. Mechanical and chemical recycling of solid plastic waste. *Waste Manag.* **2017**, *69*, 24–58. [CrossRef]
7. Peterson, J.D.; Vyazovkin, S.; Wight, C.A. Kinetics of the Thermal and Thermo-Oxidative Degradation of Polystyrene, Polyethylene and Poly(propylene). *Macromol. Chem. Phys.* **2001**, *202*, 775–784. [CrossRef]
8. Esmizadeh, E.; Tzoganakis, C.; Mekonnen, T.H. Degradation Behavior of Polypropylene during Reprocessing and Its Biocomposites: Thermal and Oxidative Degradation Kinetics. *Polymers* **2020**, *12*, 1627. [CrossRef]
9. Martinez Jothar, L.; Montes-Zavala, I.; Rivera-García, N.; Díaz-Ceja, Y.; Pérez, E.; Waldo-Mendoza, M. Thermal degradation of polypropylene reprocessed in a co-rotating twin-screw extruder: Kinetic model and relationship between Melt Flow Index and Molecular weight. *Rev. Mex. Ing. Quím.* **2021**, *20*, 1079–1091. [CrossRef]
10. Da Costa, H.M.; Ramos, V.D.; de Oliveira, M.G. Degradation of polypropylene (PP) during multiple extrusions: Thermal analysis, mechanical properties and analysis of variance. *Polym. Test.* **2007**, *26*, 676–684. [CrossRef]
11. Spicker, C.; Rudolph, N.; Kühnert, I.; Aumnate, C. The use of rheological behavior to monitor the processing and service life properties of recycled polypropylene. *Food Packag. Shelf Life* **2019**, *19*, 174–183. [CrossRef]
12. Stocchi, A.; Pettarin, V.; Izer, A.; Bárány, T.; Czigány, T.; Bernal, C. Fracture Behavior of Recyclable All-Polypropylene Composites Composed of α- and β-Modifications. *J. Thermoplast. Compos. Mater.* **2011**, *24*, 805–818. [CrossRef]
13. Martín-Alfonso, J.E.; Franco, J.M. Influence of polymer reprocessing cycles on the microstructure and rheological behavior of polypropylene/mineral oil oleogels. *Polym. Test.* **2015**, *45*, 12–19. [CrossRef]
14. González-González, V.A.; Neira-Velázquez, G.; Angulo-Sánchez, J.L. Polypropylene chain scissions and molecular weight changes in multiple extrusion. *Polym. Degrad. Stab.* **1998**, *60*, 33–42. [CrossRef]
15. Altay, L.; Sarikanat, M.; Sağlam, M.; Uysalman, T.; Seki, Y. The effect of various mineral fillers on thermal, mechanical, and rheological properties of polypropylene. *Res. Eng. Struct. Mater.* **2021**, *7*, 361–373. [CrossRef]
16. Caicedo, C.; Vázquez-Arce, A.; Ossa, O.H.; De La Cruz, H.; Maciel-Cerda, A. Physicomechanical behavior of composites of polypropylene, and mineral fillers with different process cycles. *DYNA* **2018**, *85*, 260–268. [CrossRef]
17. Techawinyutham, L.; Sumrith, N.; Srisuk, R.; Techawinyutham, W.; Siengchin, S.; Mavinkere Rangappa, S. Thermo-mechanical, rheological and morphology properties of polypropylene composites: Residual CaCO3 as a sustainable by-product. *Polym. Compos.* **2021**, *42*, 4643–4659. [CrossRef]
18. Saw, L.T.; Uy Lan, D.N.; Rahim, N.A.A.; Mohd Kahar, A.W.; Viet, C.X. Processing degradation of polypropylene-ethylene copolymer-kaolin composites by a twin-screw extruder. *Polym. Degrad. Stab.* **2015**, *111*, 32–37. [CrossRef]
19. Schyns, Z.O.G.; Shaver, M.P. Mechanical Recycling of Packaging Plastics: A Review. *Macromol. Rapid Commun.* **2021**, *42*, 2000415. [CrossRef]
20. Pfaendner, R.; Herbst, H.; Hoffmann, K. Innovative concept for the upgrading of recyclates by restabilization and repair molecules. *Macromol. Symp.* **1998**, *135*, 97–111. [CrossRef]

21. Ton-That, M.; Denault, J. *Development of Composites Based on Natural Fibers*; The Institute of Textile Science: Ottawa, ON, Canada, 2007.
22. Barbeș, L.; Radulescu, C.; Stihi, C. ATR–FTIR spectrometry characterisation of polymeric materials. *Rom. Rep. Phys.* **2014**, *66*, 765–777.
23. Karger-Kocsis, J. *Polypropylene Structure, Blends and Composites: Volume 3 Composites*; Springer Science & Business Media: Berlin/Heidelberg, Germany, 2012; ISBN 9401105235.
24. Li, X.; Bhushan, B. A review of nanoindentation continuous stiffness measurement technique and its applications. *Mater. Charact.* **2002**, *48*, 11–36. [CrossRef]
25. Oseli, A.; Prodan, T.; Susič, E.; Slemenik Perše, L. The effect of short fiber orientation on long term shear behavior of 40% glass fiber reinforced polyphenylene sulfide. *Polym. Test.* **2020**, *81*, 106262. [CrossRef]
26. Lacoste, J.; Vaillant, D.; Carlsson, D.J. Gamma-, photo-, and thermally-initiated oxidation of isotactic polypropylene. *J. Polym. Sci. Part A Polym. Chem.* **1993**, *31*, 715–722. [CrossRef]
27. Strömberg, E.; Karlsson, S. The Design of a Test Protocol to Model the Degradation of Polyolefins During Recycling and Service Life. *J. Appl. Polym. Sci.* **2009**, *112*, 1835–1844. [CrossRef]
28. Andreassen, E. Infrared and Raman spectroscopy of polypropylene. In *Polypropylene*; Karger-Kocsis, J., Ed.; Springer: Dordrecht, The Netherlands, 1999; pp. 320–328. ISBN 978-94-011-4421-6.
29. Zieba-Palus, J. The usefulness of infrared spectroscopy in examinations of adhesive tapes for forensic purposes. *Forensic Sci. Criminol.* **2017**, *2*, 1–9. [CrossRef]
30. Zdiri, K.; Elamri, A.; Hamdaoui, M.; Harzallah, O.; Khenoussi, N.; Brendlé, J. Elaboration and Characterization of Recycled PP/Clay Nanocomposites. *J. Mater. Environ. Sci.* **2018**, *9*, 2370–2378.
31. Hamad, K.; Kaseem, M.; Deri, F. Recycling of waste from polymer materials: An overview of the recent works. *Polym. Degrad. Stab.* **2013**, *98*, 2801–2812. [CrossRef]
32. Da Costa, H.M.; Ramos, V.D.; Rocha, M.C.G. Rheological properties of polypropylene during multiple extrusion. *Polym. Test.* **2005**, *24*, 86–93. [CrossRef]
33. Schmiederer, D.; Gardocki, A.; Kühnert, I.; Schmachtenberg, E. Local thermo-oxidative degradation in injection molding. *Polym. Eng. Sci.* **2008**, *48*, 717–722. [CrossRef]
34. Khak, M.; Ramazani, A.S.A. Rheological measurement of molecular weight distribution of polymers. *e-Polymers* **2013**, *13*, 1. [CrossRef]
35. Bayazian, H.; Schoeppner, V. Investigation of molecular weight distributions during extrusion process of polypropylene by rheometry experiment. *AIP Conf. Proc.* **2019**, *2065*, 30044. [CrossRef]
36. Jubinville, D.; Esmizadeh, E.; Tzoganakis, C.; Mekonnen, T. Thermo-mechanical recycling of polypropylene for the facile and scalable fabrication of highly loaded wood plastic composites. *Compos. Part B Eng.* **2021**, *219*, 108873. [CrossRef]
37. Oliver, W.C.; Pharr, G.M. An improved technique for determining hardness and elastic modulus using load and displacement sensing indentation experiments. *J. Mater. Res.* **1992**, *7*, 1564–1583. [CrossRef]
38. Flores, A.; Ania, F.; Baltá-Calleja, F.J. From the glassy state to ordered polymer structures: A microhardness study. *Polymer* **2009**, *50*, 729–746. [CrossRef]
39. Zdiri, K.; Elamri, A.; Hamdaoui, M.; Harzallah, O.; Khenoussi, N.; Brendlé, J. Reinforcement of recycled PP polymers by nanoparticles incorporation. *Green Chem. Lett. Rev.* **2018**, *11*, 296–311. [CrossRef]
40. Bourmaud, A.; Le Duigou, A.; Baley, C. What is the technical and environmental interest in reusing a recycled polypropylene–hemp fibre composite? *Polym. Degrad. Stab.* **2011**, *96*, 1732–1739. [CrossRef]
41. Ferry, J.D. *Viscoelastic Properties of Polymers*; John Wiley & Sons: New York, NY, USA, 1980.
42. Saba, N.; Jawaid, M.; Alothman, O.Y.; Paridah, M.T. A review on dynamic mechanical properties of natural fibre reinforced polymer composites. *Constr. Build. Mater.* **2016**, *106*, 149–159. [CrossRef]
43. Mazidi, M.M.; Razavi Aghjeh, M.K.; Khonakdar, H.A.; Reuter, U. Structure–property relationships in super-toughened polypropylene-based ternary blends of core–shell morphology. *RSC Adv.* **2016**, *6*, 1508–1526. [CrossRef]
44. Hidalgo-Salazar, M.; Luna Vera, F.; Correa, J. Biocomposites from Colombian sugarcane bagasse with polypropylene: Mechanical, thermal and viscoelastic properties. In *Characterizations of Some Composite Materials*; Books on Demand: Dorderstedt, Germany, 2019; ISBN 978-1-78984-912-7.
45. Hoyos, M.; Tiemblo, P.; Gómez-Elvira, J.M. The role of microstructure, molar mass and morphology on local relaxations in isotactic polypropylene. The α relaxation. *Polymer* **2007**, *48*, 183–194. [CrossRef]
46. Gitsas, A.; Floudas, G. Pressure Dependence of the Glass Transition in Atactic and Isotactic Polypropylene. *Macromolecules* **2008**, *41*, 9423–9429. [CrossRef]
47. Gaska, K.; Manika, G.C.; Gkourmpis, T.; Tranchida, D.; Gitsas, A.; Kádár, R. Mechanical Behavior of Melt-Mixed 3D Hierarchical Graphene/Polypropylene Nanocomposites. *Polymers* **2020**, *12*, 1309. [CrossRef] [PubMed]

Article

Diverted from Landfill: Reuse of Single-Use Plastic Packaging Waste

Kit O'Rourke [1], Christian Wurzer [2], James Murray [3], Adrian Doyle [4], Keith Doyle [4], Chris Griffin [5], Bernd Christensen [6], Conchúr M. Ó Brádaigh [1] and Dipa Ray [1,*]

[1] School of Engineering, Institute for Materials and Processes, The University of Edinburgh, Sanderson Building, Robert Stevenson Road, Edinburgh EH9 3FB, UK
[2] School of GeoSciences, UK Biochar Research Centre, The University of Edinburgh, Edinburgh EH9 3FF, UK
[3] Materials Research Institute, Technological University of the Shannon, University Road, Co. Westmeath, N37 HD68 Athlone, Ireland
[4] FALTECH, Ballycumber Road, Co. Offaly, R35 XR57 Clara, Ireland
[5] Johns Manville, 10, 100, West Ute Ave., Littleton, CO 80127, USA
[6] Johns Manville Europe GmbH, Werner-Schuller-Str. 1, 97877 Wertheim, Germany
* Correspondence: dipa.roy@ed.ac.uk

Abstract: Low-density polyethylene (LDPE) based packaging films mostly end up in landfill after single-use as they are not commonly recycled due to their flexible nature, low strength and low cost. Additionally, the necessity to separate and sort different plastic waste streams is the most costly step in plastics recycling, and is a major barrier to increasing recycling rates. This cost can be reduced through using waste mixed plastics (wMP) as a raw material. This research investigates the properties of PE-based wMP coming from film packaging wastes that constitutes different grades of PE with traces of polypropylene (PP). Their properties are compared with segregated individual recycled polyolefins and virgin LDPE. The plastic plaques are produced directly from the wMP shreds as well as after extruding the wMP shreds into a more uniform material. The effect of different material forms and processing conditions on the mechanical properties are investigated. The results of the investigation show that measured properties of the wMP fall well within the range of properties of various grades of virgin polyethylene, indicating the maximum possible variations between different batches. Addition of an intermediate processing step of extrusion before compression moulding is found to have no effect on the tensile properties but results in a noticeably different failure behaviour. The wMP does not show any thermal degradation during processing that was confirmed by thermogravimetric analysis. The results give a scientific insight into the adoption of wMP in real world products that can divert them from landfill creating a more circular economy.

Keywords: mixed plastic packaging waste; recycled plastics; compression moulding; mechanical testing; fracture surface; mixed polyethylenes

1. Introduction

Thermoplastic polymers are particularly appealing to use due to their recyclability over thermosetting polymers. Recently, the demand for recycled materials has increased due to government legislation and many companies switching to recycled materials to improve their brand sustainability image. In 2017, over 90% of plastics produced globally were derived from virgin material made using fossil fuels [1], however, the current climate crisis has forced a necessary movement away from the use of fossil fuels. In order to keep up with the current demand of plastics, it is necessary to find a more efficient way to utilise the plastics already in circulation. Furthermore, the current recycling rate must be increased to meet recycling targets, such as the EU target for recycling 50% of plastic packaging waste by 2025 and 55% by 2030 [2]. The necessity to separate and sort different plastic waste streams is the most labour-intensive and costly step in any plastics recycling process and one of the main barriers to increasing recycling rates.

Currently, plastic packaging consisting of high-density polyethylene (HDPE), polypropylene (PP), and polyethylene terephthalate (PET) are commonly recycled. However, a large proportion of single-use plastics packaging wastes that consists primarily of low-density polyethylene (LDPE), are not generally recycled due to their flexibility, low mechanical performance and low cost [3]. Within the UK, it is estimated that five million tonnes of plastics are used every year [4], with 67% of this consisting of plastic packaging [5]. However, LDPE films and bags are not collected with household recycling in the UK, an addition which could significantly increase recycling rates.

Waste mixed plastics (wMP) originating from single-use plastics packaging consists mostly of various grades of polyethylene including LDPE and HDPE. Such wMP have the potential to offer mechanical properties superior to LDPE alone. In addition, using such wMP can eliminate the need for separating the plastics in the recycling stream saving cost.

There is an abundance of published literature focusing on the use of recycled plastics, investigating polyolefin blends such as PE and PP (both virgin and recycled) [6–16], and studying the properties of various blends of virgin and recycled PE [17–28]. However, there is a noticeable gap in the literature characterising waste mixed plastics comprising of different grades of PEs and PPs of unknown composition.

Al-Attar [17] investigated virgin LDPE/LLDPE (vLDPE/vLLDPE) blends. The study showed that incorporating vLLDPE into vLDPE results in higher mechanical properties than vLDPE alone. This correlated with the research carried out by Cho et al. [29] who reported synergistic characteristics in vLLDPE/vLDPE blends, which exhibited higher tensile strength at yield and elongation at break than the vLLDPE, despite the lack of co-crystallisation in the blends. Similarly, Luyt et al. [20] investigated the thermal and mechanical properties of vLLDPE/vLDPE/wax ternary blends, noting that as the proportion of vLDPE in the blend increases, the mechanical properties decrease due to the decreasing crystallinity.

Rana [18] investigated the variation in properties by increasing vLLDPE content in vLLDPE/vHDPE blends. A minimal variation was revealed in mechanical properties between the blends containing between 20%–80% vLLDPE, with the plateaued values slightly lower than that of 100% vHDPE. This study is promising considering the natural material fluctuation in collected recyclable plastics; however, both vLLDPE and vHDPE have much more linearly packed polymer chains than vLDPE, which makes up a large proportion of waste plastic packaging films.

When utilising wMP, a considerable variation in material type is expected, and also variations in grades of materials, their molecular weights, and proportion of each material in each batch of mixed waste. Bai et al. [19] investigated the effect of molecular weight (MW) on the mechanical and rheological properties of HDPE blends. The high MW samples had superior impact strength, but the impact strengths of the blends were lower than the rule of mixtures theoretical values. However, the authors observed that the miscibility in the melt state and the solid-state differs and that during cooling and crystallisation there is likely to be phase separating, resulting in poor mechanical properties.

The studies investigating polyethylene blends mentioned previously have only considered the effects on virgin materials. There has been some studies on blending recycled PEs/polyolefins in different proportions. Yousif et al. [24] studied different blends of recycled LDPE (rLDPE), rHDPE, and rPP. Blending the three with different proportions was shown to improve the tensile stress up to 14–25% higher than the individual polymers alone, confirming the prospective increase in recycled plastics quality through blending.

Although many virgin PEs are shown to have better mechanical properties than their recycled counterpart, if higher properties are required, there is the potential of making blends containing both virgin and recycled PEs. For example, Cecon et al. [22] investigated the effect of adding different amounts of post-consumer recycled polyethylene (PCRPE) on virgin polyethylenes of different densities (vLDPE, vLLDPE, vMDPE, and vHDPE). Both

LDPE and LLDPE blends displayed increases up to 75% in the tensile modulus and 56% in the yield strength compared to those without PCRPE. MDPE and HDPE blends, however, presented decreases up to 70% and 56% in tensile modulus and yield strength, respectively, compared to their virgin counterparts, which is expected given the lower crystallinity of the blends.

With the exception of Cecon [22] and Yousif [23], there have not been many other studies focusing on the characterisation of recycled LDPE blends. Additionally, Cecon [22] blended recycled and virgin polymers together, which likely gives an overestimation of the expected properties of rPE blends. Mihrabi-Mazidi et al. [30] characterised rHDPE and rPP blends both of which have higher mechanical properties than LDPE, and Yousif's investigation [23] focused only on the tensile properties. These studies are not characteristic of a true recycling stream which, without segregation, could consist of LDPE, HDPE, LLDPE, PP, PET, and many more thermoplastics. Furthermore, these studies investigated uniformly mixed blends, through extrusion or injection moulding, rather than utilising the polymers as-received, which adds cost to the recycling process. The lack of comprehensive studies focusing on waste LDPE blends must be addressed if plastic packaging wastes are to become more widely recycled, given the high proportion of LDPE in plastics packaging.

The lack of comparative studies considering recycled polyolefin blends and virgin polyolefin properties, highlights the need for a benchmark recycled material database. This would be crucial in order to compare and understand the difference in using waste mixed plastics over segregated recycled plastics.

This present study characterised waste mixed plastics from packaging film wastes alongside segregated recycled and virgin plastics. The different types of plastics present in the wMP was determined using differential scanning calorimetry. The thermal stability of the materials was determined using thermogravimetric analysis to ensure the materials do not undergo thermal degradation during the processing steps. The processing was carried out using wMP shreds in the as-received condition, as well as after an intermediate extrusion step. The processing parameters were varied to understand their effects on the mechanical properties. The tensile, flexural, Izod impact tests were carried out to compare the material properties. Scanning electron microscopic (SEM) images were taken to assess the tensile fracture surfaces of the materials.

2. Experimental
2.1. Materials

The primary material of focus in this study is waste mixed plastics (wMP), washed, shredded, and supplied by PALTECH (Polymer Alloy Technology, Clara, Ireland) [31]. PALTECH takes the plastic wastes originated from food packaging from Tesco Ireland stores (Dublin, Ireland) [32]. These mixed plastics packaging wastes undergo one stage of sorting, which is float-sink separation. The wMP that float on water are collected by PALTECH, therefore these contain mainly polyolefins, having densities lower than water. In addition to the wMP, segregated individual recycled polyolefins were also investigated such as recycled low-density polyethylene (rLDPE) and recycled linear low-density polyethylene (rLLDPE), both in pellet form; recycled high-density polyethylene (rHDPE) and recycled polypropylene (rPP), both in the form of flakes; and virgin low-density polyethylene (vLDPE) in pellet form (supplied by PLASTISERVE Ltd., Leeds, UK). The segregated individually recycled polyolefins will be referred to as "recycled plastics" throughout the remainder of the paper. As this study investigates recycled plastics, the materials used are likely to be a blend of multiple grades of each segregated plastic. The materials used in this study are shown in Figure 1. A block of extrusion moulded wMP, measuring 400 × 10 × 1.5 mm, was also supplied by PALTECH to compare with the wMP shreds (Figure 1g) and this is referred as 'extruded block' in the rest of the manuscript.

Figure 1. (**a**) rLDPE pellets, (**b**) rPP flakes, (**c**) rHDPE flakes, (**d**) rLLDPE pellets, (**e**) vLDPE pellets and (**f**) shredded wMP and (**g**) extruded block.

2.2. Manufacturing

As shown above in Figure 1, the starting materials were in different forms; shredded and pellets. These materials were directly made into plaques by compression moulding. An additional plaque was produced by using the extruded block of the shredded wMP (Figure 1g). This will give a direct comparison if an intermediate extrusion step is required to melt-mix the shredded wMP into a uniform material before converting them into products by compression moulding. The elimination of any intermediate processing step could save money and facilitate the reuse and recycling of such low value plastics wastes that generally end up in landfill.

Plastic plaques were manufactured by compression moulding with a PEI lab 450 hydraulic press, using a two-part closed mould (Figure S1). The plastic plaques were of dimensions 280 mm × 280 mm × 3 mm (Figure 2). The processing cycle used in the hydraulic press was as follows:

1. Heating from 20 °C to 180 °C at a rate of 10 °C/min.
2. Holding at 180 °C for 10 min at 2 bar pressure.
3. Cooling from 180 °C to 20 °C at 10 °C/min at 2 bar pressure.

Some plaques had to undergo additional processing cycles in order to include enough material in the mould, thereby ensuring that the material quantity is consistent across all plaques. The amount of additional cycles required for each material varied due to the difference in the bulk densities of the different material forms (Table 1).

Three sets of studies were carried out with the wMP plaques. In the first set, wMP were characterised in comparison to both the recycled plastics and vLDPE. In the second set, the effect of processing pressure was studied on the quality and the mechanical properties of wMP plaques. The third set compared the mechanical properties of wMP plaques manufactured using two different processing methods; compression moulding of wMP shreds, and compression moulding of the extruded wMP block. The three sets of samples manufactured are described in Table 1.

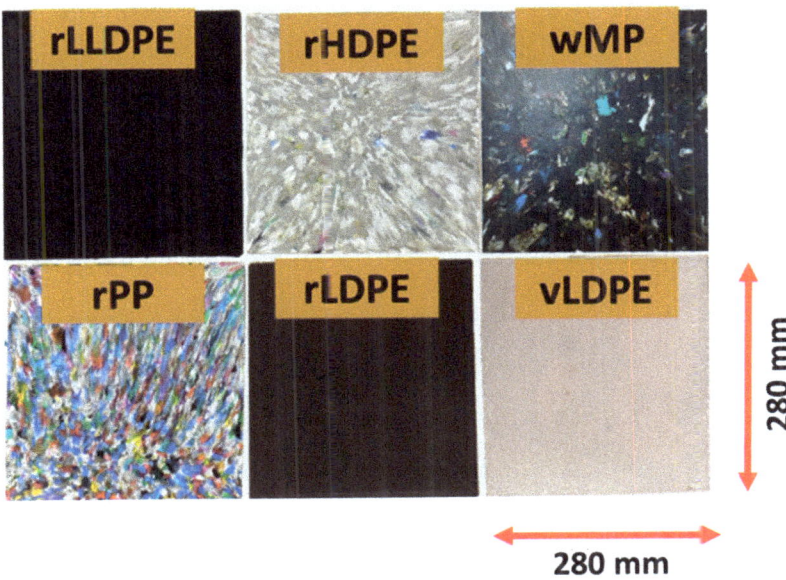

Figure 2. Manufactured plaques (280 mm × 280 mm × 3 mm) made from various recycled plastics, vLDPE, and wMP.

Table 1. Summary of recycled plastics, wMP, and vLDPE plaques manufactured by compression moulding.

Plaque	Material	Processing Pressure (bar)	No. of Processing Cycles
wMP-2 *	Waste mixed plastic	2	4
rLDPE	Recycled low-density polyethylene	2	2
rLLDPE	Recycled linear low-density polyethylene	2	5
rHDPE	Recycled high-density polyethylene	2	2
rPP	Recycled polypropylene	2	2
vLDPE	Virgin low-density polyethylene	2	4
wMP-6	Waste mixed plastic	6	4
wMP-10	Waste mixed plastic	10	4
wMP-ex	Compression moulding (from extruded block)	2	1

* wMP-2 denotes wMP plaque manufactured under 2 bar pressure.

2.3. Test Methods

2.3.1. Density

The density of each plaque was measured using an Ohaus density determination kit. Six samples were measured from each set. The following equation was used to determine the density of each sample, and an average was taken for each plaque.

$$d_{sample} = \frac{mass\ in\ air\ (g)}{mass\ in\ distilled\ water\ (g)} \times d_{distilled\ water} \left(\frac{g}{cm^3}\right)$$

2.3.2. Differential Scanning Calorimetry

Differential scanning calorimetry (DSC) analysis was performed on 5 random wMP shred samples to identify the various plastics present in it, and to determine the processing temperature. The analysis was carried out using a Perkin Elmer Pyris Thermal analyser, in a nitrogen atmosphere from 25 °C to 200 °C at a heating rate of 10 °C/min.

2.3.3. Thermogravimetric Analysis

Thermogravimetric analysis (TGA) was carried out with a Mettler Toledo, USA TGA/DSC 1 analyser between 25 °C to 600 °C at a rate of 10 °C/min in a nitrogen atmosphere to determine the thermal stability of the samples over that temperature range. TGA results were also used to understand if there is any effect from any additives present in the wMP/recycled plastics on their thermal degradation behaviour. To observe the thermal stability of the material at the processing temperature (180 °C), an isothermal TGA cycle was run from 25 °C to 200 °C in air at a rate of 10 °C/min and held at this temperature for 1 h before cooling back to 25 °C.

2.3.4. Tensile Testing

Tensile testing was used to determine the tensile strength and modulus of each type of plastic. The tests were carried out following ASTM D638, using six dumbbell-shaped specimens of Type IV from each plaque. Each sample was loaded at a speed of 5 mm/min until failure, and the values for the load and extension were recorded using the Bluehill® testing software (version 3.61).

2.3.5. Flexural Testing

Flexural testing was carried out following ASTM D790 (American Society for Testing and Materials, West Conshohocken, PA, USA), using six samples from each plaque with dimensions of 61 mm × 13 mm × 3 mm. The testing speed used was 1.3 mm/min and the span-to-thickness ratio was 16:1 following the standard.

2.3.6. Impact Testing

Izod impact testing was carried out to determine the impact strength of the plastic samples. This was carried out following ASTM D256 (American Society for Testing and Materials, West Conshohocken, PA, USA) and using a CEAST 6545 impact tester with a 5.5 J pendulum.

2.3.7. Microscopy

Scanning electron microscopy (SEM) (JEOL, Tokyo, Japan) was used to examine the tensile fracture surface of the specimens using a JEOL JSM series instrument. Each sample was prepared by gold sputtering to increase the conductivity, and the voltage used to observe the samples was 15 kV. Scanning electron microscopy was also used to see the quality of consolidation of each plastic plaque.

2.3.8. Statistical Analysis

The results were analysed using statistical methods to determine the significance of the tensile and flexural test results. As this study focuses on mixed waste plastics, the types of plastics present per batch varies. A statistical t-test can be used to determine whether there is a significant difference in the means of two groups of values [33]. Therefore, the t-test method (unequal variance) was used to determine whether the difference in the test results was due to the investigation or due to the random nature of using mixed plastics. The equation used to calculate the t-test value is as follows:

$$t = \frac{(x_1 - x_2)}{\sqrt{\frac{(s_1)^2}{n_1} + \frac{(s_2)^2}{n_2}}}$$

where:
t = t-test value
x = sample set mean
s = standard deviation
n = number of samples

Higher *t*-test values indicate that there is a significant difference between the two groups of values. A small *t*-test value indicates that the groups are similar, and there is no statistically significant difference between the means. From the *t*-test, a *p*-value is determined through comparison tables. If the corresponding *p*-value is less than a chosen value, for example $\alpha = 0.05$ for a 95% confidence interval, or $\alpha = 0.01$ for a 99% confidence interval, then the means of the two groups are statistically different.

3. Results and Discussion

The wMP samples were first characterised with DSC and TGA to identify the types of polymers present and to understand their thermal stability.

3.1. Thermal Characterisation

3.1.1. Characterisation by DSC

Five wMP samples were randomly selected and subjected to DSC. Three representative DSC curves of the wMP samples are shown in Figure 3. The melting was observed mostly between 105–135 °C, with small melting traces at 170 °C in one of the samples. These results allowed the processing temperature to be set at 180 °C.

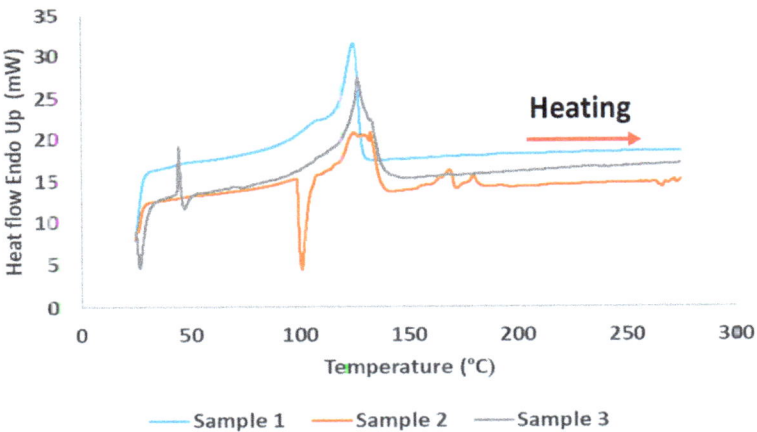

Figure 3. DSC thermograms showing the heat flow for three samples of wMP.

The melting peaks, and the typical melting ranges of different grades of polyethylenes and polypropylene observed in literature are shown in Table S1. The plastics identified in the DSC thermograms of wMP are primarily different grades of PE, with a small trace of PP in one of the samples.

The DSC data was used to calculate the crystallinity of the as-received plastic samples. The average crystallinity values are shown in Table S2. The average crystallinity of the wMP samples was calculated as around 40%. The wMP comprises of different grades of polyethylene that are commonly used as packaging films. Though some HDPE might be present, the wMP is mostly composed of different grades of LDPEs. Hence, the average crystallinity of wMP was found to be very similar to the LDPE samples, both recycled and virgin.

3.1.2. Thermogravimetric Analysis

The TGA curves in Figure 4 show that the wMP has a marginally lower degradation onset temperature than the other recycled plastics tested, at around 350 °C. This is however much higher than the processing temperature of 180 °C, and hence not likely to cause any thermal degradation during processing.

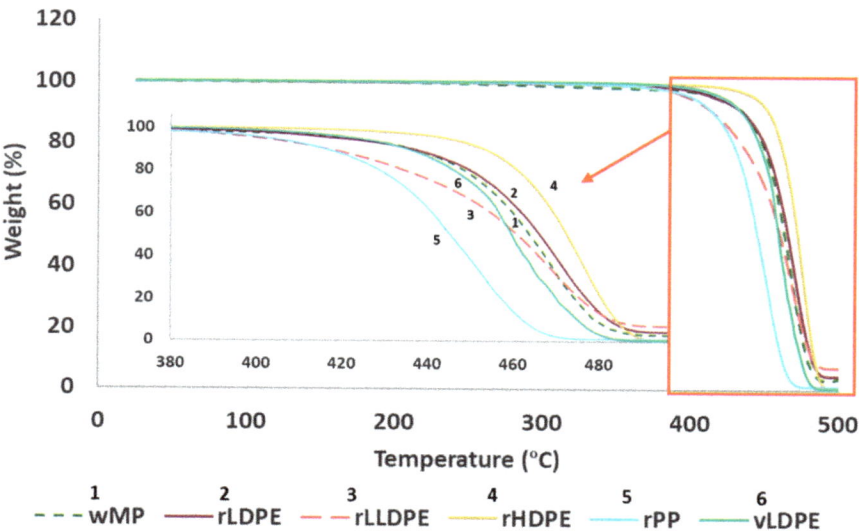

Figure 4. Thermal degradation profile of all materials from 25–600 °C in nitrogen.

The isothermal TGA results observed after holding the samples at 200 °C for 1 h in air showed no loss in weight (Figure S2). This indicates that the processing temperature at 180 °C is not likely to cause thermal degradation of any of the materials. This also indicates that there should be no effect on the mechanical properties of the plaques that experienced additional processing up to 4 cycles.

3.2. Tensile Testing

The tensile properties of the samples, described in Table 1, are given below.

3.2.1. Comparison of wMP, Recycled Plastics, and vLDPE

The representative stress–strain curves, the tensile properties, and the breaking strain of the wMP samples in comparison to recycled plastic samples and vLDPE (Table 1) are shown in Figure 5.

The tensile stress–strain curves are typical of ductile thermoplastics. The stress–strain curve of wMP-2 resembles LDPEs, but the presence of some HDPE and rPP is evident as the initial slope is higher at the start (Figure 5a).

The tensile strength and modulus of the wMP-2 and other plastics are shown in Figure 5b. The tensile strength of the wMP is found to be the lowest of all the tested materials, having a value of 7.3 ± 0.7 MPa. As mentioned before, the wMP-2 plaque was compression moulded directly from the wMP shreds and the samples were made of several shreds bonded together. As the shreds are different grades of PE, there is therefore non-uniformity across the sample, and its performance is dictated by the bonding between the different shreds. The bonding between the shreds influences the tensile strength of the material, which will be discussed later in this section. The rHDPE and rPP samples were also initially in flake form, but these were recycled materials with consistent shape and size even if the grades were different. The wMP, on the other hand, consisted of different materials, different grades of materials, different size, shape, and thickness, which created additional non-uniformity.

The tensile modulus of wMP-2 is lower than that of the rPP and rHDPE, but is higher than the LDPEs tested (Figure 5b). The increase in modulus of wMP-2 was statistically significant according to a t-test analysis with $p = 0.02$ and $\alpha = 0.05$. This higher modulus observed in wMP-2 compared to the rLDPE, rLLDPE, and vLDPE could be to be due to the effect of rHDPE and rPP present in the mix.

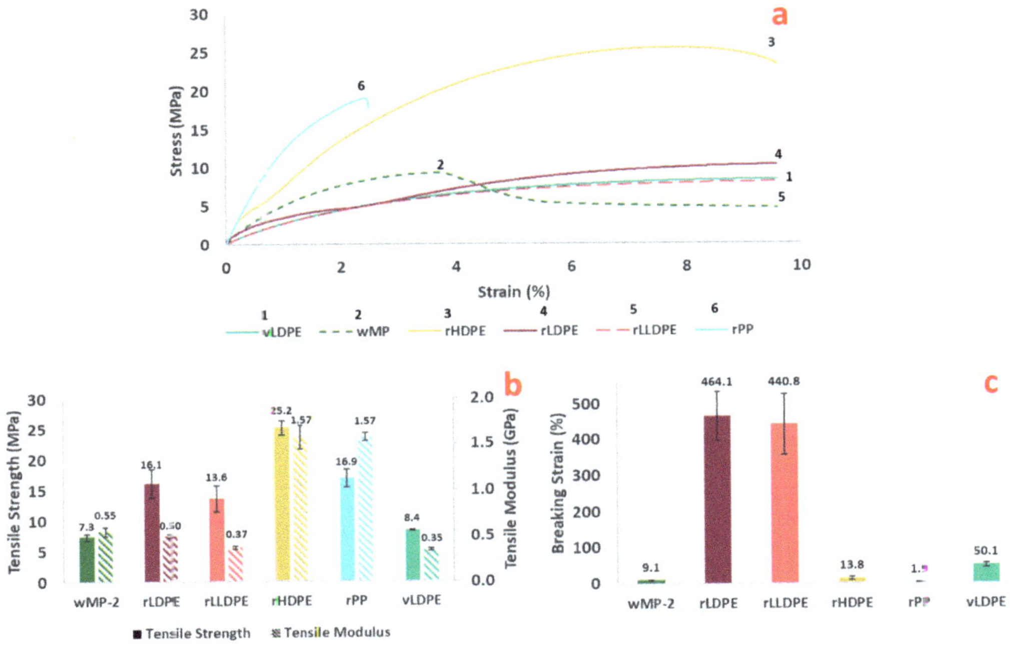

Figure 5. Graphs showing (**a**) representative tensile stress–strain curves, (**b**) tensile strength and tensile modulus, and (**c**) breaking strain of wMP-2, recycled plastics and vLDPE.

The rLDPE and rLLDPE samples exhibited much higher breaking strains than the other recycled plastics (Figure 5c). The breaking strain of vLDPE is found to be lower than that of rLDPE, which might be attributed to the fact that rLDPE contains different grades of LDPE, some of which are likely to have higher mechanical properties than the single grade of vLDPE investigated in this study. This highlights the difficulty in directly comparing virgin with recycled materials, as the recycled materials are likely to contain a mix of different grades.

LDPE and LLDPE are less crystalline than HDPE and PP and can therefore extend to higher elongations. This is because the proportion of amorphous regions are higher, which means that the polymer chains in these regions are able to detangle and uncoil much more than the crystalline regions, leading to higher elongations before breaking.

The wMP-2 had the lowest breaking strain of all the materials tested except rPP. The wMP-2 shreds were pressed together during manufacture and there was no melt mixing, so the individual plastic shreds were just adhered to one another during consolidation. During tensile testing, the force required to separate the individual plastic shreds was lower than that required to stretch the individual shreds, therefore the wMP-2 samples failed with very low elongation.

As wMP batches can vary, it is important to have an understanding about the range of variation possible in their tensile strength and tensile modulus values. Figure 6 shows the range of tensile strength and tensile modulus values that covers different grades of virgin PEs including HDPE. The experimentally measured tensile strength and tensile modulus values of wMP-2 and the recycled plastics fit well within that range, as shown in Figure 6. Any variation in the composition of wMP, originating from the PE-based packaging film wastes, are likely to vary between these two ranges.

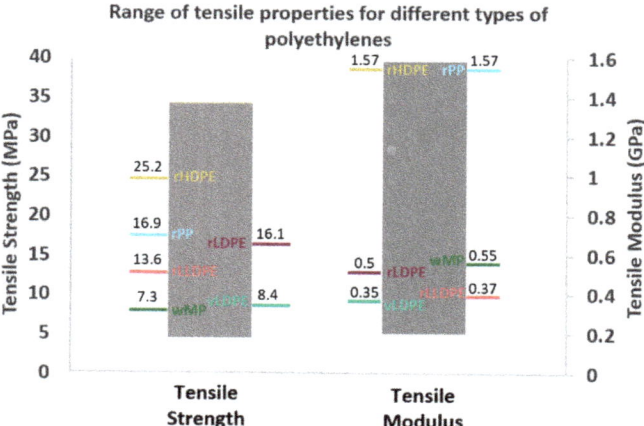

Figure 6. The possible range of variation of tensile strength and tensile modulus values of different types of recycled polyethylenes, polyethylene-based wMP as well as their measured values.

3.2.2. Comparison of wMP Compression Moulded under Different Pressures

The tensile results of the wMP samples manufactured under three different consolidation pressures are shown in Figure 7. The tensile results indicate that increasing the consolidation pressure of the shredded wMP from 2 bar to 10 bar did not influence the tensile strength of the wMP samples. Increasing the pressure was found to increase the tensile modulus by 8% (from 2 bar to 10 bar), but this is not a statistically significant result, with p = 0.26 and α = 0.05.

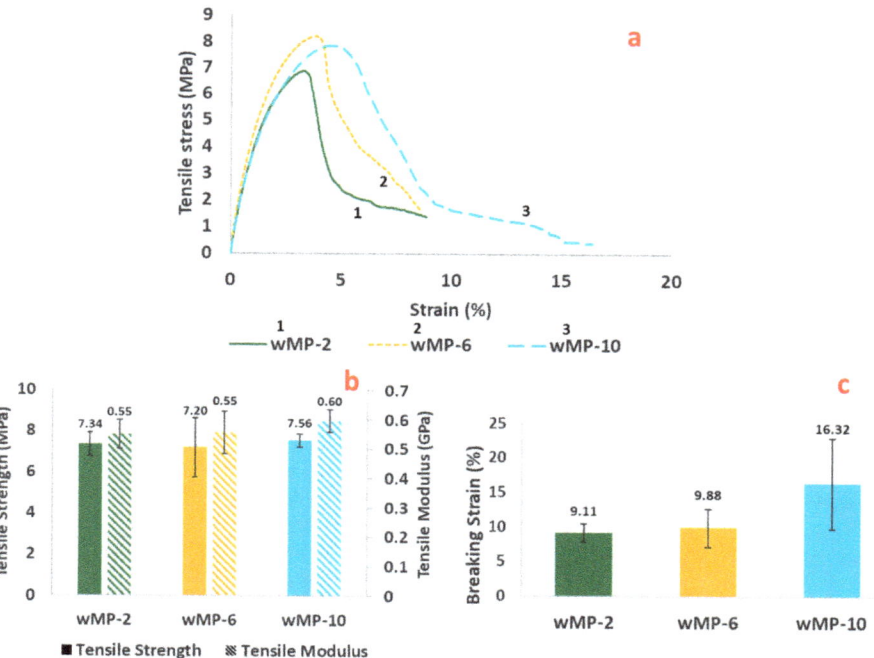

Figure 7. (**a**) representative tensile stress–strain curves, (**b**) tensile strength and tensile modulus, and (**c**) breaking strain of wMP processed by compression moulding at different pressures.

The average breaking strains of the three sets of samples show that increasing the pressure from 2 to 10 bar increases the breaking strain by 79%. This result was not statistically significant for $p = 0.06$ and $\alpha = 0.05$, and the increase might be due to the random proportion of different materials in the mix.

A schematic is shown in Figure 8 displaying the possible effects that the consolidation pressure might have on the compaction of the shredded plastics during compression moulding. This schematic is based on the measured density values of the samples, as shown in Figure 8. The higher densities indicate a higher level of compaction at higher consolidation pressures, but that did not influence the tensile properties of the wMP.

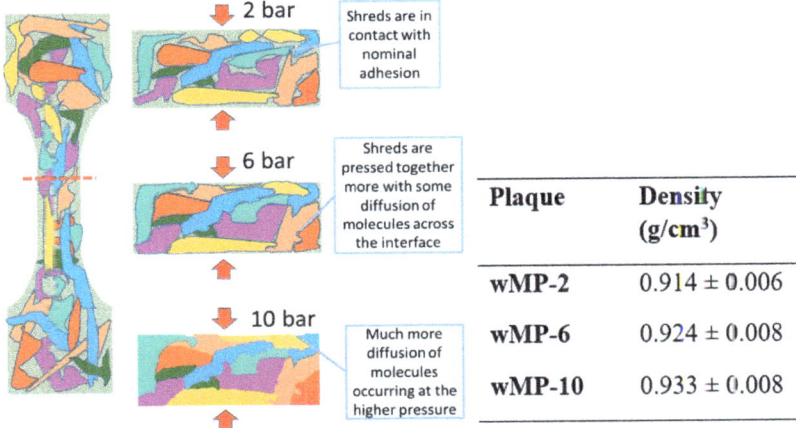

Figure 8. Different extents of adhesion between individual wMP shreds due to different consolidation pressures leading to changes in the densities.

During compression moulding, the wMP shreds melt but they do not flow as happens in the case of extrusion or injection moulding. Under heat and pressure, the molten material at the interface of two adjacent shreds fuse together. At higher pressures, it is easier for the shreds to pack together, facilitating diffusion of molecules from one molten shred to the adjacent one, allowing for higher compaction and increased density.

3.2.3. Comparison of wMP Manufactured Using Different Processing Routes

The wMP plaques reported above were manufactured by compression moulding of wMP shreds. As mentioned in Section 2.1, an equivalent wMP plaque was prepared by compression moulding a pre-extruded wMP block. This block was prepared by extrusion moulding the wMP shreds into the form of a block. A piece was cut from this extruded block and compression moulded into a plaque of equal weight and dimensions as the previous wMP plaque, and using the same processing parameters. This is shown schematically in Figure 9.

The tensile results of the two sets of samples are shown in Figure 10. The effect of different processing methods on the tensile properties of wMP are discussed below.

It is evident that the failure behaviours are different in the two samples (Figure 10a). The wMP-2 exhibited a gradual drop in load after reaching the peak, resulting in a higher breaking strain (0.0911). While the drop in load in wMP-2-ex was sudden, resulting in a lower breaking strain value (0.0591). The higher peak load and sudden drop in load in wMP-2-ex samples indicates that the sample behaved more like a uniform single material and was able to sustain a higher load before failure.

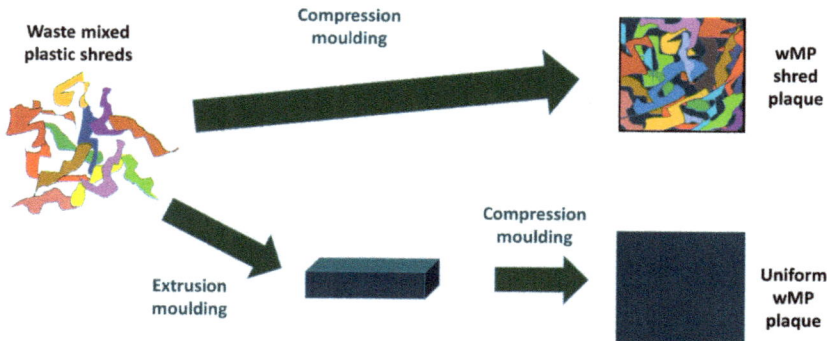

Figure 9. Schematic showing the different processing routes of manufacturing the wMP plaques.

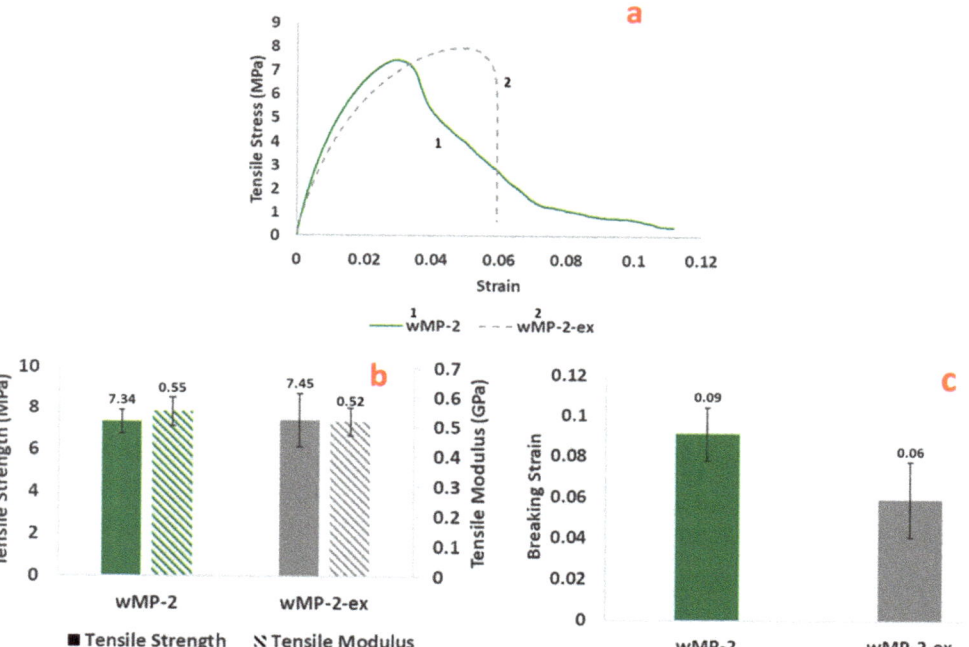

Figure 10. (**a**) Representative tensile stress–strain curves, (**b**) tensile strength and tensile modulus, and (**c**) breaking strain of wMP manufactured with different processing methods.

The tensile strength values for both the samples are quite similar, as shown in Figure 10b. The wMP-2 has a marginally higher tensile modulus, but this value is not statistically significant, with $p = 0.86$ and $\alpha = 0.05$. These results indicate that an intermediate extrusion step before compression moulding might not have any additional benefit on the tensile strength and tensile modulus values of the shredded wMP materials. Any intermediate processing step adds a cost to the process, which is not desired while reprocessing such low-cost waste materials. These results therefore indicate that shredded wMP can be directly used for producing compression moulded products. The difference in failure behaviours is explained schematically in Figure S3.

The higher strength wMP shreds in the gauge length of the tensile test specimens get stretched during tensile testing when the weaker shreds break, or the shreds are pulled out from each other. This stretching, breaking and pulling-out of individual plastic shreds one after another causes the gradual failure of the sample (Figure S3a).

On the other hand, the wMP-2-ex samples were uniformly mixed and the failure happened suddenly when the sample fractured as a single material, as shown in Figure S3b.

3.3. Flexural Testing

The flexural test results of the investigated materials described in Table 1 are given in Figure 11 below.

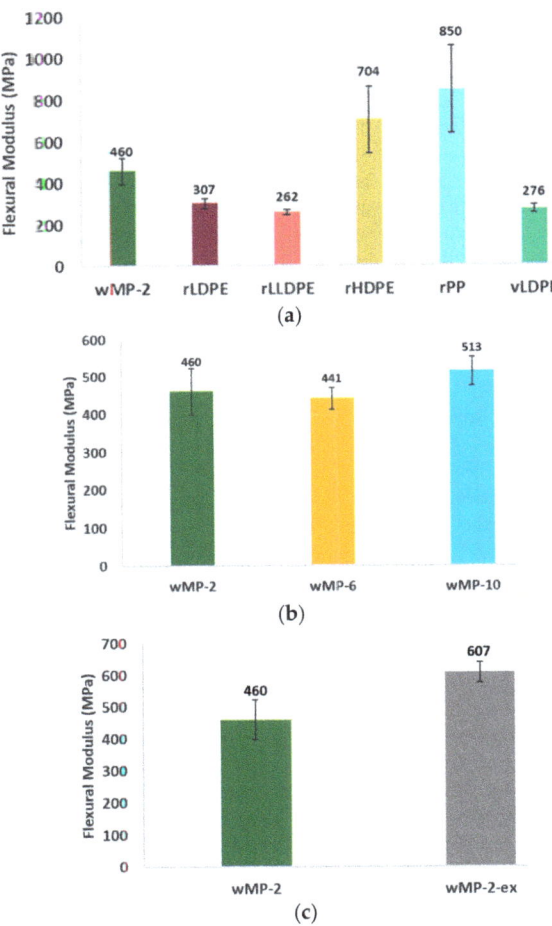

Figure 11. Flexural modulus of wMP (**a**) compared to recycled plastics and virgin LDPE, (**b**) compression moulded under different pressures, and (**c**) manufactured using different processing routes.

3.3.1. Comparison of wMP, Recycled Plastics, and vLDPE

Similarly to the tensile modulus, the flexural modulus of the wMP-2 samples is lower than rHDPE and rPP, but is noticeably higher than the other LDPEs, about 33% and 40% higher than rLDPE and vLDPE, respectively (Figure 11a). The flexural modulus of the wMP is determined by the inherent rigidity of the individual shreds. As wMP contains some proportions of HDPE and PP, which have higher crystallinity, it is expected that the wMP flexural modulus will be higher than the other LDPEs. That is evident in Figure 11a. In flexural testing, both compressive and tensile forces act on the specimens. Due to the nature of the loading, the possibility of the individual shreds being debonded from each other during testing is less pronounced here than that in the tensile testing.

3.3.2. Comparison of wMP Compression Moulded under Different Pressures

Figure 11b indicates that increasing the consolidation pressure did not affect the flexural modulus of the wMP samples. During three-point bending, the applied force is compressive on the top side while tensile on the bottom side of the sample. As mentioned above, the applied tensile force might not be large enough to separate the shreds as seen in the case of a tensile test. Therefore, the flexural modulus was not much influenced by the level of compaction and hence, the level of adhesion between the adjacent wMP shreds was not the dominant factor. This is supported by the test results as the flexural modulus values of the samples manufactured under different consolidation pressures were very similar.

3.3.3. Comparison of wMP Manufactured Using Different Processing Routes

The flexural modulus of wMP-2-ex is 32% higher than that of wMP-2, which is statistically significant for a p-value of 0.0006 and $\alpha = 0.001$. Therefore, it is clear that an intermediate extrusion step before compression moulding can significantly increase the flexural modulus. This can be attributed to the fact that the wMP-2-ex sample is much more uniform than the wMP-2. During three-point bending, the sample experiences both tension and compression, and in tensile testing there are only tensile forces. Therefore, it might be reasonable to assume that extruding the wMP is advantageous for increasing compressive properties.

3.4. Impact Testing

The impact test results of the investigated materials are given in this section.

3.4.1. Comparison of wMP, Recycled Plastics, and vLDPE

The determination of the impact strength relies on the sample failing under impact. Many of the materials tested for impact (rLDPE, vLDPE, rLLDPE) did not break with the applied 5.5 J pendulum and a comparison has been made based on the observed results.

The wMP-2 samples exhibited an impact strength 41% higher than rPP and 74% lower than rHDPE (Figure S4). It is clear from the error bars that there is a wide variation in the values measured for rHDPE, but the impact strength of rHDPE was confirmed to be statistically different from wMP-2 and rPP. The impact testing results showed that rLDPE had a higher impact strength than rHDPE as the rLDPE sample did not break during testing. As the highest proportion of polymer in wMP is LDPE, it was therefore expected that the wMP-2 sample would also have a higher impact strength than rHDPE. The average impact strength of wMP-2 was in fact lower, therefore the inclusion of other plastics in wMP-2 increased the brittleness of the sample compared to LDPE alone, reducing the impact strength. This agrees with results found in the literature, such as Shebani's investigation [28] which showed that the impact strength of vHDPE/vLDPE blends increased as the LDPE content increased.

3.4.2. Comparison of wMP Compression Moulded under Different Pressures

The impact strength comparison of wMP manufactured under different consolidation pressures is shown in Figure S5.

The impact strength of wMP samples at each pressure appears to be similar, with the average impact strength of wMP-2 only marginally lower. It was confirmed through statistical t-test that increasing the pressure has no effect on the impact strength, with $p = 0.26$ and $\alpha = 0.05$.

3.4.3. Comparison of wMP Manufactured Using Different Processing Routes

The impact strength of wMP-2 and wMP-2-ex were compared. The wMP-2-ex sample did not break during impact testing and so the impact strength could not be measured, as shown in Figure S6.

LDPE is a ductile and low modulus plastic with high toughness, so is unlikely to break during impact testing. As wMP are mainly composed of LDPE, it was expected that these samples also would not break, however this was not the case. It is possible that the wMP-2 sample only broke under impact due to the voids present at the interface of the individual plastic shreds, which created weak areas in the material. As the voids in wMP-2-ex are much smaller and more distributed, the voids did not impact the material breaking during testing. The voids are shown in SEM images in Section 3.5.2.

3.5. Fracture Surface Analysis

3.5.1. Comparison of wMP, Recycled Plastics, and vLDPE

Figure S7 shows the macroscopic tensile failure behaviour of the representative samples and Figure 12 shows the SEM micrographs of the tensile fracture surfaces of the samples.

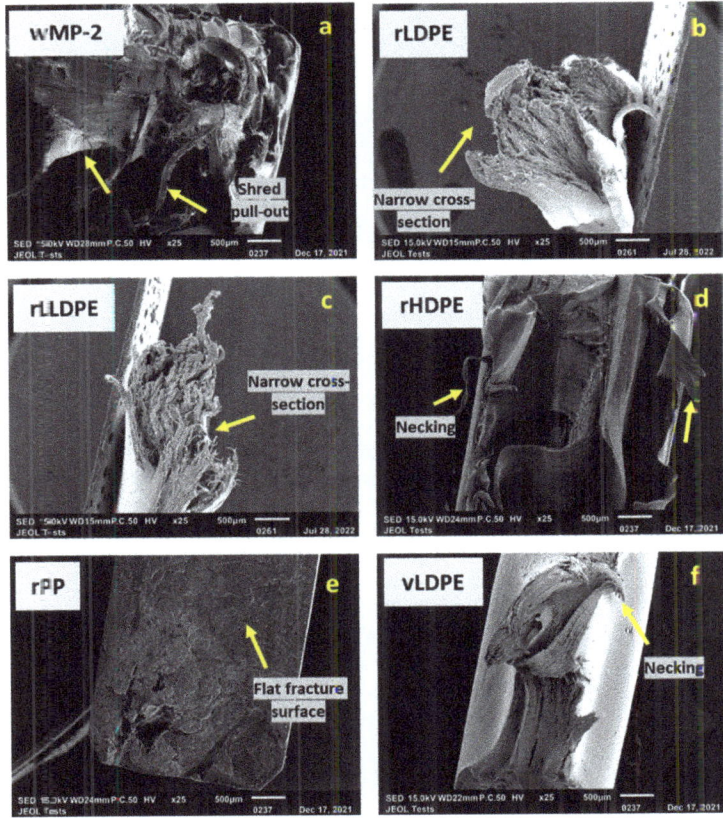

Figure 12. SEM images of the fracture surface of tested tensile specimens at ×25 magnification for (**a**) wMP-2, (**b**) rLDPE, (**c**) rLLDPE, (**d**) rHDPE, (**e**) rPP, and (**f**) vLDPE; showing varying degrees of stretching and pull-out of individual shreds in wMP, and narrow cross section of rLDPE and rLLDPE.

In wMP-2 (Figure S7a and Figure 12a), it is observed that some plastic shreds elongated more than others, which is likely due to the presence of different grades of PE shreds. Some shreds experience higher pull-outs and stretching during testing than others. Both rLDPE and rLLDPE showed high elongation during tensile testing as shown in Figure S7b,c. It can also be seen that both rLDPE and rLLDPE have much narrower cross-sections than the other samples, which is due to the large amount of necking that occurred during tensile testing (Figure 12b,c). rHDPE shows some elongation before failure, as observed in Figure S7d and Figure 12d. From Figure S7e and Figure 12e, it is clear that rPP has a nearly flat fracture surface without much elongation vLDPE exhibited some extent of necking, as shown in Figure S7f and Figure 12f, but not as high as rLDPE and rLLDPE.

The higher magnification SEM images of the tensile fracture surfaces in Figure S8a–d look quite different. Ductile deformations are clearly visible in rHDPE and varying degrees of stretching is visible in wMP. The SEM images of rLDPE and rLLDPE under higher magnification (Figure S8b,c) show that both samples appear very similar to wMP-2 rather than vLDPE (Figure S8f). Although the fracture surface of rPP appeared mostly flat in Figure S7e and Figure 12e, some micro ductility was observed in Figure S8e. The fracture surface of vLDPE exhibited a lamellar structure that is different to the other samples, which was also not observed in the rLDPE sample (Figure S8b), which appears more fibrillar. This could be due to the difference between virgin and recycled LDPEs or due to the grade of the vLDPE.

3.5.2. Comparison of wMP Manufactured Using Different Processing Routes

The tensile fracture surfaces of the wMP-2 and wMP-2-ex samples are shown in Figure 13.

The tensile fracture surface appeared more flat for wMP-2-ex compared to that of wMP-2 (Figure 13), which is characteristic of a sudden failure (Figure 10a). At a higher magnification ($\times 750$), both samples looked quite similar (Figure 13c,d). This similarity at a microscopic level could explain the comparable tensile strength despite of the contrasting failure behaviour.

There are voids visible in both wMP-2 and wMP-2-ex samples (Figure 13e,f), but the size and distribution of the voids are very different. It is evident from Figure 13e that the voids present in wMP-2 are bigger in size and mostly concentrated at the boundary between different wMP shreds, and the areas within individual shreds are relatively free of voids. This is not the case with wMP-2-ex, as shown in Figure 13f. Here, the voids are much smaller, but numerous in number and there is no area free of voids.

This difference is expected, as the voids in wMP-2 are more likely to be found between the different shreds, especially considering the different shrinkage rates of different plastics. Within the shreds themselves there is less chance of voids as each shred is composed of just one type of plastic. Conversely in the wMP-2-ex, the wMP is mixed and no large voids are present as there are no individual shred boundaries. However, in extrusion, there is still shrinkage to consider, and air pockets present, which led to the formation of smaller voids.

Using the SEM images, and the ImageJ (Fiji) version 2.5 software, the average void content of the wMP-2 and wMP-2-ex samples were measured. The void content of wMP-2-ex and wMP-2 were measured as 0.8% and 1.8%, respectively. It was therefore observed that although the void contents were not very different in wMP-2 and wMP-2-ex, the distribution of voids was different.

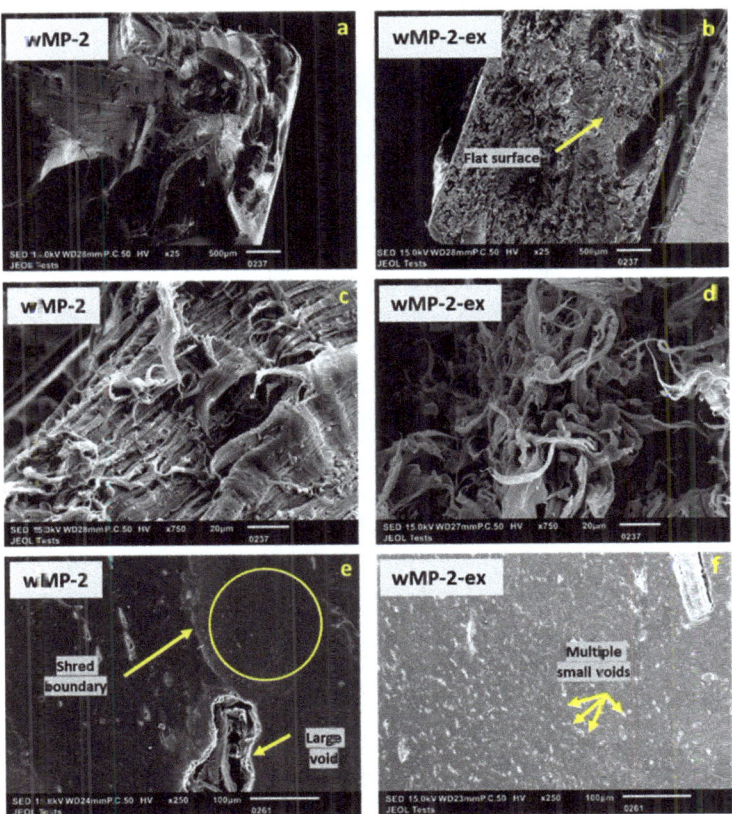

Figure 13. SEM images of the tensile fracture surfaces of wMP-2 and wMP-2-ex at (**a,b**) ×25 magnification, (**c,d**) ×750 magnification, and cross-sections of wMP-2 and wMP-2-ex at ×250 magnification (**e,f**).

4. Conclusions

This research investigated the properties of polyethylene-based waste mixed plastics (wMP) originating from plastics packaging wastes, comparing its properties to individually recycled polyolefin materials and virgin LDPE. The effects of different processing techniques and processing conditions on the mechanical properties of the processed plastic plaques were also investigated. The thermal degradation of the wMP starts at 350 °C in a nitrogen environment. The wMP did not show any thermal degradation when subjected to isothermal TGA at 200 °C for 1 h in air and this indicated that the processing or reprocessing of wMP plaques at 180 °C are not likely to cause any thermal degradation of the material. The properties of the wMP fell well within the range of properties of virgin polyethylene grades (3–33 MPa tensile strength, and 0.4–1.5 GPa tensile modulus). This also gives an idea about the variation in properties that might be seen between different batches of wMP samples. The tensile strength of the wMP is found to be closer to LDPE rather than HDPE and PP. Additionally the failure strains measured in wMP are much lower than rLDPE and rLLDPE. Therefore, the high ductility of LDPE is not observed in wMP.

The tensile modulus of the wMP is 9% higher than rLDPE, and the flexural modulus is 33% higher, therefore both of these properties improve when using a mix of polyolefins/PE grades. While processing shredded plastics directly by compression moulding, the consolidation pressure was not found to have any significant effect on the mechanical properties.

This is an important observation for converting wMP shreds directly into product forms via compression moulding. Addition of an intermediate processing step of extrusion before compression moulding was found to have no effect on the tensile properties. However, a noticeably different failure behaviour was observed. The extra processing step resulted in an increase in flexural modulus by 32%. The shredded plastics are joined together by fusion when subjected to compression moulding only. Whereas an intermediate extrusion step before compression moulding can introduce more intimate mixing between the shredded materials leading to a more uniform material. Thus, different processing routes can bring some d±ifferences in their overall performance regardless of similar property values. The failure of the samples is dictated by what type of loading is applied on them and how they are manufactured. This scientific understanding can help to decide what might be the optimum processing route for low-cost packaging wastes avoiding any additional processing cost.

Supplementary Materials: The following supporting information can be downloaded at: https://www.mdpi.com/article/10.3390/polym14245485/s1, Figure S1: Closed two-part metal mould used to manufacture plastic plaques in hydraulic press.; Figure S2: Change in weight of the samples after holding at 200 oC for 1 h in air.; Figure S3: Schematic showing the failure behaviour in (a) wMP-2 and (b) wMP-2-ex during tensile testing.; Figure S4: Izod impact strength (un-notched) of wMP-2 in comparison to recycled plastics (rPP and rHDPE).; Figure S5: Comparison of Izod impact strength (un-notched) of wMP manufactured under different pressures.; Figure S6: Comparison of Izod impact strength (un-notched) of wMP samples produced by compression moulding of wMP shreds and by compression moulding of an extruded block of wMP.; Figure S7: Representative tensile test specimens of (a) wMP-2, (b) rLDPE, (c) rLLDPE, (d) rHDPE, (e) rPP and (f) vLDPE during tensile testing with additional close up images of (g) rLDPE and (h) rLLDPE, due to the high amount of stretching.; Figure S8: SEM images of the fracture surface of tested tensile specimens at x750 magnification.; Table S1: Typical melting ranges of polyolefin materials [34–38].; Table S2: Crystallinity of wMP and segregated recycled thermoplastics calculated from DSC analysis.; Table S3: Onset and endset temperatures for thermal degradation of wMP and recycled plastics.

Author Contributions: Conceptualization, K.O. and D.R.; formal analysis, K.O.; investigation, K.O., C.W. and J.M.; resources, A.D., K.D., B.C., C.G., C.M.Ó.B. and D.R.; writing—original draft preparation, K.O.; writing—review and editing, A.D., K.D., B.C., C.M.Ó.B. and D.R.; supervision, C.M.Ó.B. and D.R.; project administration, K.O. and D.R. All authors have read and agreed to the published version of the manuscript.

Funding: This research was funded by Johns Manville (USA) and the Doctoral Training Programme (School of Engineering, University of Edinburgh. The APC was funded by Johns Manville (USA) and the Doctoral Training Programme (School of Engineering, University of Edinburgh). Funding acquisition, Y.Y.

Institutional Review Board Statement: Not applicable.

Data Availability Statement: Data is contained within the article or supplementary material.

Acknowledgments: The authors would like to thank PALTECH for providing the waste and recycled materials for this study as well as technical advice, PLASTISERVE for providing the virgin LDPE, and Johns Manville and the Doctoral Training Programme (School of Engineering, University of Edinburgh) for funding this PhD project.

Conflicts of Interest: The authors declare no conflict of interest.

References

1. *The New Plastics Economy: Rethinking the Future of Plastics & Catalysing Action*; Ellen MacArthur Foundation: Cowes, UK, 2016.
2. *Packaging Waste EU Rules on Packaging and Packaging Waste, Including Design and Waste Management*; European Commission: Brussels, Belgium, 2021.
3. Schyns, Z.O.G.; Shaver, M.P. Mechanical Recycling of Packaging Plastics: A Review. *Macromol. Rapid Commun.* **2021**, *42*, 2000415. [CrossRef]
4. Smith, L. *Plastic Waste*; House of Commons Library: London, UK, 2021.

5. Elliot, T.; Elliot, L. *A Plastic Future: Plastics Consumption and Waste Management in the UK*; World Wide Fund for Nature Inc.: Gland, Switzerland, 2018.
6. Charfeddine, I.; Majesté, J.; Carrot, C.; Lhost, O. Surface tension and interfacial tension of polyolefins and polyolefin blends. *J. Appl. Polym. Sci.* **2022**, *139*, 51885. [CrossRef]
7. Finlay, J.; Hill, M.J.; Barham, P.J.; Byrne, K.; Woogara, A. Mechanical properties and characterization of slowly cooled isotactic polypropylene/high-density polyethylene blends. *J. Polym. Sci. Part B Polym. Phys.* **2003**, *41*, 1384–1392. [CrossRef]
8. Jose, S.; Aprem, A.; Francis, B.; Chandy, M.; Werner, P.; Alstaedt, V.; Thomas, S. Phase morphology, crystallisation behaviour and mechanical properties of isotactic polypropylene/high density polyethylene blends. *Eur. Polym. J.* **2004**, *40*, 2105–2115. [CrossRef]
9. Li, J.; Shanks, R.A.; Long, Y. Mechanical properties and morphology of polyethylene-polypropylene blends with controlled thermal history. *J. Appl. Polym. Sci.* **2000**, *76*, 1151–1164. [CrossRef]
10. Madi, N.K. Thermal and mechanical properties of injection molded recycled high density polyethylene blends with virgin isotactic polypropylene. *Mater. Eng.* **2013**, *46*, 435–441. [CrossRef]
11. Sirin, K.; Balcan, M. Mechanical properties and thermal analysis of low-density polyethylene plus polypropylene blends with dialkyl peroxide. *Polym. Adv. Technol.* **2010**, *21*, 250–255.
12. Sirin, K.; Doğan, F.; Çanlı, M.; Yavuz, M. Mechanical properties of polypropylene (PP) plus high-density polyethylene (HDPE) binary blends: Non-isothermal degradation kinetics of PP+HDPE (80/20) Blends. *Polym. Adv. Technol.* **2013**, *24*, 715–722. [CrossRef]
13. Strapasson, R.; Amico, S.; Pereira, M.F.; Sydenstricker, T. Tensile and impact behavior of polypropylene/low density polyethylene blends. *Polym. Test.* **2005**, *24*, 468–473. [CrossRef]
14. Wang, J.; Dou, Q. Polypropylene/linear low-density polyethylene blends: Morphology, crystal structure, optical, and mechanical properties. *J. Appl. Polym. Sci.* **2009**, *111*, 194–202. [CrossRef]
15. Xie, M.; Chen, J.; Li, H. Morphology and mechanical properties of injection-molded ultrahigh molecular weight polyethylene/polypropylene blends and comparison with compression molding. *J. Appl. Polym. Sci.* **2009**, *111*, 890–898. [CrossRef]
16. Zhou, M.; Mi, D.; Hou, F.; Zhang, J. Tailored Crystalline Structure and Mechanical Properties of Isotactic Polypropylene/High Molecular Weight Polyethylene Blend. *Ind. Eng. Chem. Res.* **2017**, *56*, 8385–8392. [CrossRef]
17. Al-Attar, F.; Alsamhan, M.; Al-Banna, A.; Samuel, J. Thermal, Mechanical and Rheological Properties of Low Density/Linear Low Density Polyethylene Blend for Packing Application. *J. Mater. Sci. Chem. Eng.* **2018**, *6*, 32–38. [CrossRef]
18. Rana, S.K. Blend of High-Density Polyethylene and a Linear Low-Density Polyethylene with Compositional-Invariant Mechanical Properties. *J. Appl. Polym. Sci.* **2001**, *83*, 2604–2608. [CrossRef]
19. Bai, L.; Li, Y.-M.; Yang, W.; Yang, M.-B. Rheological behavior and mechanical properties of high-density polyethylene blends with different molecular weights. *J. Appl. Polym. Sci.* **2010**, *118*, 1356–1363. [CrossRef]
20. Luyt, A.S.; Hato, M.J. Thermal and mechanical properties of linear low-density polyethylene/low-density polyethylene/wax ternary blends. *J. Appl. Polym. Sci.* **2005**, *96*, 1748–1755. [CrossRef]
21. Robledo, N.; Vega, J.F.; Nieto, J.; Martínez-Salazar, J. Role of the Interface in the Melt-Rheology Properties of Linear Low-Density Polyethylene/Low-Density Polyethylene Blends: Effect of the Molecular Architecture of the Dispersed Phase. *J. Appl. Polym. Sci.* **2011**, *119*, 3217–3226. [CrossRef]
22. Cecon, V.S.; Da Silva, P.F.; Vorst, K.L.; Curtzwiler, G.W. The effect of post-consumer recycled polyethylene (PCRPE) on the properties of polyethylene blends of different densities. *Polym. Degrad. Stab.* **2021**, *190*, 109627. [CrossRef]
23. Yousif, B.; KLow, O.; El-Tayeb, N.S.M. Fabricating and Tensile Characteristics of Recycled Composite Materials. *J. Appl. Sci.* **2006**, *6*, 1380–1383. [CrossRef]
24. Patel, R.M.; Karjala, T.F.; Savargaonkar, N.R.; Salibi, P.; Liu, L. Fundamentals of structure–property relationships in blown films of linear low density polyethylene/low density polyethylene blends. *J. Plast. Film. Sheeting* **2019**, *35*, 401–421. [CrossRef]
25. Freitas, D.M.G.; Oliveira, A.D.B.; Alves, A.M.; Cavalcanti, S.N.; Agrawal, P.; Mélo, T.J.A. Linear low-density polyethylene/high-density polyethylene blends: Effect of high-density polyethylene content on die swell and flow instability. *J. Appl. Polym. Sci.* **2021**, *138*, 49910. [CrossRef]
26. Shen, G.; Shen, H.; Xie, B.; Yang, W.; Yang, M. Crystallization and fracture behaviors of high-density polyethylene/linear low-density polyethylene blends: The influence of short-chain branching. *J. Appl. Polym. Sci.* **2013**, *129*, 2103–2111. [CrossRef]
27. Lu, J.; Sue, H.-J. Morphology and mechanical properties of blown films of a low-density polyethylene/linear low-density polyethylene blend. *J. Polym. Sci. Part B Polym. Phys.* **2002**, *40*, 507–518. [CrossRef]
28. Shebani, A.; Hebani, A.; Klash, A.; Elhabishi, R.; Abdsalam, S.; Elbreki, H. The Influence of LDPE Content on the Mechanical Properties of HDPE/LDPE Blends. *Res. Dev. Mater. Sci.* **2018**, *7*. [CrossRef]
29. Cho, K.; Lee, B.H.; Hwang, K.-M.; Lee, H.; Choe, S. Rheological and mechanical properties in polyethylene blends. *Polym. Eng. Sci.* **1998**, *38*, 1969–1975. [CrossRef]
30. Mehrabi-Mazidi, M.; Sharifi, H. Post-consumer recycled high density polyethylene/polypropylene blend with improved overall performance through modification by impact polypropylene copolymer: Morphology, properties and fracture resistance. *Polym. Int.* **2021**, *70*, 1701–1716 [CrossRef]
31. PALTECH Polymer Alloy Technology. 2022. Available online: https://paltech.ie/ (accessed on 9 January 2022).
32. Tesco Ireland Partners with Paltech to Recycle Soft Plastics. 2021 [Cited 2022]. Available online: https://www.rte.ie/news/business/2021/0204/1195057-tesco-partners-with-paltech-to-recycle-soft-plastics/ (accessed on 10 January 2022).

33. Kim, T.K. T test as a parametric statistic. *Korean J. Anesthesiol.* **2015**, *68*, 540–546. [CrossRef]
34. Selke, S.E.; Hernandez, R.J. Packaging: Polymers for Containers, in Encyclopedia of Materials: Science and Technology. In *Encyclopedia of Materials: Science and Technology*; Buschow, K.H.J., Flemings, M.C., Kramer, E.J., Veyssière, P., Cahn, R.W., Ilschner, B., Mahajan, S., Eds.; Elsevier: Oxford, UK, 2001; pp. 6646–6652.
35. Jin, H.; Gonzalez-Gutierrez, J.; Oblak, P.; Zupančič, B.; Emri, I. The effect of extensive mechanical recycling on the properties of low density polyethylene. *Polym. Degrad. Stab.* **2012**, *97*, 2262–2272. [CrossRef]
36. Jordan, J.; Casem, D.T.; Bradley, J.M.; Dwivedi, A.K.; Brown, E.N.; Jordan, C.W. Mechanical Properties of Low Density Polyethylene. *J. Dyn. Behav. Mater.* **2016**, *2*, 411–420. [CrossRef]
37. Mohammadi, H.; Vincent, M.; Marand, H. Investigating the equilibrium melting temperature of linear polyethylene using the non-linear Hoffman-Weeks approach. *Polymer* **2018**, *146*, 344–360. [CrossRef]
38. Menyhárd, A.; Menczel, J.D.; Abraham, T. Polypropylene Fibers. In *Thermal Analysis of Textiles and Fibers*; Jaffe, M., Menczel, J.D., Eds.; Woodhead Publishing: Sawston, UK, 2020; pp. 205–222.

Article

Optimizing Polymer Costs and Efficiency in Alkali–Polymer Oilfield Applications

Rafael E. Hincapie [1,*], Ante Borovina [1], Torsten Clemens [1], Eugen Hoffmann [2], Muhammad Tahir [2], Leena Nurmi [3], Sirkku Hanski [3], Jonas Wegner [2] and Alyssia Janczak [1]

[1] OMV Exploration & Production GmbH, Trabrennstrasse 6-8, 1020 Vienna, Austria
[2] HOT Microfluidics GmbH, Trabrennstrasse 6-8, 1020 Vienna, Austria
[3] Kemira OYJ, Trabrennstrasse 6-8, 1020 Vienna, Austria
* Correspondence: rafaeleduardo.hincapiereina@omv.com

Citation: Hincapie, R.E.; Borovina, A.; Clemens, T.; Hoffmann, E.; Tahir, M.; Nurmi, L.; Hanski, S.; Wegner, J.; Janczak, A. Optimizing Polymer Costs and Efficiency in Alkali–Polymer Oilfield Applications. *Polymers* **2022**, *14*, 5508. https://doi.org/10.3390/polym14245508

Academic Editors: Jesús-María García-Martínez and Emilia P. Collar

Received: 13 November 2022
Accepted: 11 December 2022
Published: 15 December 2022

Publisher's Note: MDPI stays neutral with regard to jurisdictional claims in published maps and institutional affiliations.

Copyright: © 2022 by the authors. Licensee MDPI, Basel, Switzerland. This article is an open access article distributed under the terms and conditions of the Creative Commons Attribution (CC BY) license (https://creativecommons.org/licenses/by/4.0/).

Abstract: In this work, we present various evaluations that are key prior field applications. The workflow combines laboratory approaches to optimize the usage of polymers in combination with alkali to improve project economics. We show that the performance of AP floods can be optimized by making use of lower polymer viscosities during injection but increasing polymer viscosities in the reservoir owing to "aging" of the polymers at high pH. Furthermore, AP conditions enable the reduction of polymer retention in the reservoir, decreasing the utility factors (kg polymers injected/incremental bbl. produced). We used aged polymer solutions to mimic the conditions deep in the reservoir and compared the displacement efficiencies and the polymer adsorption of non-aged and aged polymer solutions. The aging experiments showed that polymer hydrolysis increases at high pH, leading to 60% higher viscosity in AP conditions. Micromodel experiments in two-layer chips depicted insights into the displacement, with reproducible recoveries of 80% in the high-permeability zone and 15% in the low-permeability zone. The adsorption for real rock using 8 TH RSB brine was measured to be approximately half of that in the case of Berea: 27 µg/g vs. 48 µg/g, respectively. The IFT values obtained for the AP lead to very low values, reaching 0.006 mN/m, while for the alkali, they reach only 0.44 mN/m. The two-phase experiments confirmed that lower-concentration polymer solutions aged in alkali show the same displacement efficiency as non-aged polymers with higher concentrations. Reducing the polymer concentration leads to a decrease in EqUF by 40%. If alkali–polymer is injected immediately without a prior polymer slug, then the economics are improved by 37% compared with the polymer case. Hence, significant cost savings can be realized capitalizing on the fast aging in the reservoir. Due to the low polymer retention in AP floods, fewer polymers are consumed than in conventional polymer floods, significantly decreasing the utility factor.

Keywords: polymer flooding; alkali–polymer flooding; chemically enhanced oil recovery; incremental oil recovery

1. Introduction

The selection of a chemically enhanced oil recovery (cEOR) agent, such as a polymer or an alkali, is critical for the correct design of technology applications. Polymer flooding has been implemented in numerous fields [1,2], although its role in reducing residual oil saturation remains a discussion topic. Due to its high viscosity, a polymer alone cannot significantly increase the capillary number; however, it can contribute to oil recovery, as its mobility can be controlled [3]. Polymer technology has been proven to be very effective in the recovery process using mobility control with multiple field applications [4].

Similarly, field applications of using an alkali alone as an EOR agent resulted in a lower oil recovery of 1–6% [5]. Although a chemical reaction between an alkali and crude oil generates in situ saponification, alkali consumption and plugging remain significant challenges [6–8]. The synergy of alkali and polymer injection can surpass the reported limitations of injecting them separately. Schumi et al. [7] and Hincapie et al. [9] reported that alkali–polymer (AP) slugs offer a higher incremental oil contribution than when injected alone (alkali/polymer/water) or as separate slugs (alkali after polymer/polymer after alkali) [10].

Furthermore, laboratory results indicated that less alkali was consumed using an AP slug than using an alkali slug [11]. One possible reason is that polymer molecules cover the rock surface to reduce alkali adsorption. Additionally, alkali increases pH and makes the rock surface charge more negative. This chemical process can hinder polymer adsorption at the rock surface [10,12]. As a result, more alkali reacts with the crude oil polar compounds to generate in situ soaps and lower interfacial tension. Furthermore, polymers hold the injected aqueous-phase viscosity and improve the sweep efficiency.

Project economics play a pivotal role for EOR technology applications. For a highly complex case such as AP involving multiple chemical agents, the business case for the project may become unfavorable or hard to justify [13]. With a proper process design, good recoveries can be observed, and it is possible that economics can be improved [9].

This work focuses on evaluating the synergies of alkali and polymers for a successful application of the technology in the Matzen field in Austria. For the optimum utility factor, aging of AP slugs is performed to optimize the project economics. This study is a continuation of our previous research [13,14], with an emphasis on AP flooding in reservoir sand packs and reservoir conditions (live oil).

The paper is organized as follows: in the next section, the approach is covered. Then, the materials and methods are covered, followed by the results and the discussion.

2. Approach

To evaluate the alkali–polymer synergies on recovery and to improve the economics, the following steps were followed:

- Fluid/rock selection and characterization: one polymer and one alkali type were selected and characterized. Oil properties were also measured.
- Aging: aging experiments were carried out to evaluate the changes in the long-term polymer performance in alkaline conditions.
- Micromodel experiments: tests were performed to evaluate the effect of AP in micromodels with a permeability contrast in a preliminary stage.
- Phase behavior and interfacial tension (IFT) experiments: these supported an understanding of the emulsion volumes generated by the fluid–fluid interactions.
- Two-phase core floods: we evaluated recovery at the core scale using various slugs; experiments were performed in real rock sand packs for live/dead oil conditions.

3. Materials

Reservoir Data: We evaluated alkali–polymers in the 8 Torton Horizon (TH) reservoir of the Matzen field in Austria in the Schoenkirchen area. As a clastic reservoir, the 8 TH reservoir is characterized by permeabilities between 150 mD and up to several darcys. The average porosity is 28–30%, the average net sand thickness is 5 m, and the reservoir temperature is 49 °C.

Oil Data: Oil from the Schoenkirchen S-85 well was used for this work. The oil is described as moderately degraded. A summary of the main data is presented in Table 1. Rupprecht [15] reports additional information on the oil. The oil is characterized by 39% saturated compounds, 42% aromatic, 16% resins, and an asphaltene content of about 3%. The oil's saponifiable acids are about 42 µmol/g and the TAN number is 2.14 mg KOH/g. Dead oil density was measured as 0.931 g/cm^3 (20 °C) and 0.891 g/cm^3 (49 °C).

Table 1. Composition of crude oils used in this work from the 8 TH reservoir.

Property	Value
Well	Schoenkirchen S-85
TAN [mg KOH/g]	2.14
Saturates [%]	39
Aromatics [%]	42
Resins [%]	16
Asphaltene [%]	3
Saponifiable Acids [µmol/g]	41 (7.5 g/L Na_2CO_3)
µ (dead oil) @ 49 °C [mPa.s]	56
µ (live oil) @ 49 °C, 150 bar [mPa.s]	19
ρ (dead oil) @ 20 °C/49 °C [g/cm^3]	0.931/0.891

Synthetic Brines: Softened water produced from the field is envisaged as the injected water [7,9]. A simplified reservoir brine was composed in g/L of 22.47 NaCl, 0.16 KCl, 0.63 $MgCl_2*6\ H_2O$, and 0.94 $CaCl_2*2\ H_2O$, here named 8 TH RSB (reservoir synthetic brine). Moreover, 8 TH RSB was softened to create a simplified injection water, resulting in a composition in g/L of 22.62 NaCl, 0.16 KCl, and 1.52 $NaHCO_3$ (buffer capacity), hereafter, Soft. 8 TH RSB. The pH for the alkali solutions is about 10.5 for 7.5 g/L Na_2CO_3 and the water viscosity at 49 °C was 0.65 mPa.s.

Alkali and Polymer: Na_2CO_3 was investigated here, as it is available at lower costs than other alkali agents. In addition, Na_2CO_3 is buffered at a pH of 10.2 for the conditions of the 8 TH reservoir and, hence, does not lead to substantial quartz dissolution (e.g., [16]) and scaling in the production wells accordingly. The high-molecular-weight anionic polyacrylamide (HPAM), a KemSweep A-5265 was used as the polymer. Three concentrations (1400 ppm, 1800 ppm, and 2000 ppm) were used, and polymer solution viscosity and concentration vary depending on the approach (aged or non-aged samples).

Outcrop Cores: Berea sandstone cores, as well as real rock material were used for the experimental evaluations. Berea sandstone is a well-sorted yellowish sandstone with approximately 87% quartz, 5% feldspar, and 2.6% clay (Table 2). Cores with similar mineral compositions were used for the evaluations. The average values for Berea cores used in the two-phase tests were a 2.96 cm diameter, 29 cm length, a porosity of 0.219, and a brine permeability of 180–220 mD. Outcrop core samples were investigated using Computed Tomography (CT) scans for inhomogeneities. According to the obtained X-ray Diffraction (XRD) data, the clay is a mix of 92% kaolinite, 7.5% chlorite, and 0.7% illite by mass. Pore walls are covered with feldspar or clay, and additional data can be found in the previous work of Scheurer et al. [17], where cores from the same block were used.

Table 2. Mineral composition of the core material used in this work. Data reported in weight percent.

Core	Quartz	Kaolinite	K-Fsp	Calcite	Illite/Mica	Smectite	Peryte	Carbonate	Dolomite	Clay Tot
Berea Outcrop	89	-	7	0.8	-	-	0.3	-	-	2.9
Real Rock Reservoir	59.0	3	12	-	2.0	3.0	-	18	-	3

K-Fsp = Potassium feldspar.

Reservoir Material: We performed core floods using sand packs made of real rock material that was crushed to a uniform sand pile. Material from the reservoir section was used (Table 2). Sand pack characteristics, such as porosity and permeability, for the specific cases are presented in a subsequent section.

4. Experimental Methods

Thermal Aging of Polymer Solutions: Polymer solutions for thermal aging studies were prepared in a glove box under nitrogen atmosphere to ensure an anaerobic environment reflecting field conditions. The method is described more in detail in [13,14]. The aging was conducted in various concentrations. A concentration of 2000 ppm KemSweep A-5265 solution in Soft. 8 TH RSB brine in the presence of alkali was used for studying the polymer long-term viscosity performance at various temperatures. In addition, 1850 ppm and 1400 ppm solutions were used for aging samples for core flood testing, representing polymer performance down in the reservoir away from the nearby wellbore area. The pH of the solutions was 10.0–10.2 throughout the aging experiments.

Static Adsorption: Crushed and sieved Berea rock material with a grain size from 125 µm to 250 µm was used for static adsorption tests. 50 g of rock was mixed with 50 g of 200 ppm polymer solution in a jar. Sets of three identical samples were stored at 49 °C for 48 h. Afterwards, the polymer solution was filtered through a 5 µm syringe filter to remove the sand, and the final polymer concentration was determined utilizing Size Exclusion Chromatography (SEC) measurements. In case of crushed Matzen rock, the grains were washed from oil by Soxhlet extraction until clean before being used for adsorption tests. A polymer concentration of 200 ppm was selected for measurement accuracy.

Phase Behavior and IFT Tests: Tests were performed using the modified version of the procedure adopted by [18,19]. For this study, tests were performed in 60 mL glass tubes. First, 30 mL of aqueous phase (brine/cEOR) was filled in a precise syringe pump, and 30 mL of dead oil was added on the top. After sealing the tube, formulations were agitated strongly up and down for fifteen minutes to ensure homogenous mixing. Note that the oil was used as dead oil (DO) or dead oil mixed with cyclohexane (DOC). Subsequently, tubes were placed on the racks with temperature control on for 23 days. During the experiments, high-quality images were taken at specific time intervals to observe changes in the micro emulsions, if produced. Changes in pH value over time were expected to provide additional information about formulation reactions between alkali and high TAN oil. Interfacial tension (IFT Upper Phase/Middle Phase) at the fluids' interface was calculated using the volume ratio between the upper phase (UP) and middle phase (MP) using the approach adopted by Liu et al. [20] and Hincapie et al. [9]. Interfacial tension data were reported in [13,14]; here, we compare results.

Micromodel Generation and Setup: Within this work, we customized and build a dual-permeability micromodel to mimic the heterogeneity of the reservoir. Furthermore, this enables us to screen the displacement efficiency of chemicals in a heterogenous formation. The micromodel dimensions are 60 × 20 mm, with a permeability contrast of 1:5 between the two zones. Micromodel generation and characterization were done by the same principle previously reported by Hincapie et al. [9]. Figure 1 displays the micromodel geometry and properties accordingly.

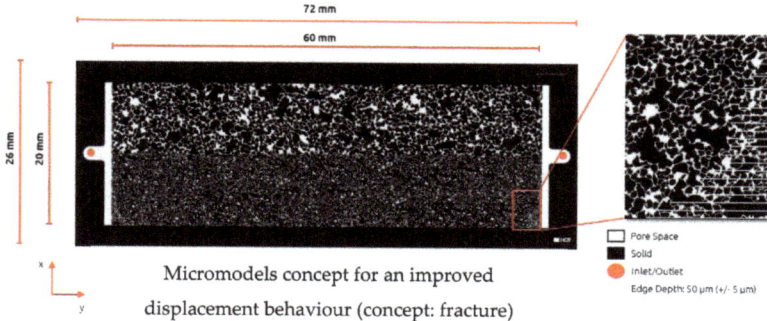

Figure 1. Illustration of the heterogenous micromodel used in this work. Top: high-permeability zone; Bottom: low-permeability zone.

All micromodel experiments were carried out in the same setup reported by Schumi et al. [7]. The setup has an automated schedule feature, which allows us to program the experimental sequence of events to carry out experiments, eliminating human error and ensuring hardware reproducibility.

Micromodel Flooding Experiments—Sequence of Events: Figure 2 shows the pore volume injected for each individual slug used in all experiments. Injection velocity was 1 ft/day Darcy velocity, which translates to 0.2 µL/min.

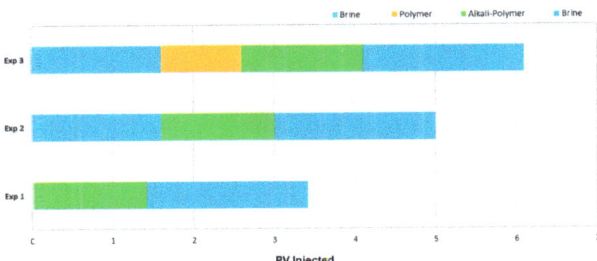

Figure 2. Summary of the injection sequences utilized for the different micromodel experiments.

Elapsed time, differential pressure, temperature at 4 positions (top, bottom, left, right), and high-resolution images were recorded during each experiment. Table 3 provides a summary of the parameters used for the experiments. Computation of saturation profiles was performed with an internal software (InspIOR Vision Pro). The software also enables computation for different sections chosen by the user. In this work, we specifically distinguished between the high and low-permeability layers as regions of interest.

Table 3. Summary of parameters in micromodel flood experiments performed in this work. Pore pressure was 1 bar and injection rate 1 ft/day in all cases.

Parameter	Unit	Experiment 1	Experiment 2	Experiment 3
Temperature	°C	49	49	49
P	-	-	-	1.85 g/L A−5265 *
Viscosity P	mPa.s	-	-	20
AP	-	7.5 g/L Na_2CO_3 + 1.85 g/L A-5265 **	7.5 g/L Na_2CO_3 + 1.85 g/L A-5265 **	7.5 g/L Na_2CO_3 + 1.85 g/L A-5265 **
Viscosity AP	mPa.s	25	25	25

* Diluted in synthetic brine 8 TH RSB (g/L): 22.53 NaCl, 0.16 KCl, 0.94 $CaCl_2$*2 H_2O, and 0.72 $MgCl_2$*6 H_2O ** Diluted in synthetic brine Soft. 8 TH RSB (g/L): 22.62 NaCl, 0.16 KCl, 1.52 $NaHCO_3$.

Core Flood Experiments in Outcrops: For cEOR slug screening, five core floods (Table 4) were performed in Berea analogue using the core flooding setup reported in previous work [7,9]. The measured petrophysical properties of Berea plugs were similar to the targeted reservoir area (shown in the results section). Standard routine core analysis (RCA) was adopted for the initialization of the core plugs. Brine permeability was measured using the Darcy approach at three injection rates, while porosity and pore volume estimation was performed using the Archimedes method. Fresh and dry core samples were loaded in the Hassler cell, with a radial confining pressure of 30 bar(g). To displace air/nitrogen from the system, CO_2 was flushed for ten minutes against the system pressure (back pressure regulator) of 10 bar(g). Following, the core was saturated with the appropriate brine (8 TH RSB (g/L)), with a system pressure of 5 bar(g). After measuring the permeability at room temperature, the system (oven) temperature was increased to the target temperature of 49 °C to validate permeability at reservoir temperature. After lowering the system temperature, the core plug was unloaded for porosity and pore volume estimation. Initial oil/brine saturations were achieved by dead oil flow through injection at a system pressure of 5 bar(g). The Dean–Stark procedure was applied to the collected fluids (oil/brine) to validate the visual volumes (saturations). Oil permeability was measured at three appropriate injection rates and the core sample was stored at target temperature for 4 weeks to age the samples and induce wettability alterations.

Table 4. Summary of parameters in Berea outcrop flood experiments performed in this work. Core orientation was vertical, pore pressure was 5 bar, injection rate was 1 ft/day, and radial pressure was 30 bar in all cases.

Parameter	Unit	Exp. 1	Exp. 2	Exp. 3	Exp. 4	Exp. 5
Temperature	°C	49	49	49	49	49
Chem. Slug 1 *	-		1.85 g/L A-5265			
η Slug 1 (7.94 s^{-1})	mPa.s	20	32	20	-	-
Chem. Slug 2 **	-	7.5 g/L Na_2CO_3 + 1.85 g/L A-5265 (1)	7.5 g/L Na_2CO_3 + 1.85 g/L A-5265 (2)	7.5 g/L Na_2CO_3 + 1.4 g/L A-5265 (1)	7.5 g/L Na_2CO_3 + 1.85 g/L A-5265 (2)	7.5 g/L Na_2CO_3 + 1.40 g/L A-5265 (1)
η Slug 2 (7.94 s^{-1})	mPa.s	38	25	24	25	24

* Diluted in synthetic brine 8 TH RSB (g/L): 22.53 NaCl, 0.16 KCl, 0.94 $CaCl_2*2 H_2O$, and 0.72 $MgCl_2*6 H_2O$
** Diluted in synthetic brine Soft. 8 TH RSB (g/L): 22.62 NaCl, 0.16 KCl, 1.52 $NaHCO_3$ (1) Solution prepared according to aging procedure. (2) Solution prepared without aging procedure.

Core Flood Experiments in Real Rock—Sand Pack Preparation: Three sand pack experiments were performed using real rock material: two with dead oil and one with live oil (Table 5). The workflow for sand packs included cleaning the sand material, screening, matching the petrophysical properties to the target reservoir, and initial saturation of fluids. Unconsolidated material was crushed to a small particle size and was packed in Soxhlet extraction apparatus. Sand cleaning was performed by cooking an 80/20 mixture of chloroform/methanol at 60 °C for four days in repeated cycles until solvents were colorless. After cleaning, sand was dried in an oven at 60 °C for two weeks, then sieved through different mesh sizes to classify different grain sizes. The mixtures of two mesh sizes of 100–200 and 200–300 were mixed with the 87.5 wt% and 12.5 wt% to achieve the target permeability. Selected sand was loaded and packed into the Viton tube using a vibrator table until complete compaction was achieved. The final length of each sand pack was 30 cm, and they had a diameter of 3 cm. Further initialization and permeability calculations were similar to that of the core flood. However, porosity/PV estimation was performed by means of Nuclear Magnetic Resonance (NMR). The sand pack was fully saturated with methanol and was displaced by the formation brine (8 TH RSB). The brine/methanol mixture was analyzed by NMR to estimate the produced methanol volume. Further procedures of oil initialization, oil permeability, and the aging process were similar to those described in the previous section on core flooding. The used setup and further details on the approaches were also presented in [9].

Table 5. Summary of parameters for sand pack flood experiments performed in this work. Core orientation was vertical and injection rate was 1 ft/day.

Parameter	Unit	Exp. 6	Exp. 7	Exp. 8 (Live)
Temperature	°C	49	49	49
Pore Pressure	Bar	5	5	125
Radial Confining Pressure	bar	30	30	150
Chem. Slug 1 *	-		1.85 g/L A-5265	
η Slug 1 (7.94 s^{-1})	mPa.s	20	21	19
Chem. Slug 2 **	-	7.5 g/L Na_2CO_3 + 1.85 g/L A-5265 (2)	7.5 g/L Na_2CO_3 + 1.4 g/L A-5265 (1)	7.5 g/L Na_2CO_3 + 1.85 g/L A-5265 (2)
η Slug 2 (7.94 s^{-1})	mPa.s	25	23	29

* Diluted in synthetic brine 8 TH RSB: 22.53 g/L NaCl, 0.16 g/L KCl, 0.94 g/L $CaCl_2*2 H_2O$, and 0.72 g/L $MgCl_2*6 H_2O$ ** Diluted in synthetic brine Soft. 8 TH RSB: 22.62 g/L NaCl, 0.16 g/L KCl, 1.52 $NaHCO_3$. (1) Solution prepared according to aging procedure. (2) Solution prepared without aging procedure.

Live Oil Preparation: One of the sand packs was re-saturated with live crude oil after the completion of the aging process with dead crude oil. Live oil preparation was performed in a high-pressure/temperature piston accumulator under reservoir conditions of 49 °C and a system pressure of 115 bar(g), adopting the procedure described in a previous study [9]. First, 150 mL of dead oil (S-85) was added to the piston accumulator; after sealing the piston, high-purity (4.5) methane gas was refilled in the piston to the target pressure of 115 bar (g). Dissolution of methane gas in dead oil was initiated by rotating the piston up and down. Gas dissolution resulted in lowering the system pressure, and the required pressure support was provided by an ICSO pump connected to the piston with a contact pressure mode. Equilibration and dissolution were established over a period of three days. In the end, the excessive amount of gas was bled off with help of a back pressure valve pre-set at 116 bar (g).

Core Flooding Experiments—Sequence of Events: Multiple fluid slugs were injected at an interstitial velocity of 1 ft/day, as shown in Figure 3. First, brine (8 TH RSB) was injected as slug 1 for all experiments for a 1.6 pore volume (PV) as a secondary mode. Second, polymer flooding was implemented as slug 2 (except Exp. 4 and Exp. 5) and was defined as a tertiary mode. Third, the alkali–polymer chemical slug was injected as a post tertiary mode (except Exp. 4 and Exp. 5) for 2 PV. Lastly, brine injection was implemented for 2 PV as a final slug for all experiments (diluted in synthetic brine 8 TH RSB). Core plugs/sand packs were unloaded after the flooding experiments to perform the Dean–Stark procedure. Final saturations (recovery factors) of the fractions' collected samples were validated with Dean–Stark volumes. Core flood effluents from core plugs/sand packs were collected in small glass fraction tubes to perform volumetrics and to generate the produced oil recovery curve versus injected PV. Moreover, pressure differential data were recorded for all injected slugs.

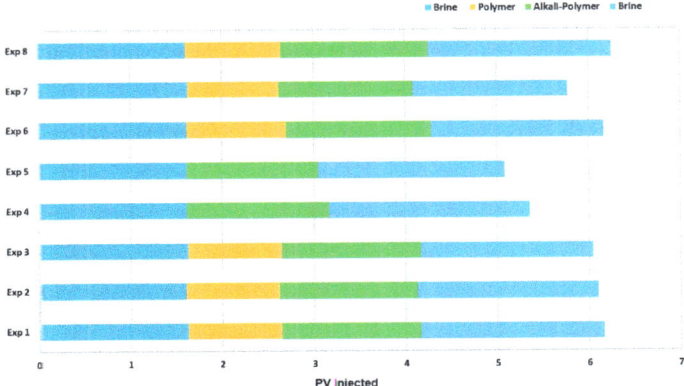

Figure 3. Injection sequences for the different core flood experiments utilized in this study.

5. Results

Thermal Aging: Long-term aging in the alkaline conditions was studied to find out the long-term polymer performance in AP applications. A rapid increase in the viscosity of the 2000 ppm solutions was observed during the first days at elevated temperatures (Figure 4A). The increase was linked to a fast change in the degree of hydrolysis providing a positive impact on viscosity in the soft brine. A more detailed description of the studies can be found in [13,14]. Even though the field temperature in Matzen is 49 °C, the aging was also carried out at 60 °C and 70 °C to accelerate the aging. We showed in [13,14] that the activation energy E_a = 110 kJ/mol, determined earlier for HPAM hydrolysis in neutral pH 6–8. Nurmi et al. [21] also applied it in the case of a high pH. With the knowledge of the activation energy, the aging results measured at one temperature can be transferred to the prediction at another temperature with the method described by Nurmi et al. [21]. Based on the activation energy, or E_a = 110 kJ/mol, the viscosity retention results in Figure 4A have been transferred to a corresponding time scale at 49 °C in Figure 4B.

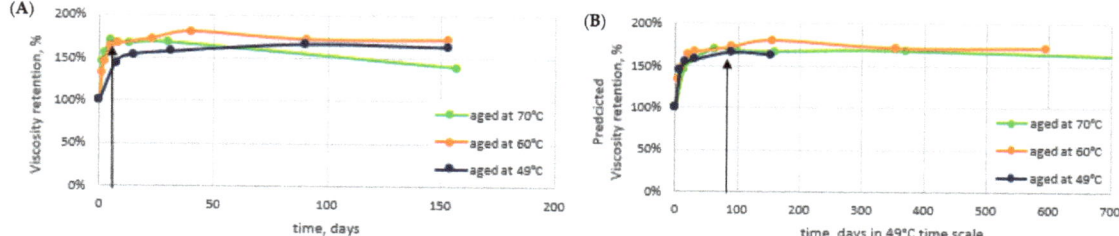

Figure 4. Aging results. Anaerobic 2000 ppm KemSweep A-5265 solutions in Soft. 8 TH RSB brine in the presence of alkali. (**A**) Aging in temperatures of 49 °C (dark blue), 60 °C (orange), and 70 °C (green). (**B**) aging results transferred to corresponding days at 49 °C. Black arrows indicate the point in time for the sampling of aged samples used in the core flood and adsorption testing.

The results suggest that the polymer gains viscosity up to 170% of the original viscosity in the present conditions within 1–2 months' time. This indicates that the concentration for the injection could be lowered from the original planned concentrations to achieve the viscosity target. The viscosity of the polymer solution increases and reaches the target viscosity in the reservoir away from the wellbore. In order to compare the performance of the polymer after aging to a fresh polymer, samples at two different concentrations, 1850 ppm and 1400 ppm, were aged. The polymers were aged for 7 days at 70 °C in fully anaerobic conditions before using them in core flood or adsorption testing. These 7 days at 70 °C correspond to approximately 90 days at 49 °C. At this point, the viscosity has reached a level at which the rate of change is already very low [13,14]. The sampling point (aged samples) is indicated with black arrows in Figure 4A,B. The initial viscosity and viscosity after aging are shown in Table 6 for all the studied concentrations. A similar increase in viscosity, ~170%, was observed irrespective of the studied concentration. The target polymer viscosity in the application was set to 20 mPas. As seen in Table 6, this viscosity is reached with 1850 ppm of fresh polymer and with 1400 ppm of polymer that has aged in the AP conditions. Our results show that the acrylic acid content (degree of hydrolysis) increases rapidly. In contrast to observations at pH 6–8 (Nurmi et al. 2018), the hydrolysis rates are clearly not constant, but instead hydrolysis rates decrease over time. The decreasing polyacrylamide hydrolysis rate is explained by auto retardation: the hydrolyzing reagent at a high pH is an anionic OH- ion which is increasingly repelled by the anionic polymer as the charge content in the polymer chain increases.

Table 6. Summary of concentrations and viscosities used here. Viscosity increases after aging, averaging 165%.

Polymer Concentration	Viscosity in Soft. 8 TH RSB + Alkali (mPas)	Viscosity after Aging in Soft. 8 TH RSB + Alkali (mPas)	Increase in Viscosity
2000	24.6	41.7	168%
1850	20.5	34.4	168%
1400	12.0	20.4	170%

Static Adsorption: Static adsorption experiments were carried to obtain insight into the adsorption phenomenon as a function of changing brine, pH, polymer charge, and rock material conditions. The results are presented in Figure 5. The trend in crushed Berea is clear. The adsorption decreases as the brine gets softer and the pH increases. Increasing the charge of the polymer in the softened brine conditions also appears to continue the positive trend. The lowest adsorption value measured was 19 µg/g. Adsorption in the actual cleaned Matzen rock material indicates even lower values. The adsorption in the 8 TH RSB brine was measured to be approximately half of that in the case of Berea: 27 µg/g vs. 48 µg/g, respectively. When applying the softer water with a pH increase in the Matzen rock, the adsorption could be expected to go even lower than with Berea, assuming a similar response to the change in conditions, as in the case of Berea. These values were not measured. Further considerations on the adsorption in the absence of oil can be found in [13,14].

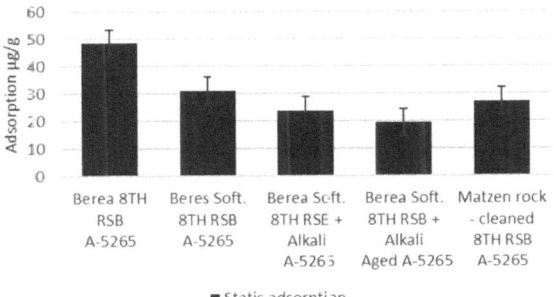

Figure 5. KemSweep A-5265 static adsorption in different conditions in crushed Berea and cleaned/crushed Matzen rock.

Phase Behavior and IFT Tests: The data gathered include samples using dead oil (DO) or dead oil mixed with cyclohexane (DOC). From Figure 6A, a slight reduction in the pH-value of all formulations can be observed for the test period. This reduction in pH value was expected due to the chemical interactions between alkali and oil polar compounds. Over time, more alkali was neutralized with oil polar compounds and, hence, resulted in lowering the pH value. Considering the lower amount of alkali required and in order to hinder the possible alkali/polymer adsorption, a 7.500A-P(DOC) formulation was the optimum one and resulted in lowering the pH value. Figure 6B presents the reduction in produced emulsions over 23 days. The reduction in volume let us infer that the produced emulsions were instable. Moreover, the 7.500A-P(DOC) formulation produced a greater emulsion volume during the initial seven days. However, during the first two days, the 15.000A-P(DO) formulation produced the highest volumes of emulsions, but these were not stable and decreased the emulsion volume rapidly. Figure 7A,B presents the reduction in IFT for the chemical formulations. The observed decline in IFT was similar to that of the reduction in emulsion volume. The IFT calculation was performed using Chun Huh equation [20] (Table 7). Compared to our previous study [9], the IFT measured using the Chun Huh equation was in line with the spinning drop tensiometer. Hence, a lower amount of alkali in the 7.500A-P(DOC) formulation also depicted a good result in lowering the IFT to an order of 10^{-2}. As can be seen from Figure 6, there was reproducibility in the lowering of the IFT in the experimental data for the phase behavior.

Micromodel Flooding Experiments: The micromodel experiments were performed stepwise to investigate the behavior of the fluid–fluid interactions. Without changing the injected fluid composition, we added additional slugs in each iteration of the experiments, as displayed in Figure 2.

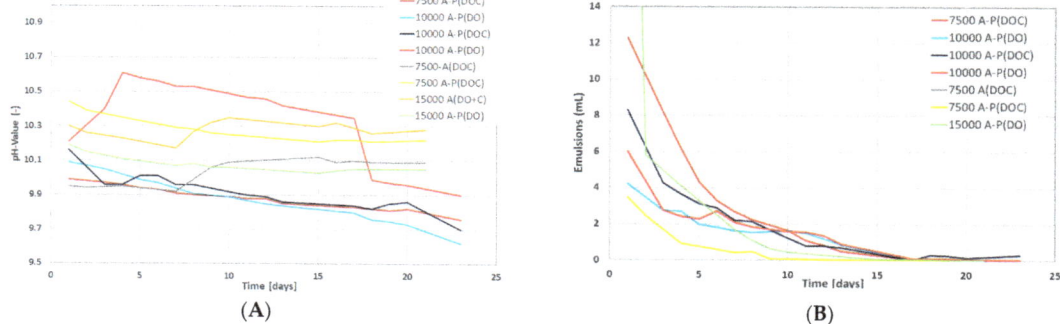

Figure 6. Change in pH value and emulsion volume over time for the phase behavior tests at different concentrations of chemicals. (**A**) Change in chemical formulation over number of days. (**B**) Change in produced emulsion volume over number of days. A refers to alkali concentration in ppm, P stands for polymer concentration of 1850 ppm, (DOC) represents dead oil with cyclohexane, and (DO) represents dead oil.

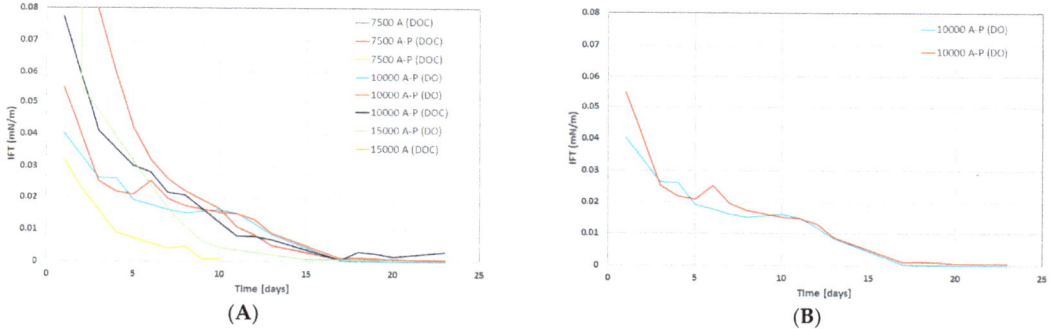

Figure 7. IFT variation over time (days) for the tested formulations in the phase behavior tests. (**A**) IFT versus time for various concentrations and (**B**) IFT versus time for 10,000 AP to show the data reproducibility. A refers to alkali concentration in ppm, P stands for polymer concentration of 1850 ppm, (DOC) represents dead oil with cyclohexane, and (DO) represents dead oil.

Table 7. Summary of volume relation from phase behavior and IFT for 7500 ppm A and AP determined using the Chun Huh equation (Liu et al. 2008) for alkali and alkali–polymer in 10 mL oil sample. IFT = Ratio Vphase/Vphase * c, c = 0.3 nM/m (Liu et al. 2008)—correction factor. Lower phase = water, middle phase = emulsion, upper phase = oil.

	Time [Days]	Lower Phase (LP) Vol. [ml/mL]	Middle Phase (MP) Vol. [ml/mL]	Upper Phase (UP) Vol. [ml/mL]	Ratio $V_{Middle\ Phase}/V_{Lower\ Phase}$ [-]	IFT $_{MP/LP}$ [mN/m]	Ratio $V_{Middle\ Phase}/V_{Lower\ Phase}$ [-]	IFT $_{UP/MP}$ [mN/m]
A	7	0.000	0.9674	0.033	5.979	1.7940	1.4705	0.4412
	10	0.091	0.5413	0.368	1.203	0.3611	0.9742	0.2922
	16	0.331	0.3987	0.409	1.189	0.3567	0.9225	0.2768
	17	0.321	0.3816	0.414	1.166	0.3499	0.8879	0.2664
	18	0.317	0.3697	0.416	1.113	0.3339	0.7543	0.2263
	21	0.294	0.3272	0.434	5.979	1.7940	1.4705	0.4412

Table 7. Cont.

	Time [Days]	Lower Phase (LP) Vol. [ml/mL]	Middle Phase (MP) Vol. [ml/mL]	Upper Phase (UP) Vol. [ml/mL]	Ratio $\frac{V_{Middle Phase}}{V_{Lower Phase}}$ [-]	IFT MP/LP [mN/m]	Ratio $\frac{V_{Middle Phase}}{V_{Lower Phase}}$ [-]	IFT UP/MP [mN/m]
AP	5	0.4137	0.0721	0.5142	0.1742	0.0523	0.1401	0.0420
	8	0.4518	0.0375	0.5108	0.0829	0.0249	0.0734	0.0220
	12	0.4755	0.0141	0.5104	0.0297	0.0089	0.0277	0.0083
	18	0.4895	0.0013	0.5092	0.0026	0.0008	0.0025	0.0007
	20	0.4902	0.0013	0.5085	0.0026	0.0008	0.0025	0.0007
	23	0.4892	0.0003	0.5105	0.0006	0.0002	0.0006	0.0002

In Table 8, the results are summarized and broken down into the obtained RF for each slug injection. Thus, we can quantify the efficiency of the injected chemical. Additionally, visual access through microfluidic technology gave us insights into the observed mechanisms during flooding. Most importantly, the impact of the cEOR fluids in the low-permeability layer and its desaturation behavior give a deeper understanding of recovery mechanisms. In the following section, the experiments are presented and discussed in detail.

Table 8. Summary of results obtained for micromodel flood experiments performed in this work. Pore pressure was 1 bar and injection rate was 1 ft/day in all cases. Temperature, 49 °C. High-permeability zone refers to 6 darcys and low-permeability zone to 1.5 darcys.

Parameter	Units	Experiment 1	Experiment 2	Experiment 3
Polymer (P)	-	-	-	1.85 g/L A-5265 *
η Polymer (7.94 s^{-1})	mPa.s	-	-	20
Alkali–Polymer (AP)	-	7.5 g/L Na_2CO_3 + 1.85 g/L A-5265 **	7.5 g/L Na_2CO_3 + 1.85 g/L A-5265 **	7.5 g/L Na_2CO_3 + 1.85 g/L A-5265 **
η AP (7.94 s^{-1})	mPa.s	25	25	25
So initial	%	82	84	81
So final	%	40	46	37
RF (Brine)	%	not applicable	21	25
Add. RF (P)	%	not applicable	not applicable	5
Add. RF (AP)	%	49	25	23
Add. RF (Brine)	%	2	1	1

* Diluted in synthetic brine 8 TH RSB: 22.53 g/L NaCl, 0.16 g/L KCl, 0.94 g/L $CaCl_2*2 H_2O$, and 0.72 g/L $MgCl_2*6 H_2O$ ** Diluted in synthetic brine Soft. 8 TH RSB: 22.62 g/L NaCl, 0.16 g/L KCl, 1.52 $NaHCO_3$.

In Exp. 1, the AP slug was injected in secondary mode to obtain a general idea about the fluid–fluid interactions between the AP and oil. The injection of the AP slug yielded an RF of 49% within the entire chip. This moderate desaturation performance is due to the remaining oil saturation (S_{or}) in the low-permeability zone. Thus, the analysis of the results focused on both regions separately.

Figure 8a,b shows the saturation, pressure differential, and RF curves for Exp. 1. Differential pressure increases until a breakthrough is reached and stabilizes during further injection. Desaturation behaves according to the pressure response. Further injection recovers additional oil over time, and after 1 PV is injected, another drop in Sor is observed.

The post flush with brine showed no effect on oil displacement; yet, we observed an increase in differential pressure. From Figure 9 we observed a "light brownish" phase that was mobilized during the brine injection. The residual AP at the boundaries that remains within the porous structure further reacts with the oil; once the brine is injected, it manages to mobilize the oil at the boundaries. This lets us infer that the post flush displacement is not as effective as it is for a pure AP solution, thus leading to a brownish-colored phase.

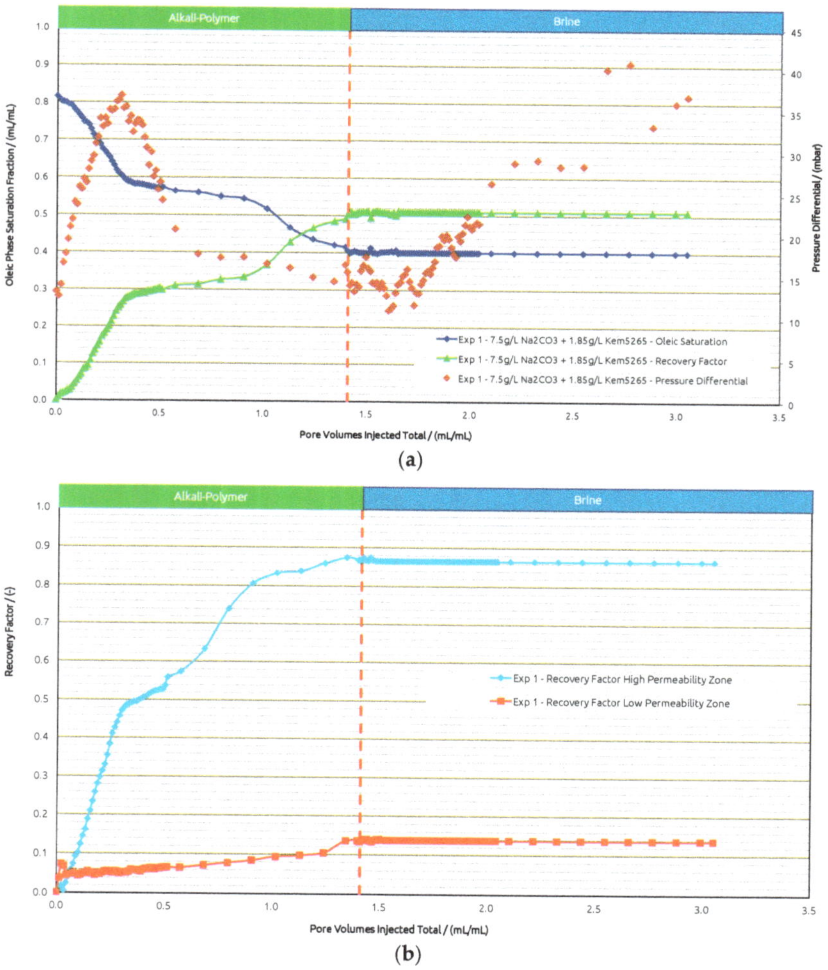

Figure 8. Summary of results obtained for Exp. 1 in micromodel experiments. (**a**) Oleic saturation, Recovery Factor (RF), and pressure differential for the entire chip and (**b**) RF of low-permeability zone vs. RF of high-permeability zone.

Figure 9. During brine injection, the remaining AP inside the porous structure reacted with the oil and the brine was displaced by an emulsion (light brown), resulting in an increase in pressure.

From the visual access, we gained insights into the displacement behavior and efficiency of the AP at different permeability layer boundaries. In Figure 10, we have selected images at different stages of injection. The images on the left show the micromodel as it was photographed during the experiment and the images are processed with highlighted areas of displacement. We see that the AP has an RF of 87% from the high-permeability region, while the RF in the low-permeability region is only 14%. The efficient recovery from the lower layer is due to the AP, which favors IFT reduction and improvement of the mobility ratio. Recovery from the low-permeability layer happens in a slow and continuous manner. Low-permeability areas that are in contact with the AP are the ones where S_{or} is reduced, namely, at the injection site and where the two layers meet in the middle.

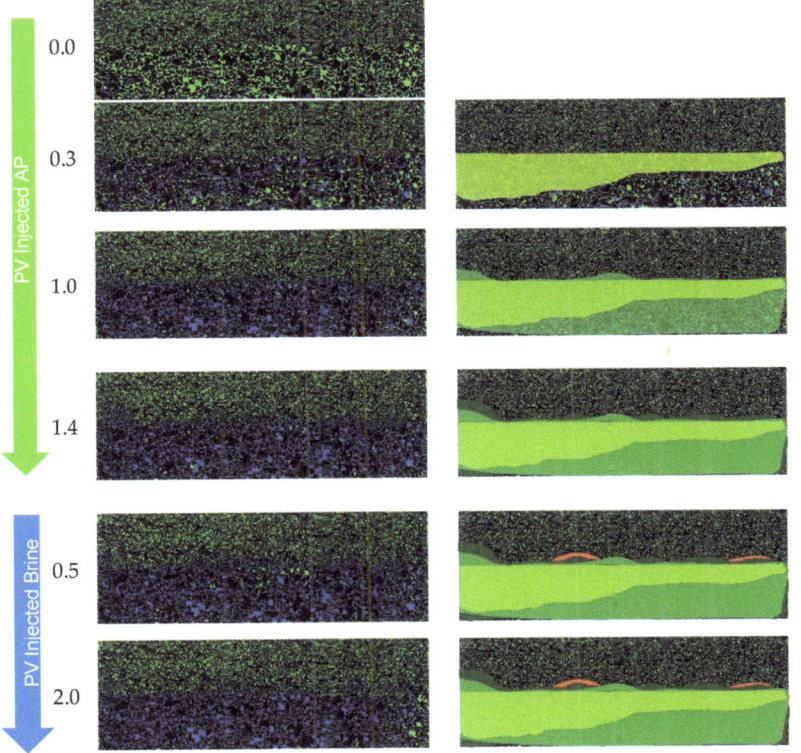

AP slug filtering through the upper part of the high-permeability layer and reaching breakthrough after 0.3 PV injected.

After 1 PV injected, AP slug has displaced most of the oil from the high-permeability layer. "Wash out" from the low-permeability layer visible.

Maximum recovery from high-permeability layer is reached and over time AP is able to recover more from the boundaries between layers.

Remaining AP resides within pore structure and post flush brine injection is able to recover slightly more oil from the low-permeability zone (red).

Further brine injection has no impact on recovery.

Figure 10. Displacement fronts during each slug injection in Exp. 1 (left: segmented images after analysis; right: post-edit images highlighting displaced area).

For Exp. 2 (Figure 11) we introduced a brine injection prior to the AP. The first brine injection yielded an RF of 21%, followed by the AP slug, which recovered an additional 25%. The pressure differential increased almost twofold when injecting the AP; yet, after all the oil was mobilized, injectivity improved and reached values even below the previous brine injection. The final post brine flush did not recover barely any additional oil (1%), and the pressure dropped slightly further. Looking at the recovery from the two layers separately, the scenario in Exp. 2 shows a slightly worse performance compared to Exp. 1. In the high-permeability layer, the AP reached an additional RF of 45% after brine, while 5% was recovered from the low-permeability layer by the AP. Overall, the injection of the AP in secondary mode has been shown to be more effective in terms of the RF, and the "wash out" from the low-permeability zone at the layer boundaries was more efficient.

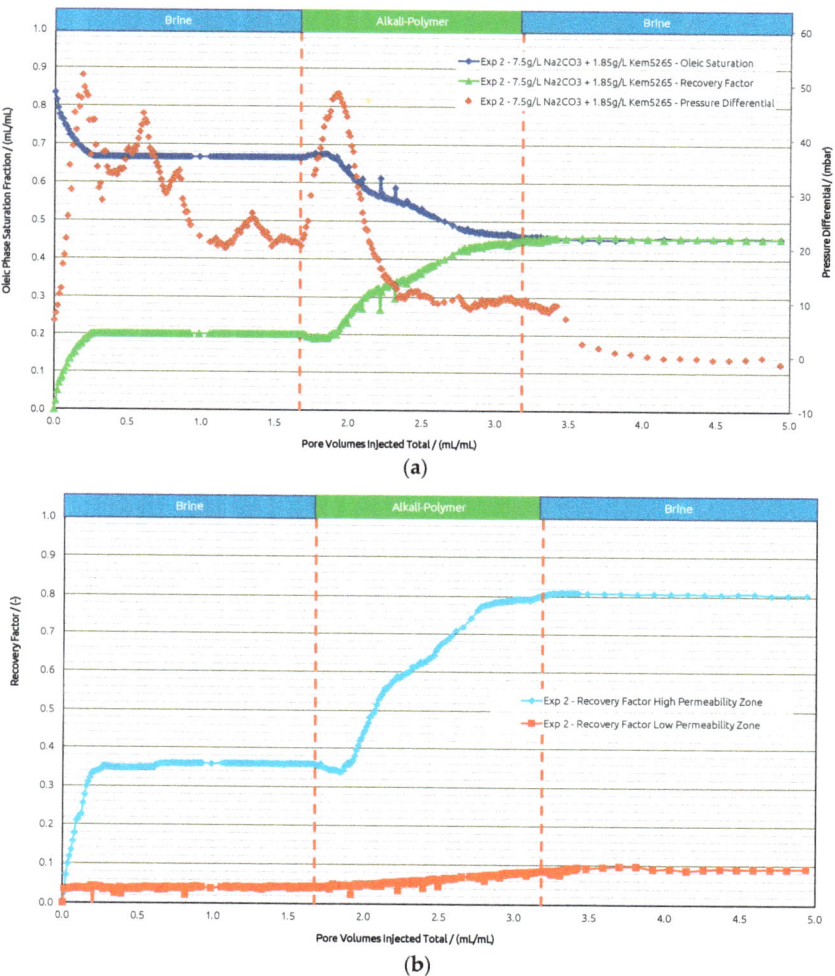

Figure 11. Summary of results obtained for Exp. 2 in micromodel experiments. (**a**) Oleic saturation, Recovery Factor (RF), and pressure differential for the entire chip and (**b**) RF of low-permeability zone vs. RF of high-permeability zone.

For comparison, in Exp. 3 (Figure 12) we targeted the experimental procedure as in Exp. 1, Exp. 2, and Exp. 3 from the core flooding section. The initial brine injection recovered 25%, followed by the polymer, which yielded only a 5% higher RF. The subsequent AP injection showed an additional 23% in RF and almost none in the post brine flush.

The pressure differential response showed an expected increase during the polymer injection. Once the oil bank was moved, the pressure declined. As soon as the AP slug entered the model, the pressure differential increased due to the mobilization of the oil; a decrease was then observed right away, thus improving injectivity. The final post brine flush slightly reduced the pressure, and the remaining AP within the pores was washed out by brine, which added a 1% increase in RF.

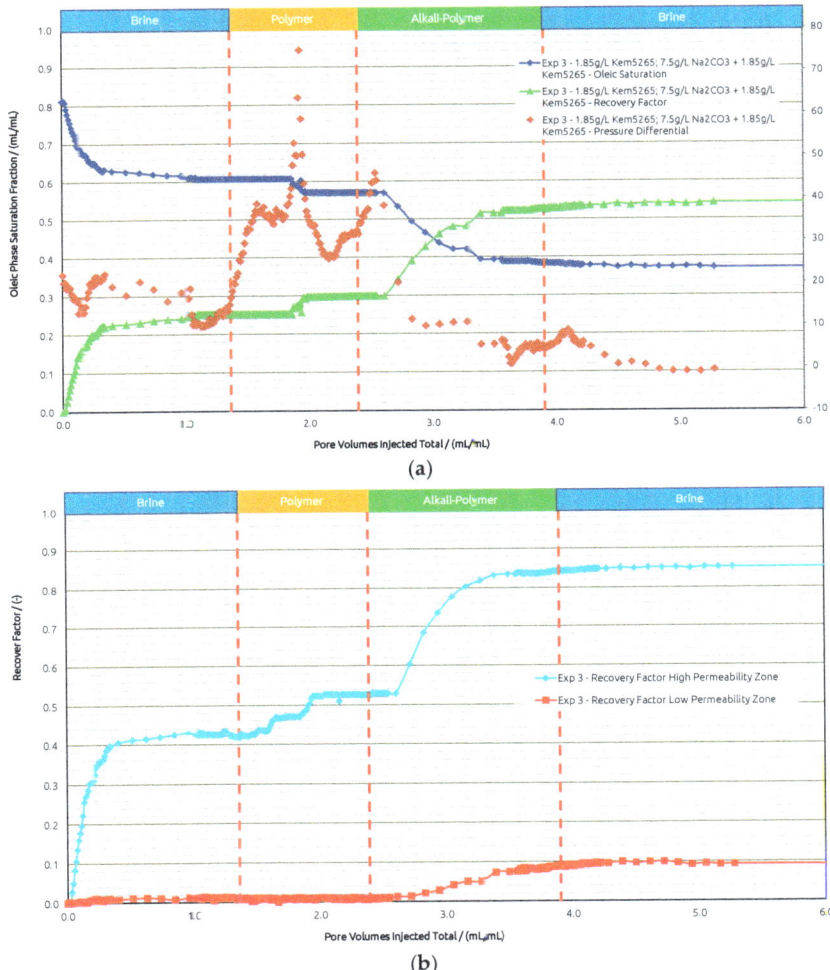

Figure 12. Summary of results obtained for Exp. 3 in micromodel experiments. (**a**) Oleic saturation, Recovery Factor (RF), and pressure differential for the entire chip and (**b**) RF of low-permeability zone vs. RF of high-permeability zone.

When looking at the layers separately, brine recovered 42% from the high-permeability layer and 1% from the lower-permeability layer. The polymer yielded an 11% RF from the higher layer and 0% from the lower-permeability layer. Only when the AP is injected, the RF from the lower-permeability layer is improved by 9%, while 32% is additionally recovered from the high-permeability zone. In summary (Figure 13), the overall RF was highest in Exp. 3 when all slugs were deployed, resulting in 54%. Secondary mode injection proved to be the most effective to displace oil from the low-permeability region and is second best in overall RF at 51%. Tertiary mode without a prior polymer injection yielded the lowest overall RF of 46%, with the least impact on recovery from the low-permeability layer.

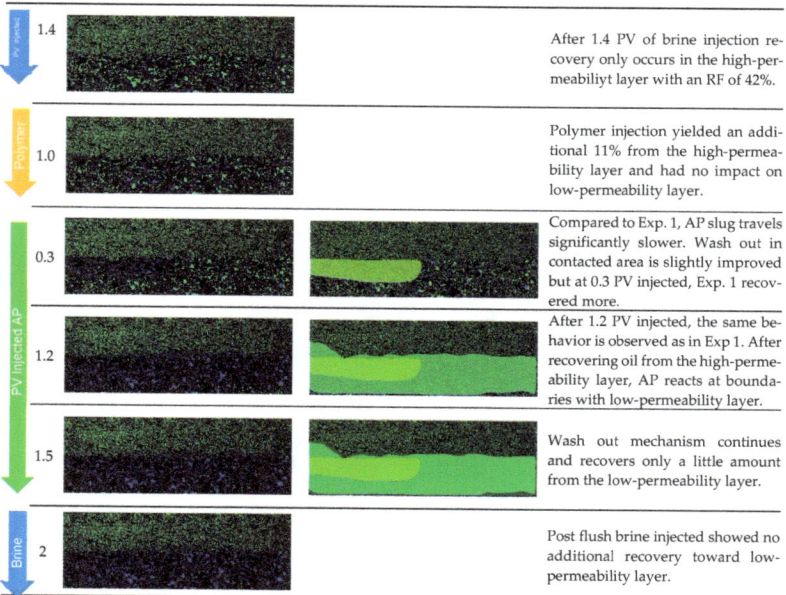

Figure 13. Displacement fronts during each slug injection in Exp. 3 (left: segmented images after analysis; right: post-edit images highlighting displaced area).

Core Flood Experiments in Outcrops: Table 9 provides the summary of the flooding experiments performed in the Berea samples. Exp. 1 and Exp. 2 compared the role of the aging process for AP formulation as shown in Figure 14a,b. The oil recovery factors from brine flood in secondary mode were the same (≈3% difference). Interestingly enough, a similar trend of oil recovery (≈0.2% difference) from polymer flood was observed. The AP slug viscosity for Exp. 1 (38 mPa.s) is significantly higher than the one in Exp. 2, due to the thermal aging of the samples, which we previously reported in [13,14]. However, there was no significant oil recovery difference from the AP slugs in both experiments.

Table 9. Summary of results obtained for Bentheimer outcrop flood experiments performed in this work. Core orientation was vertical, pore pressure was 5 bar, injection rate was 1 ft/day, and radial pressure was 30 bar in all cases. Viscosities are taken at 7.94 s^{-1}.

Parameter	Unit	Exp. 1	Exp. 2	Exp. 3	Exp. 4	Exp. 5
Temperature	°C	49	49	49	49	49
Slug 1	g/L	1.85 P *	1.85 P *	1.85 P *	-	-
η Slug 1	mPa.s	20	20	20	-	-
Slug 2 *	g/L	7.5 A + 1.85 P (aged) $^{(1)}$	7.5 A + 1.85 P	7.5 A + 1.4 P (aged) $^{(1)}$	7.5 A + 1.85 P	7.5 A + 1.4 P (aged) $^{(1)}$
η Slug 2	mPa.s	38	25	24	25	24
Length	cm	30.00	30.00	29.90	30	30.1
Diameter	cm	3.81	3.80	3.81	3.81	3.80
Bulk Volume	cm3	342.07	343.94	342.02	343.94	341.37
PV	cm3	75.10	77.54	75.97	71.67	70.38
Porosity	%	22.0	22.5	22.0	22.2	20.5
Perm. (k_w)	mD	271	277	269	369	356
Oil Sat. Init.	%	65	67	67	0.71	0.70
RF brine	%	42.5	45.6	38.8	41	43
RF Polymer	%	4.9	4.7	2.0	-	-
RFAP	%	13.3	12.4	17.6	27	26.7
RF Brine	%	0.9	1.0	1.1	2	1.4

* Diluted in synthetic brine 8 TH RSB: 22.53 g/L NaCl, 0.16 g/L KCl, 0.94 g/L CaCl$_2$*2 H$_2$O, and 0.72 g/L MgCl$_2$*6 H$_2$O. $^{(1)}$ Solution prepared according to aging procedure. P = A- 5265 A = Na$_2$CO$_3$ PV = Pore Volume.

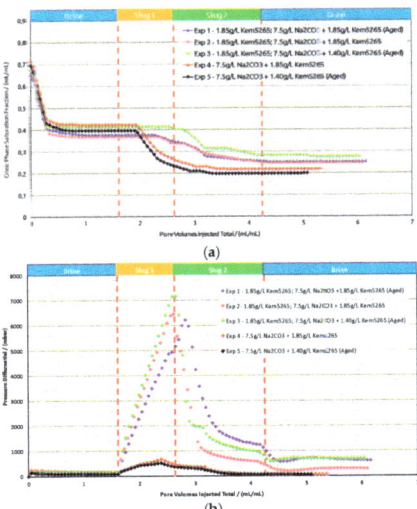

Figure 14. Oleic-phase saturation fraction/Pressure differential versus pore volumes injected total for the different slugs for core floods in Berea outcrops. (**a**) Reduction in oleic-phase saturation versus PV of slugs injected in Berea outcrops with crude dead oil initialization. (**b**) Pressure variation versus PV of injected slugs in Berea outcrops with dead oil initialization.

As expected, the pressure drop of the AP slug in Exp. 1 was higher than in Exp. 2, due to the formulation's higher viscosity. This comparison let us infer that lowering the polymer concentration for thermally aged AP slugs is required to adjust the viscosity to 25 mPa.s. Lowering the polymer concentration from 1850 ppm to 1400 ppm for the thermally aged AP slug resulted in the required viscosity value. Contrary to the first two experiments, the brine flood and polymer flood resulted in lower oil recovery for Exp. 3, but the AP injection performed better, as shown in Figure 15. One possible reason could be that more oil saturation remained in Exp. 3 before the AP injection. Furthermore, slightly higher pressure drops were observed for all slugs in Exp. 3 compared to the first two experiments and can also be the result of higher residual oil saturation; this higher saturation of the oil resulted in the increase in pressure resistance at the inlet.

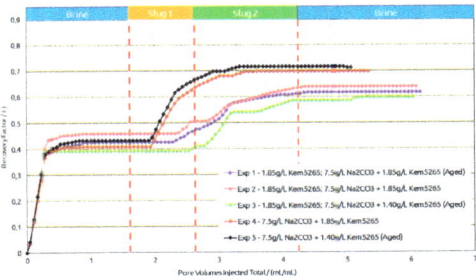

Figure 15. Recovery factor versus pore volumes injected total for the different slugs for core floods in Berea outcrops.

A comparison of Exp. 2 and Exp. 3 showed the potential in lowering the polymer concentration from 1850 ppm to 1400 ppm in the AP slug and, furthermore, to perform the slug aging process prior to injection to achieve the desired slug viscosity. Moreover, the comparison between Exp. 1 and Exp. 3 concludes that lowering the 450 ppm concentration reduces the oil recovery factor to only 2%, which falls within the marginal error of experimental repeatability. This reduction in the polymer concentration showed the potential to improve the utility factor (utility factor UF = kg of polymer injected/incremental bbl. oil produced by reducing the required number of polymers). Furthermore, lower pressure drops of the AP slug for Exp.3 were observed compared to Exp. 1, which could also improve the injectivity in field-scale applications.

Exp. 4 and Exp. 5 were only performed with an AP slug after a brine flood. The RF of the brine flood was in line with the previous experiments, but the AP formulation efficiency was better than the previous three experiments and resulted in 27% additional oil recovery. Exp. 4 and Exp. 5 concluded that, without a polymer slug prior to AP, the chemical interactions between the oil and the formulations were effective; hence, they resulted in better phase behavior interactions. Moreover, pressure drops for both experiments (4 and 5), as shown in Figure 14 were much lower than the previous experiments due to the absence of the pre-polymer slug. The oil recovery factor/reduction in oil saturation and pressure drop comparison between Exp. 4 and Exp. 5 confirmed that a good utility factor can be achieved by the aging process of the low-concentration AP slug.

Core Flood Experiments in Real Rock: Selected formulations from core flood experiments were further injected in reservoir sand packs, as shown in Figure 16 and Table 10. Contrary to the core flood experiments, the difference for the brine flood recovery factors of two sand packs (Exp. 6 and Exp. 7) was higher. The brine flood in the first two sand packs resulted in an almost 10% difference in oil recovery; however, the pressure drop data for both experiments were in line. One possible reason could be the different porosity of both sand packs. For screening purposes, the target was to match the permeability of the sand packs to the target reservoir permeability (300 mD). Similarly, regarding the AP chemical slugs of both experiments with almost the same bulk viscosity (≈24 mPa.s), the aged slug had a significantly lower oil recovery but resulted in a significantly higher pressure drop.

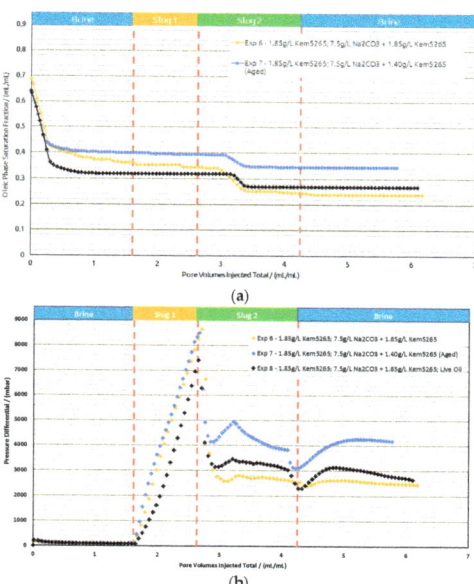

Figure 16. Oleic-phase saturation fraction/Pressure differential versus pore volumes injected total for the different slugs for core floods in real rock sand packs. Exp. 6 and 7 performed with dead oil and Exp. 8 with live oil. (**a**) Reduction in oleic-phase saturation versus PV of slugs injected in Berea outcrops with crude dead oil initialization. (**b**) Pressure variation versus PV of injected slugs in Berea outcrops with dead oil initialization.

Table 10. Summary of results obtained for sand pack flood experiments performed in this work. Core orientation was vertical and injection rate was 1 ft/day Viscosities are taken at 7.94 s^{-1}. Experiments performed at 49 °C.

Parameter	Unit	Exp. 6	Exp. 7	Exp. 8 (Live)
Pore Pressure	Bar	5	5	125
Radial Pressure	bar	30	30	150
Slug 1	g/L	1.85 P *	1.85 P *	1.85 P *
η Slug 1	mPa.s	20	21	20
Slug 2 *	g/L	7.5 A + 1.85 P	7.5 A + 1.4 P (aged) [1]	7.5 A + 1.85 P
η Slug 2	mPa.s	25	23	24
Length/Diameter	cm	29.60/2.99	30.35/2.99	30.55/2.99
Bulk Volume/PV	cm3	207.73/69.2	213.10/88.0	214.51/73
Porosity	%	33.3	41.3	34
Perm. (k_w)	mD	297	230	286
Oil Sat. Init.	%	69	64	64
RF brine	%	47.3	37.6	50
RF Polymer	%	2.5	0.9	0
RFAP	%	14.6	7.4	8
RF Brine	%	0.8	0.1	0

* Diluted in synthetic brine 8 TH WTP: 22.53 g/L NaCl, 0.16 g/L KCl, 0.94 g/L $CaCl_2$*2 H_2O, and 0 72 g/L $MgCl_2$*6 H_2O. [1] Solution prepared according to thermal aging procedure. P = A − 5265; A = Na_2CO_3; PV = Pore Volume.

Based on the promising oil recovery results, the chemical formulations of Exp. 6 were injected into the sand pack using live oil. The injection was performed under reservoir conditions presented as Exp. 8 in Figures 16 and 17. The RF from BF with live oil was higher compared to Exp. 7 but closer to Exp. 6. However, the polymer slug did not contribute any oil and the AP slug contributed to an additional 8% oil recovery. The additional oil contribution from the AP slug was similar to the thermally aged AP slug in Exp. 7. Hence, a comparison of the AP slugs between Exp. 7 and Exp. 8 let us conclude that the thermal aging could help to reduce the polymer amount required and contribute to additional oil recovery.

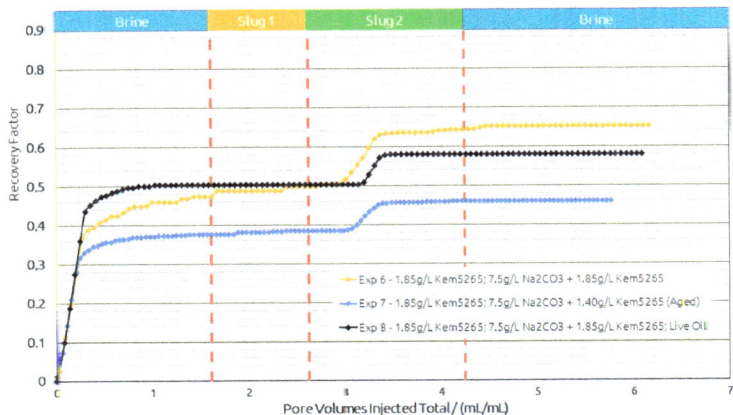

Figure 17. Recovery factor versus pore volumes injected total for the different slugs for core floods in real rock sand packs. Exp. 6 and 7 performed with dead oil and Exp. 8 with live oil.

Economic Efficiency: To evaluate the economic efficiency of the various chemical agents, utility factors (UF = kg chemicals injected/incremental oil production) can be calculated for the core flood experiments. Here, we assume the injection of 1 PV of alkali–

polymer, and the incremental oil recovery of the alkali–polymer slug observed. Note that we injected 1.5 PV in the laboratory to be sure that the displacement by the AP is complete. However, even considering dispersion, it is not necessary to inject more than 1 PV. That is why we used 1 PV for the one-dimensional EqUF calculation. Including the costs of the chemical agents, an Equivalent Utility Factor (EqUF) can be calculated:

$$EqUF = \frac{\frac{m_P * P_P + m_C * P_C + m_A * P_A + \ldots}{P_P}}{NP_{inc}} \left[\frac{kg}{bbl}\right] \quad (1)$$

where m is the injected mass in kg of the individual components, P is the price of a component in USD/kg, and NP_{inc} is the incremental oil recovery in bbl. Subscripts are P for polymer, C for co-solvent, and A for alkaline. If more components are utilized—e.g., a surfactant—the equation should be extended to n components. Here, we assume an Na_2CO_3 cost of 0.24 USD/kg and a polymer cost of 2.5 USD/kg. For the sake of simplicity, we focus here on comparing the experiments performed in outcrops.

The EqUFs for the injection of the aged and non-aged polymers with alkali for experiments 1–5 are shown in Figure 18. The EqUFs can be used to compare different chemical agents. Figure 18 shows that higher EqUFs are observed for experiments in which the aging of the polymer in an alkali solution in the reservoir was not accounted for. Reducing the polymer concentration in Experiment 3 (Figure 18) leads to a decrease in the EqUF by 40%. If the alkali–polymer is injected immediately without a prior polymer slug, then the economics are improved by 37% compared with the case (comparing Experiment 5 to Experiment 3). This is a good indication that it is beneficial to inject the alkali–polymer immediately rather than to start chemical EOR with a polymer injection. In the field test planned for the 8 TH reservoir, an area has been selected with an ongoing polymer flood for operational reasons, the roll out will be in an area of the field which is operated using waterflooding. The UFs of the polymer flood prior to the alkali–polymer are high, ranging from 9 kg/bbl. to 16 kg/bbl. for Experiments 1 to 3. The reason is the limited incremental oil observed in the experiments for polymer flooding. This shows the potential of the alkali–polymer EOR for reactive oils.

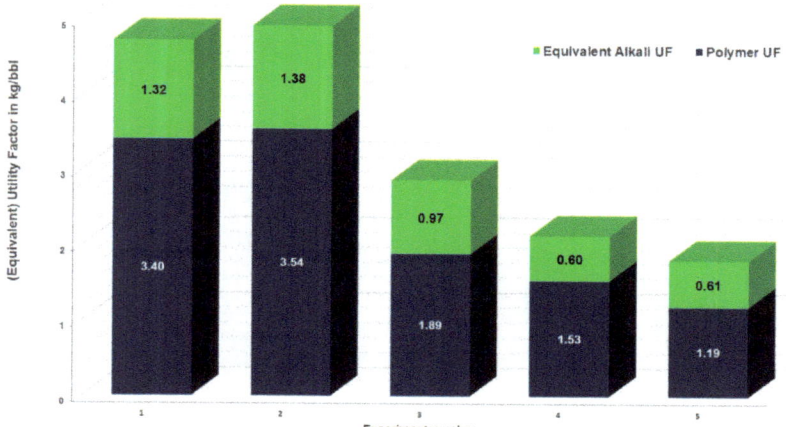

Figure 18. Equivalent utility factor (EqUF) for the experiments. The EqUF shown here reflects the incremental oil produced by the alkali–polymer slug and the mass of chemicals injected during this slug, assuming 1 PV injection. The polymer EqUF is the part of the overall EqUF related to the mass of the polymer injected during the injection of the AP slug, whereas the alkali EqUF reflects the cost-normalized mass of the alkali injected during the AP slug injection.

6. Discussion

Alkali–polymer injection is a promising technology to increase oil recovery in fields containing reactive oils. The initially generated soaps of the alkali with the reactive components of the oil can be measured to determine the amount of soap generated as a function of pH and the minimum required alkali concentration [22]. IFT measurements and phase behavior tests can be used to evaluate the presence of stable Windsor III emulsions (e.g., [23,24]) or thermodynamically instable emulsions (e.g., [9,25,26]) and the presence of salt–crude oil complexes at high alkali concentrations which are detrimental for enhanced oil recovery [27]. For the conditions of the 8 TH reservoir, a sufficient initial soap generation and the presence of instable emulsions were confirmed.

Two-layer micromodel experiments were used to investigate the displacement efficiency in heterogeneous rock. Micromodels have the advantage over sand packs or cores with two permeabilities because the boundary of the two permeability layers can be manufactured such that it neither creates a flow baffle nor a flow path. The results show that the highest incremental oil recovery over waterflooding is achieved from the high-permeability layer. Only at later stages does more crossflow occur, leading to increased recovery by AP flooding from the lower-permeability part of the micromodel. The results reveal that, to further increase recovery, polymer selection has to take the polymer molecular weight distribution and permeability distribution in the reservoir into account.

To improve the economics of AP flooding projects, the interaction of alkali with polymers over time needs to be considered. Aging polymers in alkali solution reveals the stability of the polymers but also the potential increase in polymer solution viscosity. For the conditions investigated here, the polymers were stable, and the viscosity increased with time. Displacement experiments showed that the recovery efficiency of the higher-concentration non-aged alkali–polymers is similar to the recovery efficiency of the lower-concentration but aged polymers. Hence, the polymer concentration in this AP project can be reduced compared with experiments using non-aged polymers. The polymer concentration is linked with the fracture half-length of the injection-induced fractures. The induced fractures might lead to the short circuiting of the injected fluids (e.g., [28]) and need to be included in the risk assessment for caprock integrity. Away from the wellbore, the viscosity of the polymers in alkali solution increases, leading to good sweep efficiency. Davidescu et al. [29] injected consecutive tracers and showed that the travel times for injected fluids decrease after polymer injection compared with water injection. The travel times are much longer than the required aging to increase the viscosity of the polymers in alkali. In the near wellbore, high flow velocities are observed and polymers in alkali solution are not aged and exhibit lower viscosities (neglecting visco-elastic effects) [30]. Further away from the wells, lower flow velocities (smaller arrows) are observed, and the polymer viscosity is increased. As shown in the section above, the EqUF is reduced by 40% by applying a lower polymer concentration. Furthermore, Nurmi et al. [14] showed that the adsorption of polymers on the rock is reduced in alkali solution compared to the case without alkali solution. Hence, the consumption of polymers in the reservoir is reduced, improving the economics of such projects.

7. Summary and Conclusions

Evaluating the synergies of alkali and polymers provided us with very important findings. On the one hand, polymer solutions need to be aged in alkali for alkali–polymer projects. On the other hand, the economics are much better if the alkali–polymer is injected directly. In the case studied here, we need and are planning to do the pilot in the polymer area for operational reasons.

We have shown that the performance of AP floods can be optimized by making use of lower polymer viscosities during injection but increasing polymer viscosities in the reservoir owing to the "thermal aging" of the polymers at a high pH. Furthermore, the AP conditions enable us to reduce the polymer retention in the reservoir, decreasing the utility factors (kg polymers injected/incremental bbl. produced).

The IFT measurements showed that the saponification at the oil–alkali solution interface very effectively reduces the IFT. Alkali phase experiments confirmed that emulsions are formed initially and supported the potential for residual oil mobilization. The aging experiments revealed that the polymer hydrolysis rate is substantially increased at a high pH compared to polymer hydrolysis at a neutral pH, resulting in a 60% viscosity increase in AP conditions. Micromodel experiments in two-layer chips depicted insights into the displacement. The two-phase experiments confirmed that lower-concentration polymer solutions aged in alkali show the same displacement efficiency as non-aged polymers with higher concentrations. Hence, significant cost savings can be realized, capitalizing on the fast aging in the reservoir. Due to the low polymer retention in AP floods, fewer polymers are consumed than in conventional polymer floods, significantly decreasing the utility factor (injected polymers kg/incremental bbl. produced).

We have shown that alkali/polymer (AP) injection leads to a substantial incremental oil production of reactive oils. A workflow was presented to optimize AP projects including reservoir effects. AP flood displacement efficiency must be evaluated incorporating the aging of polymer solutions. Significant cost savings and increasing efficiency can be realized in AP floods by incorporating the aging of polymers and taking the reduced polymer adsorption into account.

Author Contributions: Conceptualization, R.E.H. and A.B.; Methodology, R.E.H. and A.B.; Formal analysis, A.B., E.H., M.T., L.N., S.H., J.W. and A.J.; Investigation, R.E.H. and A.B.; Data curation, E.H. and M.T.; Writing—original draft, R.E.H.; Writing—review & editing, R.E.H.; Supervision, T.C.; Project administration, R.E.H. All authors have read and agreed to the published version of the manuscript.

Funding: This research received no external funding.

Institutional Review Board Statement: Not applicable.

Informed Consent Statement: Not applicable.

Data Availability Statement: Not applicable.

Acknowledgments: The authors would like to thank OMV E&P GmbH for granting permission to publish this paper.

Conflicts of Interest: The authors declare no conflict of interest.

Abbreviations

AP	Alkali–Polymer
ASP	Alkali–Surfactant–Polymer
cEOR	chemical Enhanced Oil Recovery
CT	Computed Tomography
DO	Dead Oil
DOC	Dead Oil with Cyclohexane
EOR	Enhanced Oil Recovery
EqUF	Equivalent Utility Factor
IFT	Interfacial Tension
HA	Saponifiable components in the oil
MP	Middle Phase
NMR	Nuclear magnetic resonance
PV	Pore Volume
RCA	Routine Core Analysis
RF	Recovery Factor
SEC	Size Exclusion Chromatography
Soft. 8 TH RSB	Softened 8 Torton Horizon Reservoir Synthetic Brine

Sor	remaining oil saturation
TAN	Total Acid Number
TH	Torton Horizon
UF	Utility Factor
UP	Upper Phase
XRD	X-ray Diffraction
8 TH RSB	8 Torton Horizon Reservoir Synthetic Brine

References

1. Gao, C. Scientific research and field applications of polymer flooding in heavy oil recovery. *J. Pet. Explor. Prod. Technol.* **2011**, *1*, 65–70. [CrossRef]
2. Anand, A.; Al Sulaimani, H.; Riyami, O.; AlKindi, A. Success and Challenges in Ongoing Field Scale Polymer Flood in Sultanate of Oman—A Holistic Reservoir Simulation Case Study for Polymer Flood Performance Analysis & Prediction. In Proceedings of the SPE EOR Conference at Oil and Gas West Asia, Muscat, Oman, 26–28 March 2018. [CrossRef]
3. Rock, A.; Hincapie, R.E.; Tahir, M.; Langanke, N.; Ganzer, L. On the Role of Polymer Viscoelasticity in Enhanced Oil Recovery: Extensive Laboratory Data and Review. *Polymers* **2020**, *12*, 2276. [CrossRef] [PubMed]
4. Delamaide, E. Is Chemical EOR Finally Coming of Age? In *SPE Asia Pacific Oil & Gas Conference and Exhibition*; OnePetro SPE: Houston, TX, USA, 2020. [CrossRef]
5. Mayer, E.; Berg, R.; Carmichael, J.; Weinbrandt, R. Alkaline Injection for Enhanced Oil Recovery—A Status Report. *J. Pet. Technol.* **1983**, *35*, 209–221. [CrossRef]
6. Al-Murayri, M.T.; Al-Mayyan, H.E.; Al-Mahmeed, N.; Muthuswamy, A.; Shahin, G.T.; Shukla, S.R. Alkali-Surfactant Adsorption and Polymer Injectivity Measurements Using Reservoir Core from a Giant High Temperature and High Salinity Clastic Reservoir to Design an ASP Pilot. In Proceedings of the SPE Kuwait Oil & Gas Show and Conference, Mishref, Kuwait, 14 October 2019. [CrossRef]
7. Schumi, B.; Clemens, T.; Wegner, J.; Ganzer, L.; Kaiser, A.; Hincapie, R.E.; Leitenmueller, V. Alkali/Cosolvent/Polymer Flooding of High-TAN Oil: Using Phase Experiments, Micromodels, and Corefloods for Injection-Agent Selection. *SPE Reserv. Eval. Eng.* **2019**, *23*, 463–478. [CrossRef]
8. Zhou, R.; Zhong, H.; Ye, P.; Wei, J.; Zhang, D.; Zhong, L.; Jiao, T. Experiment on the Scaling Mechanism of Strong Alkaline–Surfactant–Polymer in a Sandstone Reservoir. *ACS Omega* **2022**, *7*, 1794–1802. [CrossRef] [PubMed]
9. Hincapie, R.E.; Borovina, A.; Clemens, T.; Hoffmann, E.; Wegner, J. Alkali/Polymer Flooding of High-TAN Oil: Live Oil and Dead Oil Evaluation. *SPE Reserv. Eval. Eng.* **2022**, *25*, 380–396. [CrossRef]
10. Krumrine, P.; Falcone, J.J. Surfactant, Polymer, and Alkali Interactions in Chemical Flooding Processes. In *SPE Oilfield and Geothermal Chemistry Symposium*; OnePetro SPE: Houston, TX, USA, 1983. [CrossRef]
11. Sheng, J.J. Critical review of alkaline-polymer flooding. *J. Pet. Explor. Prod. Technol.* **2016**, *7*, 147–153. [CrossRef]
12. Kazempour, M.; Sundstrom, E.A.; Alvarado, V. Effect of Alkalinity on Oil Recovery During Polymer Floods in Sandstone. *SPE Reserv. Eval. Eng.* **2012**, *15*, 195–209. [CrossRef]
13. Födisch, H.; Nurmi, L.; Hincapie, R.E.; Borovina, A.; Hanski, S.; Clemens, T.; Janczak, A. Improving Alkali Polymer Flooding Economics by Capitalizing on Polymer Solution Property Evolution at High pH. In Proceedings of the SPE Annual Technical Conference and Exhibition, Houston, TX, USA, 3–5 October 2022. [CrossRef]
14. Nurmi, L.; Hincapie, R.E.; Clemens, T.; Borovina, A.; Hanski, S.; Födisch, H.; Janczak, A. Improving Alkali Polymer Flooding Economics by Capitalizing on Polymer Solution Property Evolution at High pH. In *2022 SPE Reservoir Evaluation & Engineering-Reservoir Engineering*; SPE-210043-PA; OnePetro SPE: Houston, TX, USA, 2022. [CrossRef]
15. Rupprecht, B.J. Hydrocarbon Generation and Alteration in the Vienna Basin. Ph.D. Thesis, Montan University Leoben, Leoben, Austria, 2017. Available online: https://pure.unileoben.ac.at/portal/en/publications/hydrocarbon-generation-and-alteration-in-the-vienna-basin(51a3b0dc-3ac3-4577-aa4f-feb34f51f67c).html (accessed on 6 June 2022).
16. Southwick, J.G. Solubility of Silica in Alkaline Solutions: Implications for Alkaline Flooding. *Soc. Pet. Eng. J.* **1985**, *25*, 857–864. [CrossRef]
17. Scheurer, C.; Hincapie, R.E.; Neubauer, E.; Metz, A.; Ness, D. Sorption of Nanomaterials to Sandstone Rock. *Nanomaterials* **2022**, *12*, 200. [CrossRef] [PubMed]
18. Phukan, R.; Gogoi, S.B. Tiwari, P.; Vadhan, R.S. Optimization of Immiscible Alkaline-Surfactant-Alternated-Gas/CO_2 Flooding in an Upper Assam Oilfield. In Proceedings of the SPE Western Regional Meeting, San Jose, CA, USA, 24 April 2019. [CrossRef]
19. Borji, M.; Kharrat, A.; Ott, H. Comparability of in situ crude oil emulsification in phase equilibrium and under porous-media-flow conditions. *J. Colloid Interface Sci.* **2022**, *615*, 196–205. [CrossRef] [PubMed]
20. Liu, S.; Zhang, D.L.; Yan, W.; Puerto, M.; Hirasaki, G.J.; Miller, C.A. Favorable Attributes of Alkaline-Surfactant-Polymer Flooding. *SPE J.* **2008**, *13*, 5–16. [CrossRef]
21. Nurmi, L.; Sandengen, K.; Hanski, S.; Molesworth, P. *Sulfonated Polyacrylamides—Evaluation of Long Term Stability by Accelerated Aging at Elevated Temperature*; SPE-190184-MS; Society of Petroleum Engineers: Houston, TX, USA, 2018. [CrossRef]

22. Southwick, J.; Brewer, M.; van Batenburg, D.; Pieterse, S.; Bouwmeester, R.; Mahruqi, D.; Alkindi, A.; Mjeni, R. Ethanolamine as Alkali for Alkali Surfactant Polymer Flooding—Development of a Low-Complexity Field Implementation Concept. In Proceedings of the SPE Improved Oil Recovery Conference, Virtual, 31 August–4 September 2020. [CrossRef]
23. Stoll, W.M.; al Shureqi, H.; Finol, J.; Al-Harthy, S.A.; Oyemade, S.N.; de Kruijf, A.; van Wunnik, J.; Arkesteijn, F.; Bouwmeester, R.; Faber, M.J. Alkaline/Surfactant/Polymer Flood: From the Laboratory to the Field. *SPE Reserv. Eval. Eng.* **2011**, *14*, 702–712. [CrossRef]
24. Panthi, K.; Clemens, T.; Mohanty, K.K. Development of an ASP formulation for a sandstone reservoir with divalent cations. *J. Pet. Sci. Eng.* **2016**, *145*, 382–391. [CrossRef]
25. Dezabala, E.F.; Radke, C.J. A Nonequilibrium Description of Alkaline Waterflooding. *SPE Reserv. Eng.* **1986**, *1*, 29–43. [CrossRef]
26. Sharma, M.M.; Jang, L.K.; Yen, T.F. Transient Interfacial Tension Behavior of Crude-Oil/Caustic Interfaces. *SPE Reserv. Eng.* **1989**, *4*, 228–236. [CrossRef]
27. Magzymov, D.; Clemens, T.; Schumi, B.; Johns, R.T. Experimental Analysis of Alkali-Brine-Alcohol Phase Behavior with High Acid Number Crude Oil. *SPE Reserv. Eval. Eng.* **2020**, *24*, 390–408. [CrossRef]
28. Hoek, P.J.V.D.; Al-Masfry, R.A.; Zwarts, D.; Jansen, J.-D.; Hustedt, B.; Van Schijndel, L. Optimizing Recovery for Waterflooding Under Dynamic Induced Fracturing Conditions. *SPE Reserv. Eval. Eng.* **2009**, *12*, 671–682. [CrossRef]
29. Davidescu, B.-G.; Bayerl, M.; Puls, C.; Clemens, T. Horizontal Versus Vertical Wells: Assessment of Sweep Efficiency in a Multi-Layered Reservoir Based on Consecutive Inter-Well Tracer Tests—A Comparison Between Water Injection and Polymer EOR. In Proceedings of the SPE Europec Featured at 82nd EAGE Conference and Exhibition, Amsterdam, The Netherlands, 18–21 October 2021. [CrossRef]
30. Seright, R.S.; Fan, T.; Wavrik, K.; De Carvalho Balaban, R. New Insights into Polymer Rheology in Porous Media. *SPE J.* **2010**, *16*, 35–42. [CrossRef]

Article

Plasticized Mechanical Recycled PLA Films Reinforced with Microbial Cellulose Particles Obtained from Kombucha Fermented in Yerba Mate Waste

Ángel Agüero [1,2], Esther Corral Perianes [2], Sara Soledad Abarca de las Muelas [2], Diego Lascano [1], María del Mar de la Fuente García-Soto [2,3], Mercedes Ana Peltzer [3,4,5], Rafael Balart [1] and Marina Patricia Arrieta [2,6,*]

1. Instituto de Tecnología de Materiales (ITM), Universidad Politécnica de Valencia (UPV), Plaza Ferrándiz y Carbonell 1, 03801 Alcoy, Spain
2. Departamento de Ingeniería Química Industrial y del Medio Ambiente, Escuela Técnica Superior de Ingenieros Industriales, Universidad Politécnica de Madrid (ETSII-UPM), Calle José Gutiérrez Abascal 2, 23006 Madrid, Spain
3. Grupo de Investigación: Tecnologías Ambientales y Recursos Industriales (TARIndustrial), 20006 Madrid, Spain
4. Laboratory of Obtention, Modification, Characterization, and Evaluation of Materials (LOMCEM), Department of Science and Technology, University of Quilmes, Bernal B1876BXD, Argentina
5. National Scientific and Technical Research Council (CONICET), Buenos Aires C1425FQE, Argentina
6. Grupo de Investigación: Polímeros, Caracterización y Aplicaciones (POLCA), 28006 Madrid, Spain
* Correspondence: m.arrieta@upm.es; Tel.: +34-910-677-301

Abstract: In this study, yerba mate waste (YMW) was used to produce a kombucha beverage, and the obtained microbial cellulose produced as a byproduct (KMW) was used to reinforce a mechanically recycled poly(lactic acid) (r-PLA) matrix. Microbial cellulosic particles were also produced in pristine yerba mate for comparison (KMN). To simulate the revalorization of the industrial PLA products rejected during the production line, PLA was subjected to three extrusion cycles, and the resultant pellets (r3-PLA) were then plasticized with 15 wt.% of acetyl tributyl citrate ester (ATBC) to obtain optically transparent and flexible films by the solvent casting method. The plasticized r3-PLA-ATBC matrix was then loaded with KMW and KMN in 1 and 3 wt.%. The use of plasticizer allowed a good dispersion of microbial cellulose particles into the r3-PLA matrix, allowing us to obtain flexible and transparent films which showed good structural and mechanical performance. Additionally, the obtained films showed antioxidant properties, as was proven by release analyses conducted in direct contact with a fatty food simulant. The results suggest the potential interest of these recycled and biobased materials, which are obtained from the revalorization of food waste, for their industrial application in food packaging and agricultural films.

Keywords: PLA; cellulose; yerba mate; kombucha; food packaging

Citation: Agüero, Á.; Corral Perianes, E.; Abarca de las Muelas, S.S.; Lascano, D.; de la Fuente García-Soto, M.d.M.; Peltzer, M.A.; Balart, R.; Arrieta, M.P. Plasticized Mechanical Recycled PLA Films Reinforced with Microbial Cellulose Particles Obtained from Kombucha Fermented in Yerba Mate Waste. *Polymers* **2023**, *15*, 285. https://doi.org/10.3390/polym15020285

Academic Editors: Jesús-María García-Martínez and Emilia P. Collar

Received: 13 December 2022
Revised: 30 December 2022
Accepted: 2 January 2023
Published: 5 January 2023

Copyright: © 2023 by the authors. Licensee MDPI, Basel, Switzerland. This article is an open access article distributed under the terms and conditions of the Creative Commons Attribution (CC BY) license (https://creativecommons.org/licenses/by/4.0/).

1. Introduction

Biobased and biodegradable polymers have gained attention for food packaging applications in order to reduce the consumption of non-renewable resources and prevent the accumulation of plastic waste in the environment. Among other biopolymers, poly(lactic acid) has emerged in the market as the most used biobased and biodegradable plastic due to its many advantages, such as its environmentally benign characteristics, availability in the market at a competitive cost, ease of processing by means of the current existing processing technologies for petrol-based thermoplastics (i.e., extrusion, injection molding, etc.), high transparency, and inherent biodegradability [1,2]. However, PLA also presents some disadvantages for film production which hinder its industrial exploitation in the food packaging or agricultural sectors, such as its sensitivity to thermal degradation [2], poor barrier performance [3], and inherently brittle nature [4]. Its degradability in the environment requires specific conditions (compost medium at 58 °C, a pH around 7.5, relative humidity of 60%, a C/N relationship between 20:1 and 40:1, and proper aeration) to be met, even for short periods of time [5]. Moreover, the model of the linear economy generates high levels of plastic waste and creates a dependence between economic development and the entry of new, virgin plastics into the system [6]. Therefore, the use of recycled PLA for film for food packaging or agricultural applications is gaining interest [7,8]. Cosate de Andrade et al. [8] compared the chemical recycling and mechanical recycling of PLA, and concluded that mechanical recycling generates less impact than chemical recycling due to the fact that the mechanically recycled polymers are produced using lower energy and fewer inputs than other destinations. However, bioplastic consumption is currently still low, and they can be considered contaminants in plastic recycling streams due to the fact that they can affect the mechanical performance of well-implemented mechanical recycling processes of other plastics, such as polyethylene terephthalate (PET), polypropylene (PP), and polystyrene (PS) [9–11]. Moreover, although the European Commission promotes the increase of recycled plastic in food packaging as an essential prerequisite to its strategy to introduce recycled plastics in a circular economy, the current legislation does not allow the direct use of recycled plastics coming from recycled streams for food contact materials. This is because those recycling processes originate from waste, and the legislation establishes strict requirements concerning food safety (the transfer of substances that may affect human health, or quality of the food, and microbiological safety) [12]. In this context, during the industrial production of plastic products, several parts are produced with defects, rejected from the production line, and then discarded. These rejected parts can be reprocessed and used to produce recycled pellets that do not come from waste streams and are of well-known origin. In a previous work, PLA was reprocessed up to six times, and it was observed that the main losses took place when PLA was subjected to more than four reprocessing cycles, while low degradation was found between one and three reprocessing cycles [13]. However, due to PLA's high sensitivity to hydrolysis of its ester groups at the industrial processing conditions, such as melt extrusion, the obtained recycled PLA-based products show a decrease in the polymer chain length and, thus, show lower-quality performance than PLA-based products produced with virgin PLA [14]. This is why the use of reinforcing fillers with antioxidant activity as additives have gained interest for the purpose of protecting the polymeric matrix from thermal degradation and increasing the mechanical resistance of mechanically recycled PLA [7].

Another industrial sector that generates a large amount of waste and can be introduced in the food packaging sector for the preparation of high-tech composites and/or nanocomposites is the food industry [15]. The kombucha beverage is a popular probiotic beverage typically produced by fermenting sugared tea with a symbiotic community of bacteria and yeast (SCOBY) that involves cooperative and competitive interactions [16]. While yeasts produce invertase, which releases monosaccharides to media accessible to any microbe as a carbon source, bacteria rapidly metabolize released sugars and produce organic acids that acidify the media [16]. Meanwhile, the reduction in monosaccharides increases the frequency of the invertase-producing yeast, and the ethanol produced by yeast stimulates the bacterial cellulose synthase mechanism to produce a cellulose film at the surface that acts as a physical barrier to protect from external competitors [16]. The cellulosic film is a byproduct in the kombucha tea industry, but it is very interesting for the plastic industry. Kombucha tea has been fermented in several sugared infusions (i.e., black tea, green tea, yerba mate, etc.) [16–18]. The antioxidant activity of microbial cellulose obtained from the kombucha fermentation is directly related to the high amount of bioactive compounds in the infusion used for its fermentation, such as phenolics, tannins, catechins, flavonoids, etc., which are decomposed into their simpler forms during the kombucha fermentation process [17]. In fact, it has been observed that the cellulose obtained from kombucha fermented in sugared infusions of yerba mate possesses high antioxidant activity [16]. Yerba mate (*Ilex paraguariensis*, Saint Hilaire) is a tree from the subtropical region of South America that grows in a limited zone within Argentina, Brazil, and Paraguay, where it has an important commercial purpose due to the high consumption of dried yerba mate leaves in the form of infusion, which is known as "mate" [15,19]. Its high consumption leads to a high amount of yerba mate being wasted without any kind of revalorization [15]. For instance, in 2020, the consumption of yerba mate in Argentina was over 310 million kg [20]. Thus, in this work, kombucha SCOBY was fermented in yerba mate waste.

Among other plasticizers, citrate esters such as acetyl(tributyl citrate) (ATBC) have been proven to be very effective PLA plasticizers, and are accepted for food contact applications [21]. The miscibility between PLA and ATBC has been associated with the similarity in their solubility parameters (δ) that of PLA being between 19.5 $MPa^{1/2}$ and 20.5 $MPa^{1/2}$ [2], while that of ATBC is 20.2 $MPa^{1/2}$ [22]. Likewise, to produce polymers by a solvent casting method, the selection of an effective solvent is also on the basis of a similar solubility parameter to that of the polymer. In this sense, chloroform (δ = 19 $MPa^{1/2}$) is widely used to dissolve PLA [23].

The main objective of the present research was to obtain sustainable and active films from the revalorization of plastic and food industry waste. The materials were prepared based on mechanically recycled PLA and cellulosic particles extracted from kombucha fermented in yerba mate waste. Thus, virgin PLA was subjected to three reprocessing extrusion cycles (r3-PLA) to simulate the revalorization of industrial PLA products rejected during the production line. Three reprocessing cycles were selected, since in a previous work, it was observed that between one and three reprocessing cycles, low PLA degradation occurs [13]. The decrease in the polymer chain length due to the three reprocessing cycles was investigated by measurement of the viscosity-molecular weight. On the other hand, yerba mate waste was used to obtain the sugared infusion to produce kombucha beverage from kombucha SCOBY, while the cellulosic by-product formed during its production was used to produce cellulosic particles with antioxidant activity (KMW). Another kombucha SCOBY was fermented in a sugared infusion of new yerba mate, and the cellulosic particles obtained were studied for comparison (KMN). Both particles, namely kombucha mate waste (KMW) and kombucha mate new (KMN), were used to reinforce plasticized, mechanically recycled r3-PLA with 15 wt.% of ATBC. Two reinforcing amounts were used, namely 3 wt.% and 5 wt.%, and the obtained films were characterized in terms of transparency, barrier performance against water and UV light, thermal stability, crystallization behavior, surface wettability, and mechanical performance in order to obtain information regarding the possibility of using these films as antioxidant food contact materials, such as food packaging or in the agro-industrial field.

2. Materials and Methods
2.1. Materials

PLA commercial-grade IngeoTM 2003D with a density of 1.24 $g \cdot cm^{-3}$ and a melt flow index (MFI) of 6 g/10 min (measured at 210 °C and with a load of 2.16 kg) was supplied by Natureworks (Minnetonka, MN, USA). Acetyl tributyl citrate (ATBC) (98% purity, Mw = 402 g mol^{-1}, and Tm = −80 °C), chloroform ($CHCl_3$, δ = 19 $MPa^{1/2}$), and 2,2-diphenyl-1-picrylhydrazyl (DPPH) 95% free radical were supplied by Sigma Aldrich (Madrid, Spain). The pristine yerba mate (Taragüi, Virasoro, Argentina) was used as is and called YMN, while the yerba mate waste was obtained from the residue of mate infusion after our consumption and called YMW.

2.2. Processing of Kombucha to Obtain Cellulosic Particles from Yerba Mate Waste

The native culture of kombucha was provided by Teresa Carles Manufacturing S. L. (Barcelona, Spain), and was used as the starter culture and inoculum for a new batch of kombucha fermented in an infusion of yerba mate (5 g/L) and sucrose (100 g/L). KMW was obtained from the fermentation of one kombucha SCOBY from that batch in a 2.5 L sugared infusion prepared either with 15 g of yerba mate (YMN) and/or yerba mate waste (YMW), 300 g of sucrose, and 500 mL of stock culture, which was maintained at static conditions at 22 ± 2 °C and then covered with a textile cloth for 30 days. A new floating disc was produced, and it was recovered, washed with distilled water, filtered off, and further sterilized at 121 °C and 101 kPa for 15 min in a steam autoclave. The disc was then homogenized by ultraturax at 30,000 rpm for two minutes (4 cycles of 30 s) and dried at 60 °C for 24 h. The dry matter, determined by drying at 105 °C until a constant weight was reached, showed a yield of ca. 1.3 ± 0.1%, in accordance with previous reported works [16]. Then, the obtained cellulosic paper was ground to obtain a powder and further sieved (500 μm). In Figure 1 the wall process to obtain either KMN or KMW from the SCOBY fermented in YM or YMW and convert it to the powder able to be processed by melt extrusion is schematically represented.

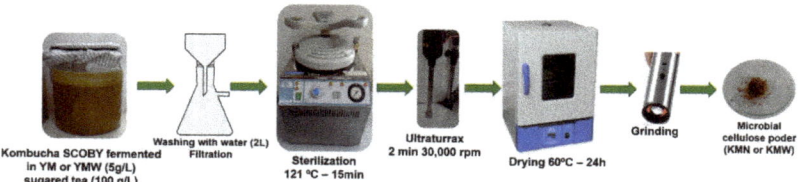

Figure 1. Schematic representation of the microbial cellulose (KMN and/or KMW) production from kombucha fermented in YM or YMW.

2.3. Processing and Reprocessing of PLA

To obtain reprocessed PLA (r3-PLA), PLA pellets were previously dried overnight to remove the residual moisture at 60 °C for 4 h in an air-circulating oven. The PLA pellets were processed 3 times in a twin-screw co-rotating extruder with a screw diameter of 30 mm, supplied by Construcciones Mecanicas Dupra, S.L. (Alicante, Spain), at a screw speed of around 22 rpm and using a temperature profile of 180 °C (feeding hopper), 185 °C, 190 °C, and 195 °C (extrusion die), on the basis of previous work [13]. After the extrusion process, the strands were cooled in air and then pelletized using an air-knife unit. They were subsequently subjected to an additional processing cycle under the same conditions, up to three times.

The capillary viscosity of virgin PLA and r3-PLA pellets was measured with a Ubbelohde viscometer (type 1C). Both pellets were diluted in $CHCl_3$ and the measurements were conducted at 25 °C using a water bath and a home-made 3D printed viscosimeter support. At least four concentrations were used. The intrinsic viscosity [η] of PLA and r3-PLA was determined to estimate the viscosity molecular weight by means of the Mark–Houwink relation (Equation (1)).

$$[\eta] = K \times M_v^a \quad (1)$$

where K and a, for PLA, are 1.53×10^{-2} and 0.759, respectively [24].

2.4. Films Preparation

KMN- and KMW-loaded r3-PLA-ATB-based materials were processed into thin films by the solvent casting method. For this purpose, 0.6 g of reprocessed PLA pellets (r3-PLA) were dissolved in 45 mL of $CHCl_3$ under continuous stirring at 1000 rpm at room temperature. ATBC was then added at 15 wt.% with respect to the polymeric matrix, on the basis of previous works [15,22,25], and named r3-PLA-ATBC. For the development of composites, the plasticized PLA films (r3PLA-ATBC) were then loaded either with kombucha mate waste (KMW) or kombucha mate new (KMN) in 1 wt.% and 3 wt.%, with respect to the r3-PLA-ATBC polymeric blend, and all films were prepared by the solvent casting method. Each suspension was cast onto a 50 mm-diameter glass mold, and then $CHCl_3$ was allowed to evaporate at 40 °C for 48 h in an oven. The obtained films are summarized in Table 1. They were dried under a vacuum to complete the drying process, ensuring the complete elimination of the solvent for about 10 h at 40 °C, prior to being characterized.

Table 1. Film formulations based on plasticized r3-PLA-ATBC.

Sample	r3-PLA (wt.%)	ATBC (wt.%)	KMN (wt.%)	KMW (wt.%)
r3-PLA	100	-	-	-
r3-PLA-ATBC	85	15	-	-
r3-PLA-ATBC-KMN1	84.15	14.85	1	-
r3-PLA-ATBC-KMN3	82.45	14.55	3	-
r3-PLA-ATBC-KMW1	84.15	14.85	-	1
r3-PLA-ATBC-KMW3	82.45	14.55	-	3

2.5. Characterization of the Films

2.5.1. UV-Visible Measurements

The transmittance of the obtained films was measured in the 800–250 nm region using a UV-Visible spectrophotometer Varian Cary 1E UV-Vis (Varian, Palo Alto CA, USA) at a scanning speed of 400 nm/min. The overall transmittance in the visible region was calculated following the ISO 13468 standard.

2.5.2. Scanning Electron Microscopy

The microstructures of films' cross-sections were observed by field emission scanning electron microscopy (FESEM) by means of a ZEISS ULTRA 55 microscope from Oxford Instruments (Abingdon, UK). The film samples were previously frozen in liquid N_2, cryofractured, and sputtered with a thin layer of gold and palladium alloy in an EMITECH sputter coating, SC7620, from Quorum Technologies, Ltd. (East Sussex, UK) to achieve a conductive surface. Then, the film samples were observed with an accelerating voltage of 2 kV. Images were taken at 10,000× magnification.

2.5.3 Differential Scanning Calorimetry

Differential scanning calorimetry (DSC) analyses were conducted in a Mettler-Toledo model 821 DSC (Schwerzenbach, Switzerland). The DSC thermal cycles were carried out under a nitrogen atmosphere. The first heating DSC scan was conducted from 30 °C to 200 °C at a rate of 10 °C/min, with the main objective of eliminating the thermal history. Then, the samples were cooled down to −50 °C at a rate of 10 °C/min. Finally, the second heating DSC scan was carried out from −50 °C to 300 °C at a rate of 10 °C/min. The degree of crystallinity (χ_c), obtained from the DSC thermograms, was calculated by Equation (2).

$$\chi_c = \frac{\Delta H_m - \Delta H_{cc}}{\Delta H_m^0} \cdot \frac{1}{W_{PLA}} 100 \qquad (2)$$

where ΔH_m is the melting enthalpy, ΔH_{cc} is the cold crystallization enthalpy, ΔH_m^0 is the melting heat associated with pure crystalline PLA (93 J g^{-1}) [26], and W_{PLA} is the weight fraction of PLA in the blend formulation.

2.5.4. Thermogravimetric Analysis

Dynamic thermogravimetric analyses were conducted in a TA Instruments TGA2050 thermobalance (TA Instruments, New Castle, DE, USA). For each measurement, around 10 mg of films were placed in a platinum crucible and heated from 30 to 800 °C at 10 °C/min, under a nitrogen atmosphere.

Isothermal thermogravimetric analyses were also conducted in a TGA/SDTA 851 thermobalance from Mettler-Toledo (Schwerzenbach, Switzerland). For each measurement, around 10 mg of films were heated at 180 °C for 20 min.

2.5.5. Tensile Test Measurements

The mechanical properties were evaluated by means of tensile test measurements using a Shimadzu AGS-X 100 N universal tensile testing machine (Shimadzu Corporation, Kyoto, Japan) equipped with a 100 N load cell, with an initial length of 30 mm and a crosshead speed of 10 mm min^{-1}. Dog-bone samples were prepared by a JBA electrohydraulic cutter (Instruments J. Bot SA) for tensile specimen 1BB, according to ISO 527-2. The Young modulus (E), tensile strength (TS), and average percentage elongation at break (ε%) were calculated from the obtained stress–strain curves, and the media of at least five specimens were reported.

2.5.6. Static Contact Angle Measurements

Surface wettability of the films was studied through static water contact angle (WCA) measurements by using a standard goniometer (EasyDrop-FM140, KRÜSS GmbH, Hamburg, Germany) equipped with a camera and Drop Shape Analysis SW21; DSA1 software. Drops of ~15 µL distiller water were placed onto the films' surfaces with the aid of a syringe, and approximately ten contact angle measurements were taken for each sample, with the films in random positions.

2.5.7. Water Vapor Transmission Rate

The water vapor transmission rate (WVTR) measurements of the films were determined by gravimetry, using silica gel as a desiccant agent. Films were placed in permeability cups with an exposed area (A) of 10 cm^2, filled with 2 g of previously dried silica gel, and further placed in a desiccator at 23 ± 1 °C with a saturated KNO$_3$ solution, obtaining a relative humidity of 85 ± 4%. The cups were weighed every hour for 7 h, and then again after 24 h. The mass increase in the cups was plotted against time, with slope n. WVTR (g/day cm^2) was determined through Equation (3):

$$\text{WVTR} = \frac{n}{A} \quad (3)$$

Because the water vapor transmission is dependent on the film thickness, the WVTR values were normalized to 100 µm [27].

2.5.8. Specific Migration Test and Antioxidant Activity

Double-sided total immersion migration tests were performed by total immersion of films in a glass vial containing a fatty food simulant (Simulant D1 = ethanol 50% v/v) at 40 °C for 10 days (area-to-volume ratio = 6 dm^2/L) [28]. After 10 days, films were removed and the food simulant was used to determine their antioxidant ability, which was measured by determination of the radical scavenging activity (RSA) through the DPPH method. This was accomplished by the determining the reduction in the absorbance at 517 nm by means of a UV-Vis Varian Cary spectrophotometer. The radical scavenging activity (RSA) was determined using Equation (4).

$$\text{RSA (\%)} = \frac{A_{control} - A_{sample}}{A_{control}} \times 100\% \quad (4)$$

where $A_{control}$ is the absorbance of 2,2-difenil-1-picrylhydrazyl (DPPH) in ethanolic solution and A_{sample} the absorbance of DPPH after 15 min in contact with each food simulant sample.

3. Results

3.1. Reprocessed PLA Characterization

The materials developed herein were prepared with mechanically recycled PLA, which was processed three times by melt extrusion using a temperature profile from feeding to hopper of 180 °C, 185 °C, 190 °C, and 195 °C, based on previous work [13], to simulate the revalorization of industrial waste produced during the production line, in which some parts are rejected. The viscosity–molecular weight (M_v) relationship of the reprocessed PLA (3r-PLA) and PLA pellets was determined in order to obtain insights into the degradation of the polymeric matrix as a consequence of the reprocessing procedure. The obtained results of the estimated M_v of PLA and r3-PLA were 181,770 ± 3370 g/mol and 115,410 ± 5080 g/mol, respectively (a reduction of around 36%). A reduction in the intrinsic viscosity ([η]) has already been reported in PLA samples subjected to a simulated mechanical recycling process in which one melt extrusion reprocessing cycle was applied, showing a reduction of around 14% with respect to samples prepared with virgin PLA [29]. In the present work, a reduction in the intrinsic viscosity ([η]) due to the three reprocessing melt extrusion cycles was around 30%, leading to the aforementioned reduction in the M_v. This reduction in the M_v is due to chain scission, produced by thermal degradation during each thermal processing cycle as a consequence of a hydrolysis process which is augmented by the heating [13].

3.2. UV-Visible Measurements

The transmittance of the obtained films was measured by means of a UV-Visible spectrophotometer, and the absorption spectra of the films are shown in Figure 2. Neat PLA film was also analyzed for comparison. From the spectra, it could be seen that although all formulations based on recycled PLA (r3-PLA) resulted in less transparent materials than PLA, they were mostly transparent in the visible region of the spectra (400–700 nm) allowing the films to be seen through, which is one of the most important requirements for food packaging due to consumers' acceptance [30]. It is also very important that films intended for agricultural applications should not only protect crops, but also permit the photosynthesis process to occur [1,31]. Among reprocessed films, r3-PLA film was the most transparent film, showing the highest transmission along the visible region of the spectra (400–700 nm). The incorporation of ATBC slightly affected the transparency of r3-PLA, as was already observed with the addition of ATBC to a virgin PLA matrix [22]. The transparency was slightly reduced with the incorporation of KMN and/or KMW. Absorption measurements were conducted in the range of 540–560 nm (see zoom image in Figure 2) of the visible region of the spectra, and it can be seen that the materials resulted as highly transparent (between 81% and 87% of transmittance).

When comparing r3-PLA with neat PLA film, a slight UV-light absorption in the 260 to 290 nm region can be observed in the 3r-PLA sample, which has already been observed in recycled PLA [7,14]. It has been reported that recycled PLA leads to a reduction in the UV light transmission in the 260 to 290 nm region of the spectrum, ascribed to the formation of –COOH chain end groups in PLA as a consequence of the chain scission (carbonyl carbon-oxygen bond cleavage) during thermal processing [14]. Nevertheless, it should be highlighted that the UV light transmission reduction is less marked than in the post-consumer, mechanically recycled PLA bottles studied by Chariyachotilert et al. They observed higher UV light transmission reduction, which can be related not only to the thermal degradation during reprocessing, but also with the degradation of PLA products during service, as well as under the conditions typically used for cleaning PET (85 °C, 1 wt.% NaOH and 0.3 wt.% Triton® X-100 surfactant for 15 min) [14]. In this sense, in a previous work, Agüero et al. studied the mechanical recyclability of injected molded PLA parts in more depth, performing between one and six reprocessing melt extrusion cycles, and showed that low degradation takes place between one and three reprocessing cycles in [10]. Thus, this means that less degradation had taken place after three reprocessed melt extrusion cycles than in post-consumed, washed, and further reprocessed PLA, highlighting the viability of mechanical recyclability of rejected PLA parts from the production line.

3.3. Scanning Electron Microscopy

FESEM investigations were conducted to study the microstructure of the films, and the micrographs of the cross-fractured surface are shown in Figure 3. The r3-PLA film (Figure 3a) showed the typical regular and smooth fracture of PLA films based on semi-crystalline virgin PLA [9,32]. An increased ductile fracture was observed in r3-PLA-ATBC film (Figure 3b), with more plastic behavior and no apparent phase separation, demonstrating the plasticizing effect of ATBC on the reprocessed PLA matrix. The ternary composites for both formulations with 1 wt.% (Figure 3c,e) and those with 3 wt.% (Figure 3d,f), showed a uniform dispersion of both KMN and KMW into the r3-PLA matrix. In the case of the higher reinforcing amount used here (3 wt.%) (Figure 3d,f), it seems that there was an increase in surface roughness. However, the formulations had very similar surface patterns to those reinforced with lower amounts of kombucha particles (1 wt.%), suggesting that cellulosic particles are well-distributed in the reprocessed PLA matrix. It has been observed that plasticizers such as ATBC improve the dispersion of cellulosic particles into the PLA matrix [25].

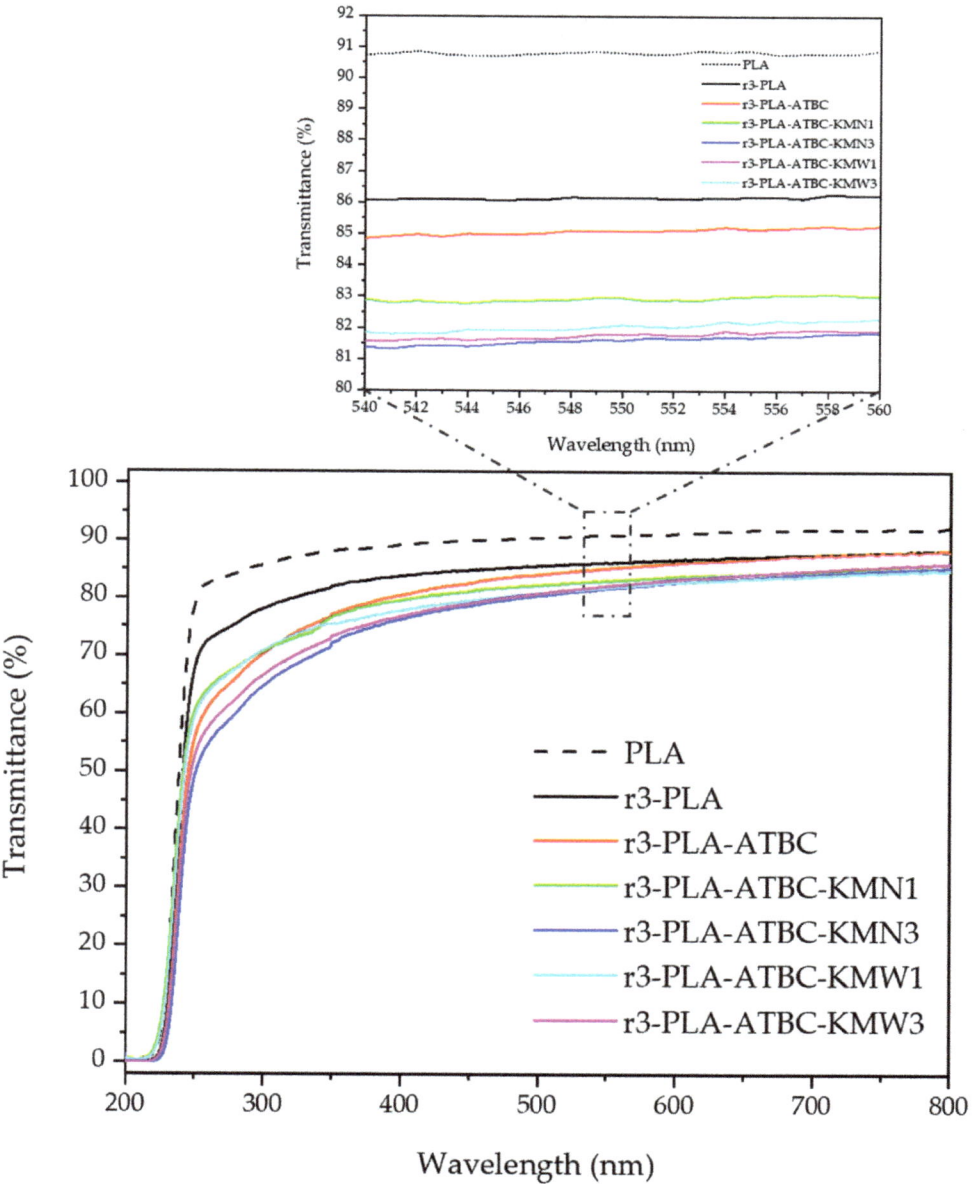

Figure 2. UV-vis spectra of films 200–800 nm and zoom image 540–560 nm.

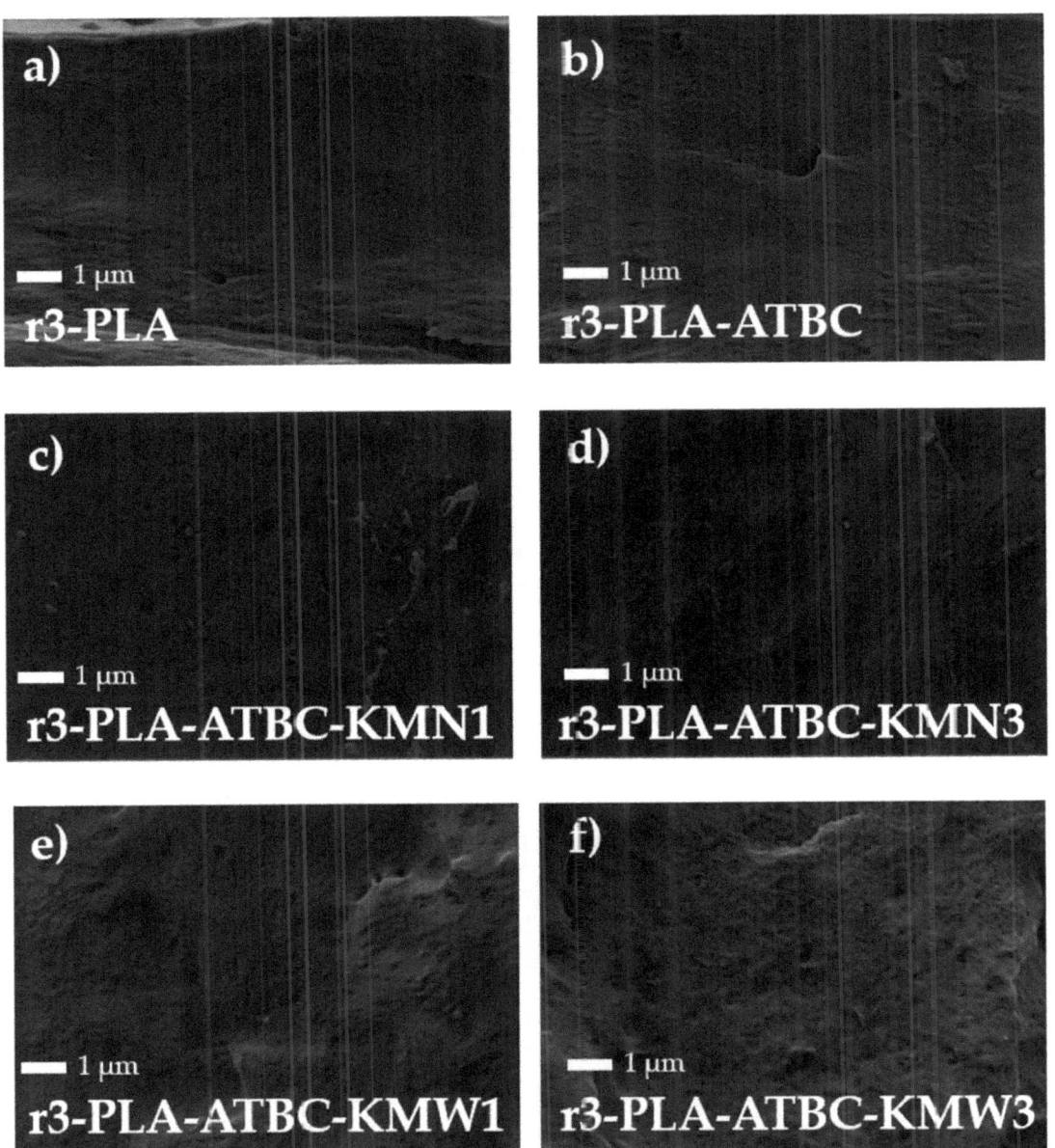

Figure 3. FE-SEM observations at 10,000× of 3r-PLA-ATBC based films: (**a**) r3-PLA, (**b**) r3-PLA-ATBC, (**c**) r3-PLA-ATBC-KMN1, (**d**) r3-PLA-ATBC-KMW1, (**e**) 3r-PLA-ATBC-KMN3, and (**f**) r3-PLA-ATBC-KMW3.

3.4. Differential Scanning Calorimetry

DCS analysis was conducted and used to investigate the glass transition (T_g), cold crystallization (T_{cc}), melting temperatures (T_m), and crystallinity (χ_c) of plasticized 3r-PLA-ATBC films, and the obtained DSC curves are shown in Figure 4 while the obtained results are summarized in Table 2. The r3-PLA film showed the T_g at a lower value than the

PLA samples which were processed three times by melt extrusion and further processed by injection molding (T_g = 64 °C [13]), due to the presence of a residual solvent, as was demonstrated by Yang et al. Their study compared PLA-based composites processed by extrusion with those processed by solvent casting method, and concluded that limited variations in the DSC parameters were observed for samples processed with the two different processing techniques (melt extrusion and solvent casting method) [33]. In the r3-PLA film, a cold crystallization peak appeared which was not present in the virgin PLA pellet [13], and this has been related to the fact that the shorter PLA chains formed during the reprocessing cycles, such as oligomers, showed higher mobility levels and promoted the crystallization of PLA [13,29]. In fact, it has been observed that PLA plasticized with oligomeric lactic acid (OLA) showed a reduction in cold crystallization temperature, which was further reduced by increasing the OLA content [34,35]. Similarly, the incorporation of the ATBC plasticizer produced a decrease in the T_g and T_{cc}, as well as in the T_m, which is ascribed to the ability of the ATBC plasticizer to increase the free volume between the polymer chains. Accordingly, their mobility was also decreased, enhancing the slow crystallization rate [22,36]. On the other side, the combination of the ATBC plasticizer and the microbial cellulose particles onto a plasticized 3r-PLA matrix produced higher T_g and T_{cc} values and higher crystallinity degrees, suggesting that the segmental motion of PLA matrix may have been affected by the presence of KMN and KMW [33]. Moreover, the synergic effect on the crystallization of PLA as a consequence of a potential nucleating agent in a presence of citrate ester plasticizers has been already reported [25,37]. The DSC thermograms show a double melting behavior, which has already been observed in mechanically recycled PLA [38,39] and PLA plasticized with OLA [34]. This behavior, in PLA-based materials, is ascribed to the presence of different crystalline structures with different levels of perfection and thermodynamic stability. The melt PLA crystallizes at temperatures higher than 120 °C in an ordered form (α form) [40]. In the present work, all samples were crystallized at temperatures below 120 °C. This is related to the ability of shorter polymer chains (oligomers), produced as a consequence of the PLA degradation during reprocessing steps, to promote the aforementioned crystallization of PLA. The reduction in the cold crystallization temperature was particularly marked in plasticized 3r-PLA-ATBC samples. When PLA crystallized below 110 °C, less stable crystals appeared, known as α' crystals [40]. From the cold crystallization peak in the DSC thermogram, it can be observed that disorder (crystals with α' form) to order (crystals with α form) phase transition took place, suggesting that a great fraction of the polymer was in an amorphous state due to the DSC cooling scan applied, as this was already observed in plasticized PLA-ATBC samples [15,41]. An increase in the T_{cc} values of KMN- and KMW-loaded films was observed with respect to the r3-PLA-ATBC film, suggesting that somewhat fewer disordered crystals (α') are present in composite materials. A different crystallization degree was observed for PLA-ATBC-KMN-based films with respect to PLA-ATBC-KMW-based films. A higher crystallinity degree was found for those particles obtained from the fermentation of kombucha in yerba mate waste (KMW), and could be directly related to the superior dispersion of KMW particles into the plasticized PLA-ATBC matrix, which are able to promote a higher nucleation effect [25].

Table 2. DSC thermal properties of 3r-PLA-ATBC-based films.

Sample	T_g (°C)	T_{cc} (°C)	ΔH_{cc} (J g^{-1})	T_{mI} (°C)	T_{mII} (°C)	ΔH_m (J g^{-1})	χ_c (%)
r3-PLA	49.1	106.1	20.0	144.4	151.3	23.2	3.4
r3-PLA-ATBC	32.1	95.0	21.2	135.6	146.8	23.4	2.8
r3-PLA-ATBC-KMN1	39.1	99.5	19.2	139.7	148.9	19.8	0.7
r3-PLA-ATBC-KMN3	36.8	99.9	19.5	138.3	148.2	20.2	1.0
r3-PLA-ATBC-KMW1	37.3	98.4	20.3	137.7	148.1	22.8	3.1
r3-PLA-ATBC-KMW3	33.5	99.3	19.0	138.3	147.8	21.5	3.2

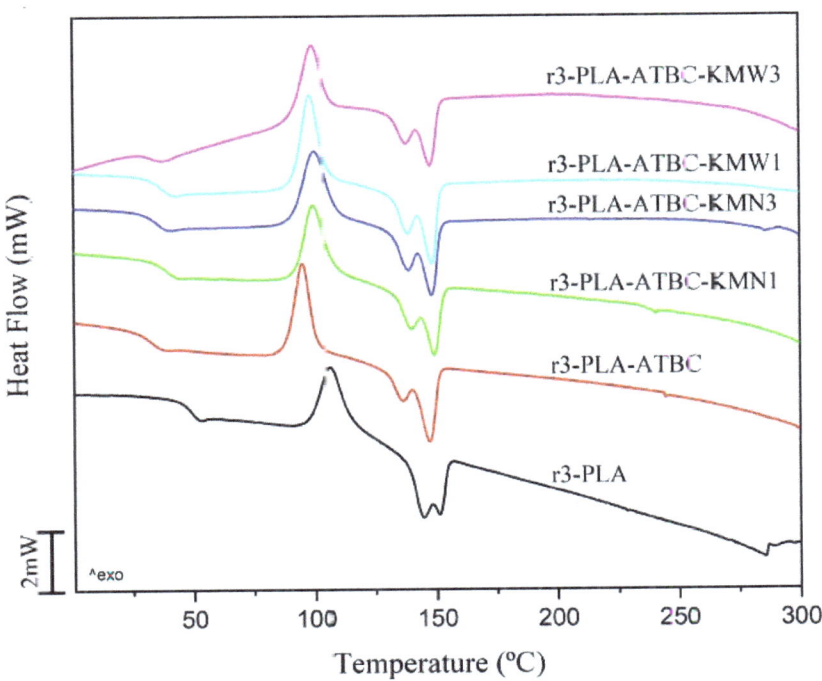

Figure 4. DSC second heating scan of 3r-PLA-ATBC-based films.

3.5. Thermogravimetric Analysis

The thermal stability of the materials was studied under isothermal mode at 180 °C to ensure enough thermal stability for the typical melt-processing temperature of PLA (Figure 5a). During the first minute under TGA isothermal conditions, films experienced a quick weight loss, probably due to the evaporation of the remaining chloroform. Then, all the materials showed a mass loss of around 1 or 2% in 10 min, which allows enough time to process the PLA-based materials. 3r-PLA-ATBC-KMW1 showed very similar thermal stability to 3r-PLA-ATBC, whereas 3r-PLA-ATBC-KMW3 presented less thermal stability. This could be related to the fact that when higher amounts than 3 wt.% of KMW reinforce 3r-PLA, some part of the PLA matrix is non-stabilized due to the deficient particle dispersion. Nevertheless, it should be highlighted that all the materials showed enough thermal stability for melt extrusion purposes.

The thermal degradation parameters obtained by TGA are described in Table 3. Plasticization of PLA produced a decrease in $T_{5\%}$ and $T_{10\%}$, due to the decomposition of the plasticizer [42,43], and a slight decrease in T_{max} was observed, since the plasticization could make the polymer chains available to thermal degradation, as was already observed on plasticized PLA with citrate esters [15,22,30]. The addition of 1% of KM increased $T_{5\%}$ and $T_{10\%}$ in the case of KMW, compared with plasticized r3-PLA. However, the addition of 3 wt.% of KM decreased $T_{5\%}$ and $T_{10\%}$ due to the low thermal stability of bacterial cellulose [44]. Regarding T_{max}, it seemed that this value was enhanced by the addition of KMW and KMN at 1 wt.%, since the value was close to the unplasticized r3-PLA film, but the addition of a higher amount of KM showed a significant decrease for 3 wt.% KMN. However, no modifications were observed for KMW. An overall conclusion for this study is that the addition of KMW enhanced the thermal properties of r3-PLA-ABTC better than the KMN. This could be due to the higher crystallinity of r3-PLA-ABTC-KMW1 and r3-PLA-ABTC-KMW3 composites, as shown by the DSC results.

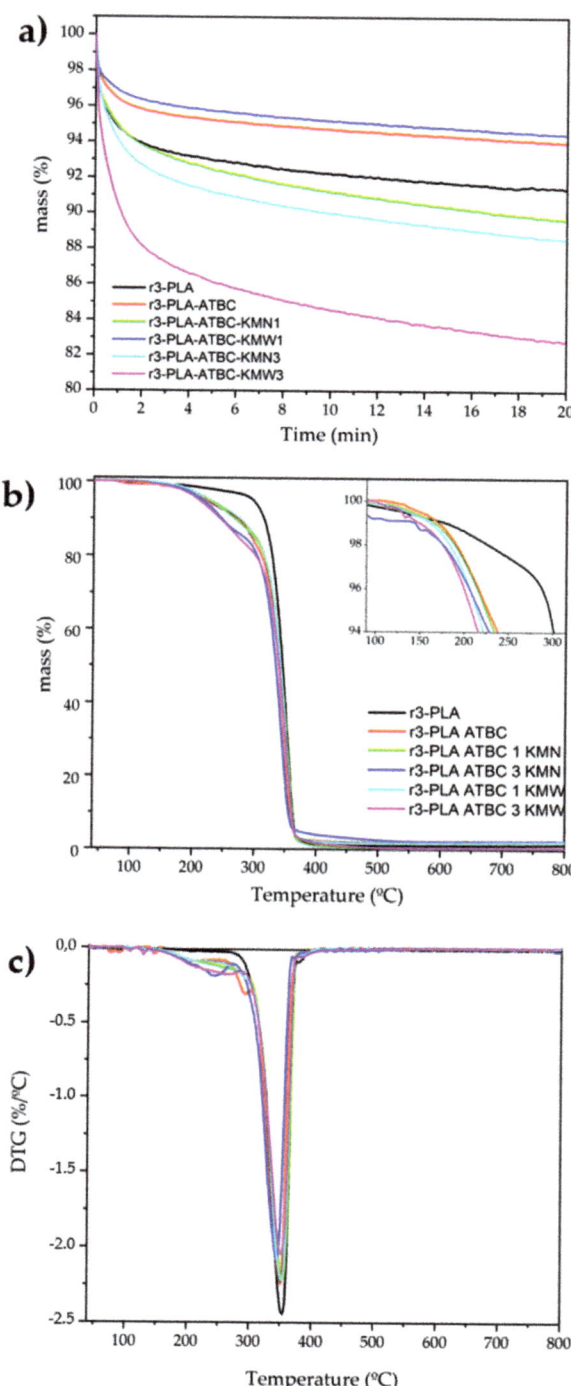

Figure 5. (**a**) Isothermal TGA analysis at 180 °C and Dynamic TGA (**b**) and DTG (**c**) of 3r-PLA-ATBC-based films.

Table 3. TGA thermal properties of 3r-PLA-ATBC-based films.

Sample	$T_{5\%}$ (°C)	$T_{10\%}$ (°C)	T_{max} (°C)	Residual Mass (%)
r3-PLA	296.1	313.52	354.4	0.4
r3-PLA-ATBC	217.4	272.28	351.1	0.8
r3-PLA-ATBC-KMN1	224.8	266.74	354.1	0.5
r3-PLA-ATBC-KMN3	214.7	246.21	344.7	0.7
r3-PLA-ATBC-KMW1	227.2	274.62	352.4	0.5
r3-PLA-ATBC-KMW3	208.1	243.19	351.8	0.5

Figure 5 also shows the TGA (Figure 5b) and DTG (Figure 5c) curves. Thermal degradation of the composites presented two steps of degradation. Firstly, the evaporation/degradation of the plasticizer overlaps with the initial degradation of KM, then in the second step of degradation, a PLA matrix was observed. For samples with higher concentrations of KM, a higher percentage of the mass was lost in the first step, confirming that in this event, the KM was starting to degrade. It is important to note that the composites were thermally stable at the processing temperatures usually used for PLA.

3.6. Tensile Test

The mechanical performance of the r3-PLA and plasticized r3-PLA based films was analyzed by tensile test, the results of which, in terms of Young Modulus (MPa), Tensile Strength (MPa), and Elongation at Break (%), are represented in Figure 6. Firstly, it is worth noting the decreased tensile parameters obtained for the r3-PLA films analyzed herein, which were processed by solvent casting, compared with in other works centered around PLA samples, which used melt extrusion [13]. As was mentioned previously, this effect of the processing condition on the mechanical behavior of PLA films has been studied by Yang et al. [33], who related this reduction in the tensile values, especially in terms of modulus, to the presence of captured residual chloroform, which acts as a plasticizer. Otherwise, with the addition of ATBC (r3-PLA-ATBC), an enhanced in ductility was observed, as was expected due to the proven effectiveness of this citrate ester as a PLA plasticizer [22,25,37,45]. Specifically, r3-PLA-ATBC showed a notable modulus reduction with respect to r3-PLA, from ~1750 MPa to ~1550 MPa, as well as an increment in the elongation at break from 12% to 14%, while a very small drop in the tensile strength was observed.

Regarding the addition of KM, at low content (1 wt.%), both KMN and KMW induce a very similar effect on the modulus and tensile strength, showing r3-PLA-ATBC-KMN1 and r3-PLA-ATBC-KMN1 values around 750–850 MPa and 15–17 MPa, respectively, for these parameters. In composite materials, when the particle dispersion is not homogenous enough, the transfer load does not occur appropriately, which caused a reduction in modulus and strength. This drawback of the introduction of cellulose-derived particles in PLA films has already been reported by other authors [46]. However, a significant difference between KMN and KMW was observed for the results of the elongation at break. In this sense, r3-PLA-ATBC-KMN1 showed a slight enhancement with respect to unloaded r3-PLA-ATBC (up to 17%) while the results of the r3-PLA-ATBC-KMW1 samples were closer to the non-plasticized r3-PLA. This difference in terms of ductility is directly correlated with the crystallinity values obtained in the thermal analysis. When the KM content was increased, the r3-PLA-ATBC film loaded with 3 wt.% of KMN showed a somewhat higher modulus and tensile strength, approaching the r3-PLA-ATBC, while the elongation at break was shown to be lower. Lignocellulosic and cellulosic materials have been widely studied as fillers of PLA-based composites due to their high weight/strength ratio, and can act as reinforcement when the interface contact area is adequate [47,48]. Instead, with the addition of 3 wt.% of KMW, mechanical reinforcement does not occur properly. This behavior can be explained by the more marked plasticizing effect produced by the KMW as a consequence of the possible degradation of some phenolic compounds (with less –OH able to establish hydrogen bonding interaction between them) present in pristine yerba mate, due to the hydrothermal extraction process used to prepare the infusion. Thus, permitting better interaction between PLA and ATBC allowed for a higher elongation at break, in good accordance with the T_g value, close to that of r3-PLA-ATBC (Table 2).

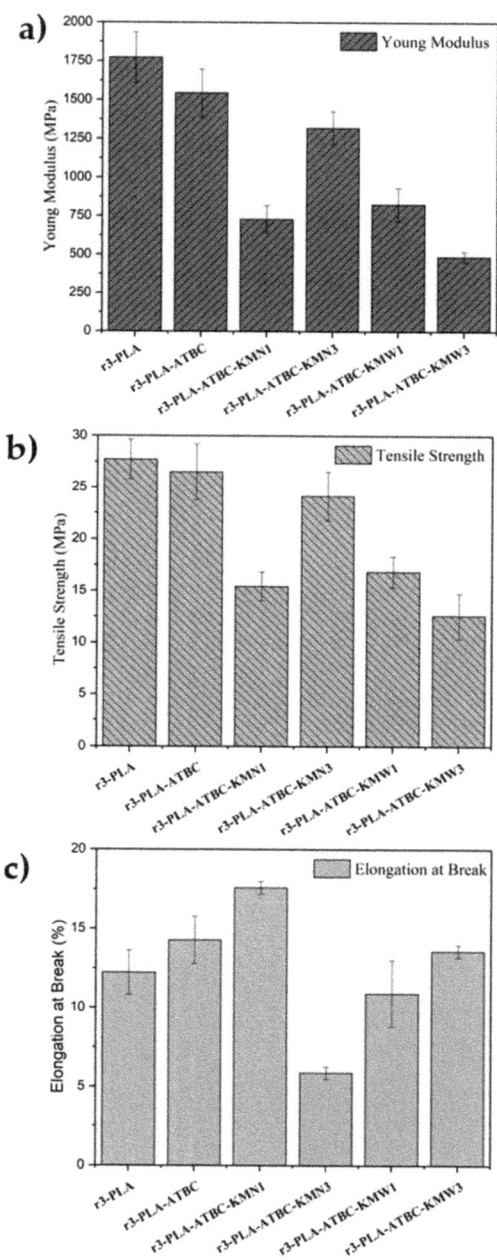

Figure 6. Tensile test measurements of 3r-PLA-ATBC-based films: (**a**) Young modulus, (**b**) Tensile Strength and (**c**) Elongation at break.

3.7. Release Studies and Antioxidant Ability

Specific migration tests were conducted to evaluate the potential antioxidant activity of the films, as microbial cellulose obtained from kombucha fermentation showed that the tea used for its production had antioxidant properties [18]. In this work, kombucha was fermented in a yerba mate-sugared infusion. The antioxidant activity of yerba mate is well-known, and mainly arises from its composition in phenolic compounds [19]; even yerba mate waste is still able to provide antioxidant activity [15]. The radical scavenging activity (RSA) of each food simulant D1 sample after 10 days of contact at 40 °C, considered by the current legislation the worst foreseeable conditions for intended use [28], was determined by means of the DPPH method [41], and the results are shown in Figure 7.

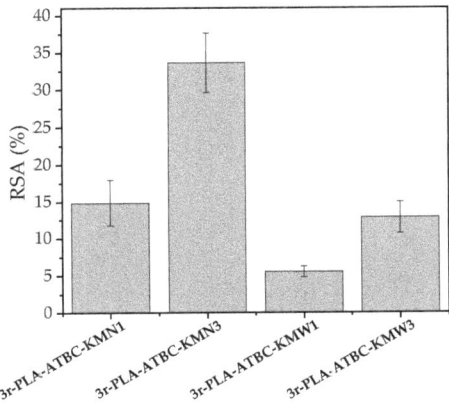

Figure 7. Radical scavenging activity of 3r-PLA-ATBC-based films.

As expected, the r3-PLA film did not show any antioxidant activity. The KMN and KMW were processed into thin films and also subjected to the food simulant for comparison. The neat KMN film showed high RSA activity of 70.2 ± 1.1% and 55.2 ± 11.7%. Meanwhile, the films loaded with KMN or KMW showed some antioxidant activity, as microbial cellulose obtained from kombucha fermentation possesses natural and remarkable antioxidant activity, which is directly related to the infusion used for the fermentation [18]. Films loaded with KMN showed higher antioxidant activity than those loaded with KMW, as it is known that yerba mate waste possesses a low polyphenol content due to the hydrothermal extraction process used during infusion preparation (temperature higher than 80 °C) before obtaining the waste. The scavenging effect obtained here for 3r-PLA-ATBC-based materials loaded with microbial cellulose kombucha fermented in yerba mate or yerba mate waste (between 1 wt.% and 3 wt.%) are low, but it should be highlighted that they still possess some antioxidant activity and, thus, the results interesting for food crops and food packaging. A high level of antioxidant activity has been observed in materials containing yerba mate extract. For instance, Deladino et al. studied corn starch-loaded materials with yerba mate extract at a concentration of around 10 wt.% with respect to the starchy matrix, and found between 40% and 60% RSA [19]. The values obtained in the present work are in the range of other antioxidant materials based on tri-layer recycled PLA/sodium caseinate (SC)/recycled PLA-based materials reinforced with 1 and 3 wt.% of nanoparticles obtained from yerba mate waste (YMN) (RSA (%) of rPLA/SC/rPLA-YMN1 = 6.4 ± 0.1 and RSA (%) of rPLA/SC/rPLA-YMN3 = 11.0 ± 0.2) [39].

Besides the biobased origin, biodegradability, and recyclability of PLA, the modification of PLA-based composite films through the incorporation of cellulosic nanoparticles has already shown improvements in thermal, barrier, and mechanical properties, and the possibility to provide additional antioxidant properties makes these films highly interesting for food packaging or agricultural applications [15,49].

3.8. Water Contact Angle and Water Vapor Transmission Rate

The surface hydrophilic/hydrophobic properties of films were determined by the measurement of the static water contact angle (WCA) and the results are reported in Figure 8a. Meanwhile, the water vapor transmission rate (WVTR) values of plasticized r3-PLA-ATBC-based films are reported in Figure 8b. The PLA film, after three cycles of melt extrusion (3r-PLA), showed a water contact angle higher than 65°, which was ascribed to hydrophobic surfaces (θ lower than 65° are ascribed to hydrophilic surfaces) [50]. The plasticized 3r-PLA sample (3r-PLA-ATBC) showed a lower WCA value, as was observed in an already reported work on PLA and PLA-ATBC [25]. This was also in agreement with the WVTR value, in which r3-PLA andr3-PLA-ATBC showed similar WVTR properties, despite being slightly higher the WVTR of 3r-PLA-ATBC. The plasticizing effect of ATBC into the r3-PLA matrix influenced the diffusion process as a consequence of the increased polymer chain mobility. The presence of either KMN or KMW in the 3r-PLA-ATBC matrix produced a decrease in the surface wettability of the films (Figure 8a), leading to values higher than 3r-PLA-ATBC and, thus, higher hydrophobicity. This unexpected increment of the hydrophobicity of the film surface, even when hydrophilic cellulosic particles were added, can be related to the changes in the topographical properties as a consequence of the presence of cellulose particles. However, it should be highlighted that the composite films were still more hydrophilic than 3r-PLA. The WVTR showed increased values with the presence of either KMN or KMW in the 3r-PLA-ATBC matrix (Figure 8b), particularly in the case of KMN, probably due to the high amount of active compounds.–OH groups were able to interact with water, increasing the water diffusion through the film. Meanwhile, the KMW, which showed less antioxidant activity (Figure 7) and, consequently, a lower amount of bioactive compounds within the polymeric matrix, allowed less water vapor to be transmitted through the film.

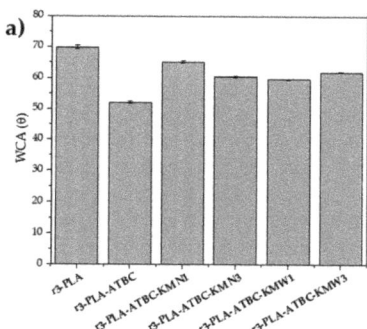

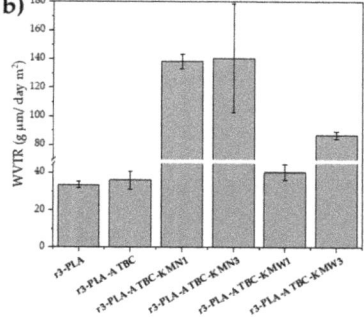

Figure 8. (a) Static water contact angle measurements and (b) water vapor transmission rate of 3r-PLA-ATBC-based films.

4. Conclusions

Microbial cellulose particles were successfully obtained from kombucha beverage fermented in both infusion, pristine yerba mate, and yerba mate waste. They were further used to reinforce a plasticized PLA matrix subjected to three extrusion cycles (r3-PLA), aiming to simulate the revalorization of PLA from industrial PLA products rejected during the production line. The r3-PLA-based biocomposites, reinforced with KMN and KMW, were effectively prepared by solvent casting, and the effect of yerba mate starting material (pristine or waste), as well as the amount added (1 and 3 wt.%) into the plasticized r3-PLA-ATBC, were deeply investigated.

All films resulted to be optically transparent, and FESEM micrographs revealed a good dispersion of microbial cellulose particles in the reprocessed polymeric matrix.

DSC analysis showed a crystallinity increase in r3-PLA-ATBC composites reinforced with KMW, indicating that it favored the crystal growth and nucleation effects, while for the tensile test, measurements showed a more marked plasticization effect. Moreover, the materials reinforced with KMW showed less WVTR, indicating improved barrier properties against water than the materials reinforced with pristine KMN. However, KMN-reinforced r3-PLA-ATBC-KMN-based films showed higher antioxidant activity, although it should be highlighted that r3-PLA-ATBC-KMW-based films still showed antioxidant activity.

The reprocessed PLA (r3-PLA) in combination with ATBC and KM particles offers a promising perspective to produce transparent and flexible films with good water barriers and mechanical properties that are suitable as antioxidant films for food packaging or agricultural mulch films.

Author Contributions: Conceptualization, M.d.M.d.l.F.G.-S. and M.P.A.; methodology, Á.A., D.L. and M.P.A.; software, Á.A., E.C.P. and D.L. validation, Á.A., E.C.P., S.S.A.d.l.M., D.L., M.d.M.d.l.F.G.-S., M.A.P., R.B. and M.P.A.; formal analysis, Á.A., E.C.P., S.S.A.d.l.M., D.L., M.d.M.d.l.F.G.-S., M.A.P. and M.P.A.; investigation, Á.A., E.C.P., S.S.A.d.l.M., D.L., M.d.M.d.l.F.G.-S., M.A.P., R B. and M.P.A.; resources, M.d.M.d.l.F.G.-S., M.P.A. and R.B.; data curation, Á.A., E.C.P., S.S.A.d.l.M. and D.L.; writing—original draft preparation, Á.A., D.L., M.d.M.d.l.F.G.-S., M.A.P. and M.P.A ; writing—review and editing, M.A.P., R.B. and M.P.A.; visualization, Á.A., E.C.P., S.S.A.d.l.M., D.L., M.d.M.d.l.F.G.-S., M.A.P., R.B. and M.P.A.; supervision, M.d.M.d.l.F.G.-S. and M.P.A.; project administration, M.d.M.d.l.F.G.-S., R.B. and M.P.A.; funding acquisition, M.A.P., R.B. and M.P.A. All authors have read and agreed to the published version of the manuscript.

Funding: This research was funded by the Spanish Ministry of Science and Innovation (MICINN) through PID-AEI project (grants PID2020-116496RB-C22 and PID2021-123753NA-C32) and TED-AEI project (grants TED2021-129920A-C43 and TED2021-131762A-I00) funded by MCIN/AEI/10.13039/5011C0011033 and by ERDF "A way of making Europe" by the "European Union". It was also funded by Comunidad de Madrid-CAM ("Ayudas de Estímulo a la Investigación de Jóvenes Doctores de la Universidad Politécnica de Madrid", APOYO-JOVENES-21-VUI9G6-56-I7GABC), Generalitat Valenciana-GVA grant numbers AICO/2021/025 and CIGE/2021/094 and National University of Quilmes (UNQ, Argentina) through the R&D program (Expediente 1300/19). Ph.D. Ángel Agüero and Ph.D. Diego Lascano acknowledge the Margarita Salas postdoctoral grants from the Ministerio de Universidades, Spain, funded by the European Union—Next Generation EU.

Institutional Review Board Statement: Not applicable.

Informed Consent Statement: Not applicable.

Data Availability Statement: The data presented in this study are available on request from the corresponding author.

Acknowledgments: Á.A., E.C.P., S.S.A.d.l.M. and M.P.A. also thank the "The Circular and Regenerative Campus" community and M.d.M.d.l.F.G.-S. thanks "Water in an Era of Change" community from the EELISA European University Alliance.

Conflicts of Interest: The authors declare no conflict of interest.

References

1. Garcia-Garcia, D.; Carbonell-Verdu, A.; Arrieta, M.P.; López-Martínez, J.; Samper, M. Improvement of PLA film ductility by plasticization with epoxidized karanja oil. *Polym. Degrad. Stab.* **2020**, *179*, 109259. [CrossRef]
2. Auras, R.; Harte, B.; Selke, S. An Overview of Polylactides as Packaging Materials. *Macromol. Biosci.* **2004**, *4*, 835–864. [CrossRef] [PubMed]
3. Hernández-García, E.; Vargas, M.; Chiralt, A. Effect of active phenolic acids on properties of PLA-PHBV blend films. *Food Packag. Shelf Life* **2022**, *33*, 100894. [CrossRef]
4. Arrieta, M.P.; Samper, M.D.; Aldas, M.; López, J. On the Use of PLA-PHB Blends for Sustainable Food Packaging Applications. *Materials* **2017**, *10*, 1008. [CrossRef]
5. Arrieta, M.P. Influence of plasticizers on the compostability of polylactic acid. *J. Appl. Res. Technol. Eng.* **2021**, *2*, 1–9. [CrossRef]
6. Gil Muñoz, V.; Muneta, L.M.; Carrasco-Gallego, R.; de Juanes Marquez, J.; Hidalgo-Carvajal, D. Evaluation of the Circularity of Recycled Pla Filaments for 3d Printers. *Appl. Sci.* **2020**, *10*, 8967. [CrossRef]
7. Beltrán, F.; Arrieta, M.; Antón, D.E.; Lozano-Pérez, A.; Cenis, J.; Gaspar, G.; de la Orden, M.; Urreaga, J.M. Effect of Yerba Mate and Silk Fibroin Nanoparticles on the Migration Properties in Ethanolic Food Simulants and Composting Disintegrability of Recycled PLA Nanocomposites. *Polymers* **2021**, *13*, 1925. [CrossRef]
8. de Andrade, C.M.; Souza, F.P.; Cavalett, O.; Morales, A.R. Life Cycle Assessment of Poly(Lactic Acid)(PLA): Com-parison between Chemical Recycling, Mechanical Recycling and Composting. *J. Polym. Environ.* **2016**, *24*, 372–384. [CrossRef]
9. Samper, M.D.; Arrieta, M.P.; Ferrandiz, S.; López, J. Influence of Biodegradable Materials in the Recycled Polystyrene. *J. Appl. Polym. Sci.* **2014**, *131*, 41161. [CrossRef]
10. Samper, M.D.; Bertomeu, D.; Arrieta, M.P.; Ferri, J.M.; López-Martínez, J. Interference of Biodegradable Plastics in the Polypropylene Recycling Process. *Materials* **2018**, *11*, 1886. [CrossRef]
11. Aldas, M.; Pavon, C.; De La Rosa-Ramírez, H.; Ferri, J.M.; Bertomeu, D.; Samper, M.D.; López-Martínez, J. The Impact of Biodegradable Plastics in the Properties of Recycled Polyethylene Terephthalate. *J. Polym. Environ.* **2021**, *29*, 2686–2700. [CrossRef]
12. European Commission. *Commission Regulation (Eu) 2022/1616 of 15 September 2022 on Recycled Plastic Materials and Articles Intended to Come into Contact with Foods, and Repealing Regulation (Ec) No 282/2008 (Text with Eea Relevance)*; European Commission: Brussels, Belgium, 2022.
13. Agüero, A.; Morcillo, M.d.C.; Quiles-Carrillo, L.; Balart, R.; Boronat, T.; Lascano, D.; Torres-Giner, S.; Fenollar, O. Study of the Influence of the Reprocessing Cycles on the Final Properties of Polylactide Pieces Obtained by Injection Molding. *Polymers* **2019**, *11*, 1908. [CrossRef] [PubMed]
14. Chariyachotilert, C.; Joshi, S.; Selke, S.; Auras, R. Assessment of the properties of poly(L-lactic acid) sheets produced with differing amounts of postconsumer recycled poly(L-lactic acid). *J. Plast. Film Sheeting* **2012**, *28*, 314–335. [CrossRef]
15. Arrieta, M.; Peponi, L.; López, D.; Fernández-García, M. Recovery of yerba mate (*Ilex paraguariensis*) residue for the development of PLA-based bionanocomposite films. *Ind. Crop. Prod.* **2018**, *111*, 317–328. [CrossRef]
16. Tapias, Y.A.R.; Di Monte, M.V.; Peltzer, M.A.; Salvay, A.G. Bacterial cellulose films production by Kombucha symbiotic community cultured on different herbal infusions. *Food Chem.* **2021**, *372*, 131346. [CrossRef] [PubMed]
17. Ashrafi, A.; Jokar, M.; Nafchi, A.M. Preparation and characterization of biocomposite film based on chitosan and kombucha tea as active food packaging. *Int. J. Biol. Macromol.* **2018**, *108*, 444–454. [CrossRef] [PubMed]
18. Tapias, Y.A.R.; Peltzer, M.A.; Delgado, J.F.; Salvay, A.G. Kombucha Tea By-product as Source of Novel Materials: Formulation and Characterization of Films. *Food Bioprocess Technol.* **2020**, *13*, 1166–1180. [CrossRef]
19. Deladino, L.; Teixeira, A.; Navarro, A.; Alvarez, I.; Molina-García, A.; Martino, M. Corn starch systems as carriers for yerba mate (*Ilex paraguariensis*) antioxidants. *Food Bioprod. Process.* **2015**, *94*, 463–472. [CrossRef]
20. INYM—Instituto Nacional de la Yerba Mate. 2022. Available online: https://Inym.Org.Ar/Noticias/Estadisticas/79445-En-2020-El-Consumo-De-Yerba-Mate-Totalizo-311-7-Millones-De-Kilos (accessed on 9 October 2022).
21. EFSA Panel of Food Contact Materials; Enzymes and Processing Aids. Scientific Opinion on Flavouring Group Evaluation 10, Revision 3 (Fge. 10rev3): Aliphatic Primary and Secondary Saturated and Unsaturated Alcohols, Aldehydes, Acetals, Carboxylic Acids and Esters Containing an Additional Oxygenated Functional Group and Lactones from Chemical Groups 9, 13 and 30. *EFSA J.* **2012**, *10*, 2563.
22. Arrieta, M.P.; Samper, M.D.; Lopez, J.; Jiménez, A. Combined Effect of Poly(hydroxybutyrate) and Plasticizers on Polylactic acid Properties for Film Intended for Food Packaging. *J. Polym. Environ.* **2014**, *22*, 460–470. [CrossRef]
23. Arrieta, M.P.; López, J.; López, D.; Kenny, J.M.; Peponi, L. Development of Flexible Materials Based on Plasticized Electrospun PLA–PHB Blends: Structural, Thermal, Mechanical and Disintegration Properties. *Eur. Polym. J.* **2015**, *73*, 433–446. [CrossRef]
24. Van den Oever, M.J.A.; Beck, B.; Müssig, J. Agrofibre Reinforced Poly (Lactic Acid) Composites: Effect of Moisture on Degradation and Mechanical Properties. *Compos. Part A Appl. Sci. Manuf.* **2010**, *41*, 1628–1635. [CrossRef]
25. Arrieta, M.; Fortunati, E.; Dominici, F.; López, J.; Kenny, J. Bionanocomposite films based on plasticized PLA–PHB/cellulose nanocrystal blends. *Carbohydr. Polym.* **2015**, *121*, 265–275. [CrossRef] [PubMed]
26. Turner, J.F.; Riga, A.; O'Connor, A.; Zhang, J.; Collis, J. Characterization of drawn and undrawn poly-L-lactide films by differential scanning calorimetry. *J. Therm. Anal.* **2004**, *75*, 257–268. [CrossRef]
27. Trifol, J.; Quintero, D.C.M.; Moriana, R. Pine Cone Biorefinery: Integral Valorization of Residual Biomass into Lignocellulose Nanofibrils (LCNF)-Reinforced Composites for Packaging. *ACS Sustain. Chem. Eng.* **2021**, *9*, 2180–2190. [CrossRef]

28. European Commission. No. 10, 2011 of 14, on Plastic Materials and Articles Intended to Come into Contact with Food. *Off. J. Eur. Union* **2011**, *L12*, 1–89.
29. Beltrán, F.R.; Arrieta M.P.; Gaspar, G.; de la Orden, M.U.; Martínez Urreaga, J. Effect of lignocellulosic Nanoparticles Extracted from Yerba Mate (*Ilex Paraguariensis*) on the Structural, Thermal, Optical and Barrier Properties of Mechanically Recycled Poly(Lactic Acid). *Polymers* **2020**, *12*, 1690. [CrossRef]
30. García-Arroyo, P.; Arrieta, M.P.; Garcia-Garcia, D.; Cuervo-Rodríguez, R.; Fombuena, V.; Mancheño, M.J.; Segura, J.L. Plasticized Poly(Lactic Acid) Reinforced with Antioxidant Covalent Organic Frameworks (COFs) as Novel Nanofillers De-signed for Non-Migrating Active Packaging Applications. *Polymer* **2020**, *196*, 122466. [CrossRef]
31. Molinaro, S.; Cruz-Romero, M.; Boaro, M.; Sensidoni, A.; Lagazio, C.; Morris, M.; Kerry, J. Effect of nanoclay-type and PLA optical purity on the characteristics of PLA-based nanocomposite films. *J. Food Eng.* **2013**, *117*, 113–123. [CrossRef]
32. Armentano, I.; Fortunati, E.; Burgos, N.; Dominici, F.; Luzi, F.; Fior., S.; Jiménez, A.; Yoon, K.; Ahn, J.; Kang, S ; et al. Processing and Characterization of Plasticized Pla/Phb Blends for Biodegradable Multiphase Systems. *Express Polym. Lett.* **2015**, *9*, 583–596. [CrossRef]
33. Yang, W.; Fortunati, E.; Dominici, F.; Kenny, J.; Puglia, D. Effect of processing conditions and lignin content on thermal, mechanical and degradative behavior of lignin nanoparticles/polylactic (acid) bionanocomposites prepared by melt extrusion and solvent casting. *Eur. Polym. J.* **2015**, *71*, 126–139. [CrossRef]
34. Burgos, N.; Martino, V.P.; Jiménez, A. Characterization and ageing study of poly(lactic acid) films plasticized with oligomeric lactic acid. *Polym. Degrad. Stab.* **2013**, *98*, 651–658. [CrossRef]
35. Arrieta, M.P.; Perdiguero, M.; Fiori, S.; Kenny, J.M.; Peponi, L. Biodegradable electrospun PLA-PHB fibers plasticized with oligomeric lactic acid *Polym. Degrad. Stab.* **2020**, *179*, 109226. [CrossRef]
36. Labrecque, L.V.; Kumar, R.A.; Gross, R.A.; McCarthy, S.P. Citrate esters as plasticizers for poly(lactic acid). *J. Appl. Polym. Sci.* **1997**, *66*, 1507–1513. [CrossRef]
37. Courgneau, C.; Ducruet, V.; Avérous, L.; Grenet, J.; Domenek, S. Nonisothermal Crystallization Kinetics of Poly (Lactide)—Effect of Plasticizers and Nucleating Agent. *Polym. Eng. Sci.* **2013**, *53*, 1085–1098. [CrossRef]
38. Beltrán, F.; Lorenzo, V.; Acosta, J.; de la Orden, M.; Urreaga, J.M. Effect of simulated mechanical recycling processes on the structure and properties of poly(lactic acid). *J. Environ. Manag.* **2018**, *216*, 25–31. [CrossRef]
39. Arrieta, M.P.; Beltran, F.; Abarca de las Muelas, S.S.; Gaspar, G.; Sanchez Hernandez, R.; de la Orden, M.U.; Martinez Urreaga, J. Development of Tri-Layer Antioxidant Packaging Systems Based on Recycled PLA/Sodium Caseinate/Recycled PLA Rein-forced with Lignocellulosic Nanoparticles Extracted from Yerba Mate Waste. *Express Polym. Lett.* **2022**, *16*, 881–900. [CrossRef]
40. Zhang, J.; Tashiro, K.; Tsuji, H.; Domb, A.J. Disorder-to-Order Phase Transition and Multiple Melting Behavior of Poly(l-lactide) Investigated by Simultaneous Measurements of WAXD and DSC. *Macromolecules* **2008**, *41*, 1352–1357. [CrossRef]
41. Arrieta, M.P.; Castro-Lopez, M.M.; Rayón, E.; Barral-Losada, L.F.; López-Vilariño, J.M.; López, J.; Gon-zález-Rodríguez, M.V. Plasticized Poly(Lactic Acid)–Poly (Hydroxybutyrate)(PLA-PHB) Blends Incorporated with Catechin In-tended for Active Food-Packaging Applications. *J. Agric. Food Chem.* **2014**, *62*, 10170–10180. [CrossRef]
42. Tee, Y.B.; Talib, R.A.; Abdan, K.; Chin, N.L.; Basha, R.K.; Md Yunos, K.F. Comparative Study of Chemical, Mechanical, Thermal, and Barrier Properties of Poly (Lactic Acid) Plasticized with Epoxidized Soybean Oil and Epoxidized Palm Oil. *BioResources* **2016**, *11*, 1518–1540. [CrossRef]
43. Arrieta, M.P.; López, J.; Ferrándiz, S.; Peltzer, M.A. Characterization of PLA-Limonene Blends for Food Packaging Applications. *Polym. Test.* **2013**, *32*, 760–768. [CrossRef]
44. Threepopnatkul, P.; Sittattrakul, A.; Supawititpattana, K.; Jittiarpon, P.; Raksawat, P.; Kulsetthanchalee, C. Effect of bacterial cellulose on properties of poly(lactic acid). *Mater. Today: Proc.* **2017**, *4*, 6605–6614. [CrossRef]
45. Maiza, M.; Benaniba, M.T.; Quintard, G.; Massardier-Nageotte, V. Biobased additive plasticizing Polylactic acid (PLA). *Polimeros* **2015**, *25*, 581–590. [CrossRef]
46. Geng, S.; Wei, J.; Aitomäki, Y.; Noël, M.; Oksman, K. Well-dispersed cellulose nanocrystals in hydrophobic polymers by in situ polymerization for synthesizing highly reinforced bio-nanocomposites. *Nanoscale* **2018**, *10*, 11797–11807. [CrossRef]
47. Iwatake, A.; Nogi, M.; Yano, H. Cellulose nanofiber-reinforced polylactic acid. *Compos. Sci. Technol.* **2008**, *68*, 2103–2106. [CrossRef]
48. Luzi, F.; Torre, L.; Kenny, J.M.; Puglia, D. Bio- and Fossil-Based Polymeric Blends and Nanocomposites for Packaging: Structure–Property Relationship. *Materials* **2019**, *12*, 471. [CrossRef]
49. Khalid, M.Y.; Arif, Z.U. Novel biopolymer-based sustainable composites for food packaging applications: A narrative review. *Food Packag. Shelf Life* **2022**, *33*, 100892. [CrossRef]
50. Hambleton, A.; Fabra, M.J.; Debeaufort, F.; Dury-Brun, C.; Voilley, A. Interface and Aroma Barrier Properties of Io-ta-Carrageenan Emulsion–Based Films Used for Encapsulation of Active Food Compounds. *J. Food Eng.* **2009**, *93*, 80–88. [CrossRef]

Disclaimer/Publisher's Note: The statements, opinions and data contained in all publications are solely those of the individual author(s) and contributor(s) and not of MDPI and/or the editor(s). MDPI and/or the editor(s) disclaim responsibility for any injury to people or property resulting from any ideas, methods, instructions or products referred to in the content.

Article

Effect of Innovative Bio-Based Plastics on Early Growth of Higher Plants

Ewa Liwarska-Bizukojc

Institute of Environmental Engineering and Building Installations, Lodz University of Technology, Al. Politechniki 6, 90-924 Lodz, Poland; ewa.liwarska-bizukojc@p.lodz.pl; Tel.: +48-42-631-35-22

Abstract: Plastic particles are widespread in the environment including the terrestrial ecosystems. They may change the physicochemical properties of soil and subsequently affect plant growth. In recent decades, traditional, petroleum-derived plastics have been increasingly replaced by more environmentally friendly bio-based plastics. Due to the growing role of bio-based plastics it is necessary to thoroughly study their impact on the biotic part of ecosystems. This work aimed for the assessment of the effect of five innovative bio-based plastics of different chemical composition and application on the early growth of higher plants (sorghum, cress and mustard). Each bio-based plastic was tested individually. It was found that the early stages of growth of monocotyledonous plants were usually not affected by any of plastic materials studied. At the same time, the presence of some kinds of bio-based plastics contributed to the inhibition of root growth and stimulation of shoot growth of dicotyledonous plants. Two PLA-based plastics inhibited root growth of dicotyledonous plants more strongly than other plastic materials; however, the reduction of root length did not exceed 22% compared to the control runs. PBS-based plastic contributed to the stimulation of shoot growth of higher plants (sorghum, cress and mustard) at the concentrations from 0.02 to 0 095% w/w. In the case of cress shoots exposed to this plastic the hormetic effect was observed. *Lepidium sativum* turned out to be the most sensitive plant to the presence of bio-based plastic particles in the soil. Thus, it should be included in the assessment of the effect of bio-based plastics on plant growth.

Keywords: bio-based plastics; plant growth; seed germination; terrestrial ecosystem

Citation: Liwarska-Bizukojc, E. Effect of Innovative Bio-Based Plastics on Early Growth of Higher Plants. *Polymers* **2023**, *15*, 438. https://doi.org/10.3390/polym15020438

Academic Editors: Jesús-María García-Martínez and Emilia P. Collar

Received: 28 November 2022
Revised: 9 January 2023
Accepted: 10 January 2023
Published: 13 January 2023

Copyright: © 2023 by the author. Licensee MDPI, Basel, Switzerland. This article is an open access article distributed under the terms and conditions of the Creative Commons Attribution (CC BY) license (https://creativecommons.org/licenses/by/4.0/).

1. Introduction

Soil plays many important ecological functions such as accumulation and filtration of water and nutrients, transformation of chemicals, biomass production and carbon storage [1]. The presence of chemical contaminants in soil reduces these functions and aggravates soil properties. Micro- and nanoplastic particles became one of the most common pollutants in the terrestrial ecosystems in the last decades [2]. It was demonstrated that they changed soil properties, including soil aggregation, bulk density and water holding capacity [3,4]. In order to evaluate soil quality and vitality, apart from the determination of chemical and physical indicators, the use of soil biota is recommended [5,6].

Plants are one of the most important functional groups of organisms in the terrestrial ecosystems [6]. They produce oxygen and food for other living creatures, as well as they are regulators of key ecosystem processes and services, e.g., carbon dynamics and sequestration, nutrient dynamics and soil structural stability [7]. Thus, various plant species are used as biological indicators in the soil ecotoxicity tests. With regard to the evaluation of the effects of microplastics on higher plants, the two following species have been frequently assessed: *Lepidium sativum* representing dicotyledonous plants [8–10] and *Triticum aestivum* being a monocotyledonous plant [11–13]. In addition, other plants such as *Daucus carota* [14] and *Allium fistulosum* [15] were used for testing the potential impacts of plastic particles on the organisms representing producers in the food chain. However, it should be emphasized that the number of studies concerning the assessment of the effects of plastic particles on higher plants that have been published so far is very limited.

It was found that microplastics usually did not affect the seed germination processes of either mono- or dicotyledonous plants [8,12,13]. This concerns both petroleum-derived and bio-based plastic particles. At the same time, plastic particles might act on the early growth of plants stimulating or inhibiting it [9,16]. Balestri et al. [9] observed that a significant number of seedlings exposed to leachates from high-density polyethylene (HDPE) or the bio-based biodegradable plastic Materbi® showed developmental abnormalities or seedling growth reduction. Polylactide (PLA) and polyhydroxybutyrate (PHB) microparticles at the concentration 11.9% w/w contributed to the inhibition of root growth of *Sinapsis alba* and *Lepidium sativum* [16]. At the same time, Lozano et al. [14] reported about the increase in the growth of *Daucus carota* cultivated for 28 days in the soil containing from 0.1 to 0.4% (w/w) plastic microparticles made of polyester (PES), polyamide (PA), polypropylene (PP), low-density polyethylene (LDPE), poly(ethylene terephthalate) (PET), polyurethane (PU), polystyrene (PS) and polycarbonate (PC). Huerta Lwanga et al. [13] did not observe any effect of PLA on *Triticum aestivum* growth.

In this work, five newly synthesized bio-based plastics were tested with regard to their potential impact on seed germination and the early growth of plants. All bio-based plastics were obtained with the cooperation of the project Bio-plastic Europe (Horizon 2020, grant agreement no. 860407). It was hypothesized that the materials tested would not affect seed germination but they might inhibit or stimulate the early growth of roots and/or shoots. In order to verify these hypotheses, a series of early growth tests using three different higher plants as model organisms were carried out.

2. Materials and Methods

2.1. Bio-Based Plastics

Five innovative bio-based plastic materials subjected to study were received due to the realization of Bio-plastic Europe Project (Horizon 2020, grant agreement no. 860407). Three of them were provided by NaturePlast SAS (NP, Mondeville, France) and these were the following compounds: BPE-AMF-PLA (Bio-Plastic Europe—Agriculture Mulch Film—PolyLactic Acid), BPE-T-PHBV (Bio-Plastic Europe—Toys—Poly Hydroxy Butyrate Valerate), BPE-SP-PBS (Bio-Plastic Europe—soft Packaging—PolyButylene Succinate), while the other two, i.e., BPE-C-PLA (Bio-Plastic Europe—Rigid Packaging—PolyLactic Acid), BPE-RP-PLA (Bio-Plastic Europe—Rigid Packaging—PolyLactic Acid) were provided by Arctic Biomaterials OY Ltd. (ABI, Tampere, Finland). All bio-based plastics were supplied by the manufacturers in the form of microparticles. The abbreviations of tested bio-based plastics were the same as it was assumed in the project nomenclature. The characteristics of the plastic materials tested are shown in Table 1.

Table 1. Data on the bio-based plastics tested (provided by the manufacturers).

Acronym of Bio-Based Plastic	APPLICATION	Desired Properties	Material Type	Density g cm^{-3}	Size of Granules	Innovation	Material Details	Manufacturer
BPE-AMF-PLA	Mulch film	Bio-based and both recyclable and bio-degradable, degrades in controlled fashion	PLA-based	1.26	Length 3 mm; diameter 2.5 mm	Blending of PLA and polyhydroxy butyrate-hydroxyvalerate (PHBV) for controlled degradation, fertilizer added for controlled release	PLA blended with 15% polybutylene adipate terephthalate (PBAT) and <5% process additives, intended to be used for extrusion application	NaturePlast SAS (Mondeville, France)

Table 1. Cont.

Acronym of Bio-Based Plastic	APPLICATION	Desired Properties	Material Type	Density g cm^{-3}	Size of Granules	Innovation	Material Details	Manufacturer
BPE-T-PHBV	Toys	Recyclable, industrially compostable rheology thermal stability, melt viscosity, resistance to hydrolysis, migration	PHBV-based	1.24	Length 3 mm; diameter 2.5 mm	Blending of PLA, and soft unsaturated PHAs, first mechanical, thermal characterization (smallest scale); in vitro (enzymatic) degradation	PHVBV blended with <15% additives (mostly impact modifier), intended to be used for injection application	NaturePlast SAS (Mondeville, France)
BPE-SP-PBS	Soft Packaging	Recyclable, industrially compostable rheology, thermal stability, melt viscosity, resistance to hydrolysis, barrier properties	PBS-based	1.26	Length 3 mm; diameter 2.5 mm	Improve processing and hydrolysis resistance, testing of recyclability, blending with PLA to increase mechanical properties (soft packaging to more rigid packaging)	PBS blended with <15% additives (mostly mineral filler), intended to be used for thermoforming or injection application	NaturePlast SAS (Mondeville, France)
BPE-C-PLA	Cutlery	Reusable cutlery with good mechanical properties and heat resistance	PLA-based	1.40	Length 3 mm; diameter 2.5 mm	Thermal stability, processing, resistance to hydrolysis, suitable for dishwasher cleaning, environmental degradation, ecotoxicology	PLA-based compound filled with 20% of degradable glass fiber	Arctic Biomaterials OY Ltd. (Tampere, Finland)
BPE-RP-PLA	Rigid packaging	Water and oxygen barrier, bio-based and biodegradable	PLA-based	1.50	Length 3 mm; diameter 2.5 mm	Cold mold, fast cycle time, good heat resistance, food grade	PLA-based mineral filled compound (food grade) for injection molding and potentially sheets for thermoforming	Arctic Biomaterials OY Ltd. (Tampere, Finland)

2.2. Methods of Evaluation of Phytotoxicity

Impacts of bio-based plastics on plants were assessed in agreement with ISO Standards 18763 [17] using the commercial toxicity bioassay—Phytotoxkit Solid Samples provided by Microbiotests (Ghent, Belgium). This assay enables for the determination of the number of germinated seeds and the growth of roots and shoots of selected higher plants exposed to the contaminated matrix (the reference soil containing plastic particles) compared with the controls (the reference soil only). Five following concentrations of plastic particles in the reference OECD soil were tested: 0.02, 0.095, 0.48, 2.38 and 11.9% w/w. The tests for each particle concentration were made in three replications for each plastic material and each plant, whereas the control tests were made in nine replications for each plant. The monocotyledonous plant *Sorghum saccharatum* (sorghum, series no. SOS041019) and two dicotyledonous plants *Lepidium sativum* (garden cress, series no. LES260820) and *Sinapis alba* (mustard, series no. SIA020719) were used as model organisms in these experiments. The appropriately prepared soil and ten seeds of one of the higher plants were placed in the special test plate dedicated to this assay. All these materials (the reference soil, seeds and test plates) were delivered by Microbiotests (Ghent, Belgium). Then, all test plates were incubated for 72 h at 25 ± 1 °C in the darkness in the acclimation chamber FITO 700 (Biogenet, Józefów, Poland). After incubation the number of germinated seeds was recorded for each test and control plate and germination index was calculated [16]. Additionally, a digital picture of each plate was made and then subjected to image analysis using the NIS ELEMENTS AR software (Nikon, Japan). As a result, the lengths of roots and shoots were measured. The composition of the OECD reference soil and more detailed description of the phytotoxicity assay used in this work are presented elsewhere [16].

2.3. Statistical Analysis

The results of measurements were subjected to statistical elaboration. It comprised basically the calculation of mean values, standard deviation and goodness of normal distribution. The latter was checked with the use of the Kolmogorov–Smirnov test. In addition, one-way analysis of variance (ANOVA) at statistical significance $\alpha = 0.05$ was applied to evaluate whether the lengths of roots or shoots of plants exposed to one of the plastics tested were statistically equal or different than those that were not exposed to the plastics. As the null hypothesis it was assumed that they were equal. The statistical elaboration of results was performed with the use of MS Excel (Analysis ToolPak) software and OriginPro 9.0 (OriginLab, Northampton, MA, USA).

3. Results and Discussion

Seed germination is a key process in the seed plant life cycle that influences total biomass yield and quality [18,19]. This process depends on both intrinsic (e.g., seed dormancy and available food store) and extrinsic (e.g., temperature, light, relative humidity, chemicals) factors [19,20]. Thus, the presence of plastic particles in the soil may influence it. Bio-based plastics studied in this work did not hamper the seed germination processes of any of three higher plants used as model organisms. The values of GI in the tests with the addition of plastic particles were approximately at the same level as those determined in the control tests (Figure 1). This was observed for each plant, i.e., monocotyledonous *S. saccharatum* as well as dicotyledonous *L. sativum* and *S. alba*. The results of one-way ANOVA confirmed that there were not statistically significant differences between the germination efficiency in the tests with and without plastic particles in the soil. The *p*-values were in the range from 0.3319 to 0.9515 for sorghum, while for the cress and mustard they were from 0.05039 to 0.1245 and from 0.05004 to 0.9478, respectively. Consequently, they were less than or equal to the significance level $\alpha = 0.05$. Other studies on this subject also showed that plastic microparticles did not affect the germination of wheat [12] or cress [9,16]. This proves that germination of seed plants is relatively uninfluenced by the soil composition [9].

The analysis of the effect of bio-based plastics on early growth of higher plants comprised both the development of roots and shoots. The results of measurements of plant roots and shoots varied widely, and they were difficult to use for interpretation in spite of being subjected to statistical elaboration. Depending on the compound tested and its concentration in the soil, inhibition and/or stimulation of root/shoot growth or no impact were found. A high degree of variability of results of phytotoxicity tests was also observed in other studies concerning petroleum-derived and/or bio-based plastics [11,14,21].

Roots are regarded to be the first organ exposed to the impact of toxic compounds present in the soil. As a consequence, roots react to these stress conditions mainly by growth inhibition [22]. The changes of root length of each of higher plant used as bioindicators in this study are depicted in Figure 2. The growth of sorghum roots was not inhibited by any of bio-based plastic tested irrespective of its concentration in the soil. This was statistically confirmed with the help of one-way ANOVA ($p > 0.05$) (Table 2). Only in the case of BPE-SP-PBS, was the stimulation of root growth observed at the two lowest concentrations tested ($p < 0.05$) as seen in Figure 2, and Table 2. At the same time the cress root growth was inhibited by three out of five bio-based plastics tested, BPE-AMF-PLA, BPE-T-PHBV and BPE-RP-PLA. This was found for each concentration (BPE-AMF-PLA) or for four out of five (BPE-T-PHBV, BPE-RP-PLA) concentrations of bio-based plastic particles in the soil (Table 2). The presence of the other two materials (BPE-C-PLA, BPE-SP-PBS) in the soil did not affect the root growth of cress significantly (Figure 2, Table 2). With regard to mustard, the inhibition of root growth was revealed in the tests with BPE-AMF-PLA, BPE-RP-PLA and BPE-SP-PBS. However, this was not found in every case over the entire range of concentrations of plastic particles in the soil. In the case of BPE-AMF-PLA, this was found in three out of five concentrations of this material in the soil, while in the case of BPE-RP-PLA or BPE-SP-PBS, the inhibition of root growth was statistically confirmed at the two highest concentrations (Figure 2, Table 2).

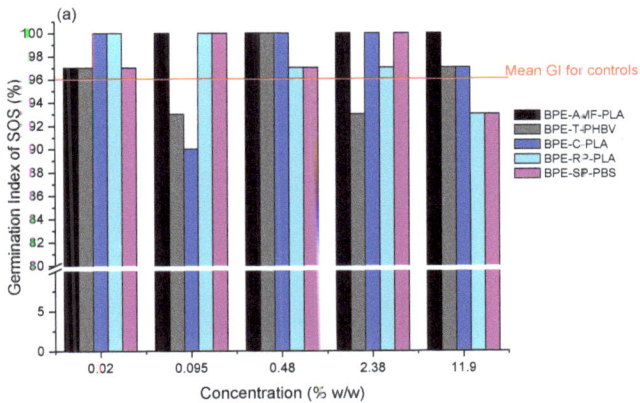
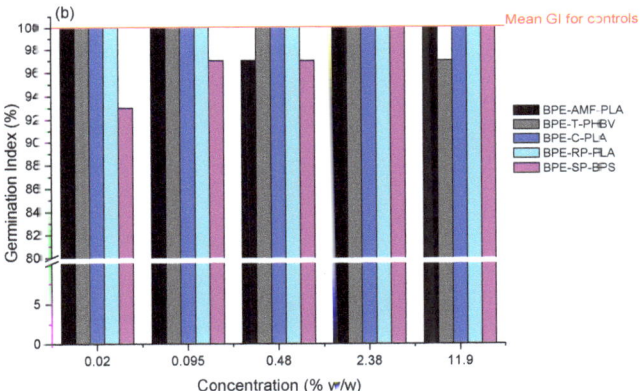
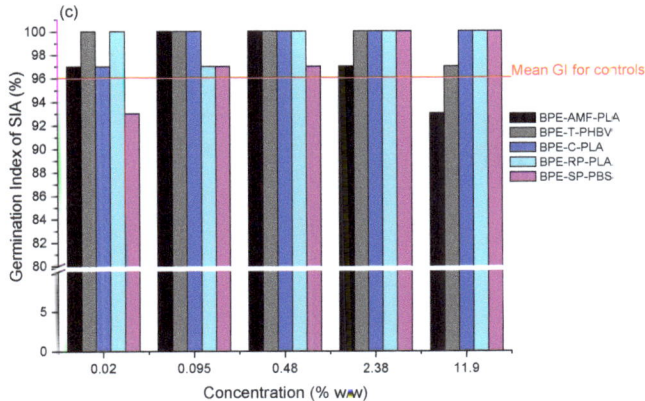

Figure 1. Effect of bio-based plastics on seed germination of higher plants: (**a**) *Sorghum saccharatum* (SOS); (**b**) *Lepidium sativum* (LES); (**c**) *Sinapsis alba* (SIA).

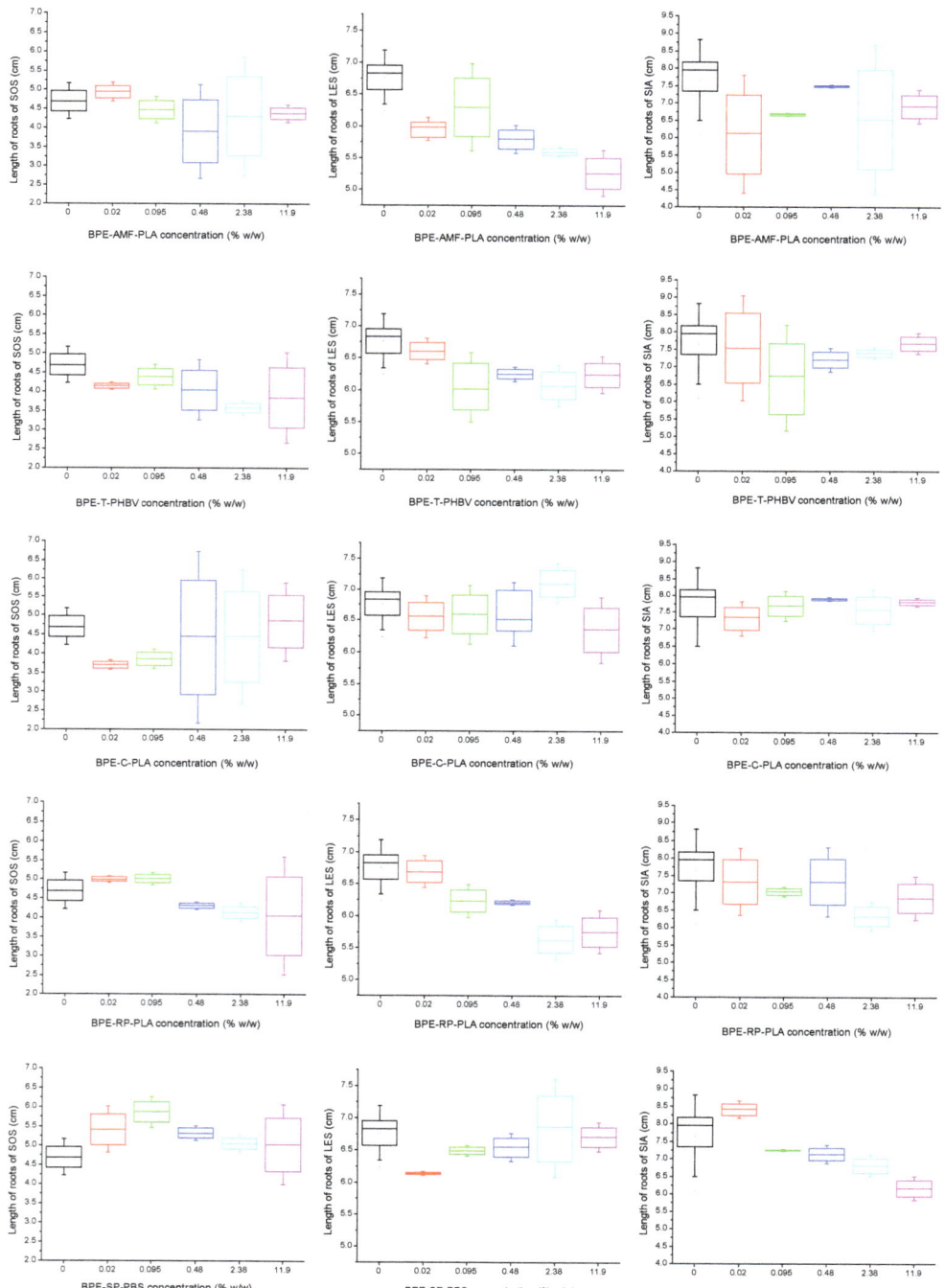

Figure 2. Effect of bio-based plastics on root growth of higher plants: *Sorghum saccharatum* (SOS), *Lepidium sativum* (LES) and *Sinapsis alba* (SIA).

Table 2. Results of one-way ANOVA for S. saccharatum (SOS), L. sativum (LES) and S. alba (SIA).

Tested compound	Exposed plant organ	p-values for SOS				
		Concentrations (% w/w)				
		0.02	0.095	0.48	2.38	11.9
BPE-AMF-PLA	roots	0.6231	0.5702	0.0678	0.3654	0.6693
	shoots	0.9034	0.6639	0.01781 (I)	0.3142	0.7238
BPE-T-PHBV	roots	0.1868	0.4567	0.6671	0.01371(I)	0.7165
	shoots	0.1638	0.4523	0.3511	0.00735(I)	0.9482
BPE-C-PLA	roots	0.02192(I)	0.06123	0.7624	0.2606	0.7531
	shoots	0.2482	0.5049	0.1488	0.07073	0.2089
BPE-RP-PLA	roots	0.5214	0.5082	0.3626	0.1977	0.1625
	shoots	0.6457	0.1868	0.6723	0.3267	0.4432
BPE-SP-PBS	roots	0.03223(S)	0.008538(S)	0.2049	0.4288	0.8969
	shoots	0.001491(S)	$1.78 \cdot 10^{-5}$(S)	0.2255	0.03400(S)	0.5318
Tested compound	Exposed plant organ	p-values for LES				
		Concentrations (% w/w)				
		0.02	0.095	0.48	2.38	11.9
BPE-AMF-PLA	roots	$2.78 \cdot 10^{-5}$(I)	0.0371(I)	$3.72 \cdot 10^{-6}$(I)	$7.13 \cdot 10^{-8}$(I)	$7.21 \cdot 10^{-12}$(I)
	shoots	0.639	0.716	0.378	0.389	0.136
BPE-T-PHBV	roots	0.4633	0.000132(I)	0.00855(I)	0.00195(I)	0.0471(I)
	shoots	0.000159(S)	0.477	0.277	0.570	0.000662 (I)
BPE-C-PLA	roots	0.386	0.499	0.253	0.0335 (S)	0.0507 (I)
	shoots	0.000834(S)	$2.09 \cdot 10^{-5}$(S)	0.00311(S)	$7.19 \cdot 10^{-8}$(S)	0.194
BPE-RP-PLA	roots	0.826	0.00488(I)	0.00329(I)	$9.25 \cdot 10^{-8}$(I)	$1.46 \cdot 10^{-8}$(I)
	shoots	$3.90 \cdot 10^{-6}$(S)	0.0166(S)	$2.14 \cdot 10^{-6}$(S)	0.825	0.458
BPE-SP-PBS	roots	0.0103 (I)	0.234	0.352	0.513	0.822
	shoots	$9.28 \cdot 10^{-5}$(S)	$2.19 \cdot 10^{-6}$(S)	0.000194(S)	0.0214(S)	$4.01 \cdot 10^{-6}$ (I)
Tested compound	Exposed plant organ	p-values for SIA				
		Concentrations (% w/w)				
		0.02	0.095	0.48	2.38	11.9
BPE-AMF-PLA	roots	0.000545(I)	0.00654(I)	0.612	0.00626(I)	0.0812
	shoots	0.959	0.596	0.843	0.528	0.0142(I)
BPE-T-PHBV	roots	0.737	0.0196(I)	0.223	0.465	0.993
	shoots	0.0584	0.454	0.509	0.517	0.000444(I)
BPE-C-PLA	roots	0.375	0.969	0.599	0.791	0.677
	shoots	0.987	0.00235(S)	0.0133(S)	0.0632	0.979
BPE-RP-PLA	roots	0.394	0.116	0.349	0.000938(I)	0.0332(I)
	shoots	0.577	0.659	0.634	0.224	0.0955
BPE-SP-PBS	roots	0.0506	0.311	0.214	0.0416(I)	0.000111(I)
	shoots	0.00984(S)	0.0345(S)	0.329	0.214	0.110

(I)—inhibition; (S)—stimulation.

The results of root length obtained for three plant species showed that dicotyledons were more sensitive to the presence of bio-based plastic particles in the soil than monocotyledons. At the same time, in the case of the stress factor induced by stimuli other than plastic particle soil contaminations (e.g., metals), the response of monocotyledons might be the same or stronger in comparison to dicotyledons [22,23]. Out of three plants used as bioindicators, cress appeared to be the most sensitive organism in the assessment of the effect of bio-based plastic materials on the early stages of root growth. Comparing to what extent each bio-based plastic affected root length, it was observed that two bio-based plastics (BPE-AMF-PLA and BPE-RP-PLA) acted more strongly on root growth than other plastic materials did. In particular, it affected the root growth of cress and mustard. The presence of BPE-AMF-PLA or BPE-RP-PLA in the soil contributed to the decrease of cress roots from 6.3 to 21.8% or from 0.6 to 16.4%, respectively. In the case of mustard roots, it was from 2.1 to 19.8% in the tests with BPE-AMF-PLA and from 4.3 to 17.4% in the tests with BPE-RP-PLA (Figure 2). The reduction of root length of any plant did not exceed 21.8% irrespective of the type of bio-based plastic added to the soil (Figure 2).

Shoot growth, just like root growth, was affected by bio-based plastics in different ways depending mainly on the plant used as the indicator, the type of bio-based plastic and its concentration in the soil (Figure 3). In the case of shoot growth, the stimulation was more often observed than it was noted for roots (Table 2). The stimulatory effect of low doses of toxic substances on growth of plants or animals is a known phenomenon called as hormesis [24–26]. The growth of sorghum shoots was generally not inhibited by the presence of plastic material in the soil (Table 2). At the same time, BPE-SP-PBS contributed to the stimulation of growth of sorghum shoots at the lower concentrations 0.02–0.095% w/w of plastic particles in the soil (Figure 3, Table 2).

Shoot development of cress was the most often stimulated in the tests with the addition of the bio-based plastic particles to the soil (Figure 3, Table 2). This phenomenon was observed at almost all concentrations tested in the case of three out five bio-based plastics tested, namely, BPE-C-PLA, BPE-RP-PLA and BPE-SP-PBS.

The inhibition of cress shoot growth was found only at the highest concentrations of bio-based plastic particles in the soil. It concerned the tests with BPE-SP-PBS and BPE-T-PHBV. At the same time, the presence of BPE-AMF-PLA did not affect shoot growth of cress at all (Figure 3, Table 2). The impact of bio-based plastic particles on mustard shoot development was weaker than that with regard to cress. The inhibition was observed at the highest concentration (11.9% w/w) in the case of tests with BPE-AMF-PLA or BPE-T-PHBV, while the presence of BPE-SP-PBS or BPE-C-PLA stimulated the growth of mustard shoots at the concentrations 0.02–0.095% w/w or 0.095–0.48% w/w, respectively (Figure 3, Table 2).

As was the case with the roots, cress turned out to be the most sensitive model organism in the evaluation of the effect of bio-based plastic particles on shoot growth. It is in line with the previous results concerning PLA, PHB and polypropylene (PP), which indicated that *L. sativum* was the most sensitive plant in the tests with these three plastics [16]. Therefore, this plant should be used as one of bioindicators in the assessment of potential phytotoxicity of plastics, in particular bio-based plastic materials.

Comparing the effect of the individual plastics on shoot growth, it was easy to see that BPE-SP-PBS stimulated the growth of shoots of all higher plants tested, when it was present in the soil at concentrations from 0.02 to 0.095% w/w. In the case of cress at lower concentrations of plastic particles in the soil, the stimulation of cress shoot growth occurred, while at the highest concentration (11.9% w/w) their growth was inhibited (Figure 3, Table 2). This response of the biological parameter (shoot growth) in the presence of BPE-SP-PBS in the soil at various concentrations (stimulation at lower concentrations, inhibition at higher concentrations) corresponded with the description of the hormetic effect [25,26]. Thus, the hormetic effect of bio-based plastic BPE-SP-PBS on shoot growth of cress is very probable.

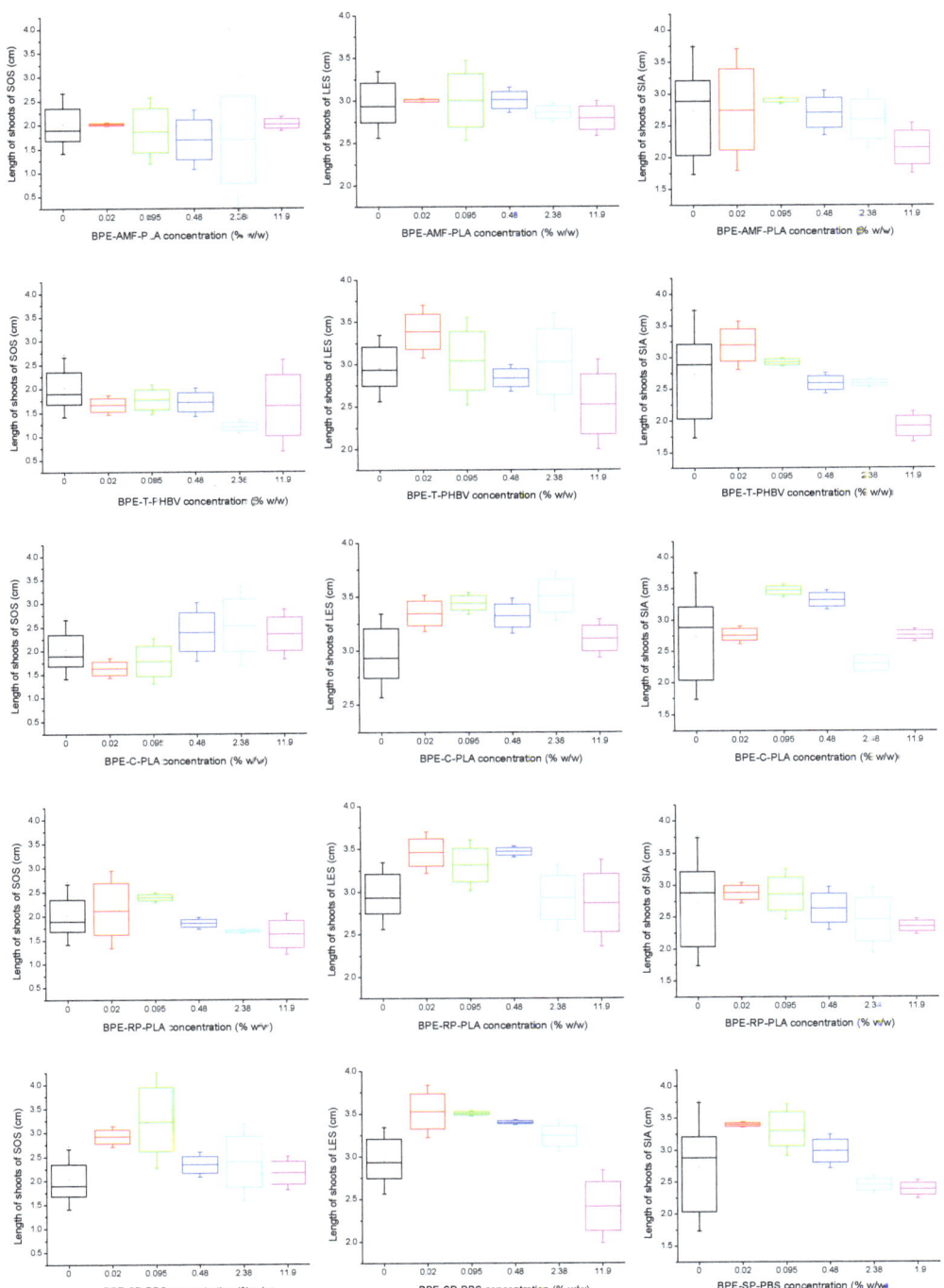

Figure 3. Effect of bio-based plastics on shoot growth of higher plants: *Sorghum saccharatum* (SOS), *Lepidium sativum* (LES) and *Sinapsis alba* (SIA).

4. Conclusions

In this work, five bio-based plastics of different chemical composition and various potential applications were studied towards their impact on early growth of higher plants.

None of bio-based plastics even at relatively high concentration in the soil (11.9% w/w) interferes with the germination of plant seeds. This is the case of monocotyledonous and dicotyledonous plants.

The growth of roots of *S. saccharatum* (monocotyledonous plant) is not inhibited irrespective of the type of bio-based plastic and its concentration in the soil. At the same time, the growth of roots of dicotyledonous plants is inhibited by some kinds of bio-based plastics. The inhibition of the development of *L. sativum* roots is statistically confirmed for two PLA-based plastics and one PHBV-based plastic. In the case of roots of *S. alba* it is confirmed for the same two PLA-based plastics as for cress and additionally for PBS-based plastic.

Two PLA-based plastics, BPE-AMF-PLA and BPE-RP-PLA, inhibit root growth of dicotylodonous plants more strongly than other plastic materials. Nevertheless, the shortening of roots does not exceed 22% in comparison to the control runs in any case.

Shoot growth of the monocotyledonous plant is usually not affected by the presence of bio-based plastics in the soil. With regard to the dicotyledonous plants, in particular cress shoots, stimulation of growth often occurs. This is statistically confirmed for the growth of *L. sativum* shoots exposed to BPE-C-PLA, BPE-RP-PLA or BPE-SP-PBS. The presence of PBS-based plastic in the soil contributes to the stimulation of shoot growth of higher plants (sorghum, cress and mustard) at the concentrations from 0.02 to 0.095% w/w. In the case of cress shoots exposed to PBS-based plastic, the hormetic effect is observed.

This work confirms that *L. sativum* is the most sensitive plant out of three bioindicators used with regard to the presence of bio-based plastic particles in the soil. It is recommended to include *L. sativum* in the evaluation of the effect of bio-based plastics on the early growth of higher plants.

Funding: This work was supported by the European Union's Horizon 2020—Research and Innovation Framework Programme through the research project BIO-PLASTICS EUROPE (Grant agreement No. 860407).

Institutional Review Board Statement: Not applicable.

Informed Consent Statement: Not applicable.

Data Availability Statement: The data presented in this study are available on request from the corresponding author.

Conflicts of Interest: The author declares no conflict of interest.

References

1. Wong, J.W.Y.; Hitzfeld, B.; Zimmermann, M.; Werner, I.; Ferrari, B.J.D. Current developments in soil ecotoxicology and the need for strengthening soil ecotoxicology in Europe: Results of a stakeholder workshop. *Environ. Sci. Eur.* **2018**, *30*, 49. [CrossRef] [PubMed]
2. Rillig, M.C.; Lehmann, A.; de Souza Machado, A.A.; Yang, G. Microplastic effects on plants. *New Phytol.* **2019**, *223*, 1066–1070. [CrossRef] [PubMed]
3. de Souza Machado, A.A.S.; Lau, C.W.; Till, J.; Kloas, W.; Lehmann, A.; Becker, R.; Rillig, M.C. Impacts of microplastics on the soil biophysical environment. *Environ. Sci. Technol.* **2018**, *52*, 9656–9665. [CrossRef]
4. Wan, Y.; Wu, C.; Xue, Q.; Hui, X. Effects of plastic contamination on water evaporation and desiccation cracking in soil. *Sci. Total Environ.* **2019**, *654*, 576–582. [CrossRef] [PubMed]
5. Lavelle, P.; Decaens, T.; Aubert, M.; Barot, S.; Blouin, M.; Bureau, F.; Margerie, P.; Mora, P.; Rossi, J.P. Soil invertebrates and ecosystem services. *Eur. J. Soil Biol.* **2006**, *42*, S3–S15. [CrossRef]
6. Bogum, G. (Ed.) *Ecotoxicology*; InTech: Rijeka, Croatia, 2012; ISBN 978-953-51-0027-0.
7. Faucon, M.-P.; Houben, D.; Lambers, H. Plant Functional Traits: Soil and Ecosystem Services. *Trends Plant Sci.* **2017**, *22*, 385–394. [CrossRef]

8. Arcos-Hernandez, M.V.; Laycock, B.; Pratt, S.; Donose, B.C.; Nikolić, M.A.L.; Luckman, P.; Werker, A.; Lant, P.A. Biodegradation in a soil environment of activated sludge derived polyhydroxyalkanoate (PHBV). *Polym. Degrad. Stab.* **2012**, *97*, 2301–2312. [CrossRef]
9. Balestri, E.; Menicagli, V.; Ligorini, V.; Fulignati, S.; Galletti, A.M.R.; Lardicci, C. Phytotoxicity assessment of conventional and biodegradable plastic bags using seed germination test. *Ecol. Indic.* **2019**, *102*, 569–580. [CrossRef]
10. Bosker, T.; Bouwman, L.J.; Brun, N.R.; Behrens, P.; Vijver, M.G. Microplastics accumulate on pores in seed capsule and delay germination and root growth of the terrestrial vascular plant Lepidium sativum. *Chemosphere* **2019**, *226*, 774–781. [CrossRef]
11. Qi, Y.; Yang, X.; Pelaez, A.M.; Lwanga, E.H.; Beriot, N.; Gertsen, H.; Garbeva, P.; Geissen, V. Macro- and micro- plastics in soil-plant system: Effects of plastic mulch film residues on wheat (Triticum aestivum) growth. *Sci. Total Environ.* **2018**, *645*, 1048–1056. [CrossRef]
12. Judy, J.D.; Williams, M.; Gregg, A.; Oliver, D.; Kumar, A.; Kookana, R.; Kirby, J.K. Microplastics in municipal mixed-waste organic outputs induce minimal short to long-term toxicity in key terrestrial biota. *Environ. Pollut.* **2019**, *252*, 522–531. [CrossRef] [PubMed]
13. Huerta-Lwanga, E.; Mendoza-Vega, J.; Ribeiro, O.; Gertsen, H.; Peters, P.; Geissen, V. Is the Polylactic Acid fiber in green compost a risk for Lumbricus terrestris and Triticum aestivum? *Polymers* **2021**, *13*, 703. [CrossRef] [PubMed]
14. Lozano, Y.M.; Lehnert, T.; Linck, L.T.; Lehmann, A.; Rillig, M.C. Microplastic shape, polymer type, and concentration affect soil properties and plant biomass. *Front. Plant. Sci.* **2021**, *12*, 616645. [CrossRef] [PubMed]
15. de Souza Machado, A.A.; Lau, C.W.; Kloas, W.; Bergmann, J.; Bachelier, J.B.; Faltin, E.; Becker, R.; Görlich, A.S.; Rillig, M.C. Microplastics can change soil properties and affect plant performance. *Environ. Sci. Technol.* **2019**, *53*, 6044–6052. [CrossRef]
16. Liwarska-Bizukojc, E. Phytotoxicity assessment of biodegradable and non-biodegradable plastics using seed germination and early growth tests. *Chemosphere* **2022**, *289*, 133132. [CrossRef]
17. ISO 18763:2016; Soil Quality—Determination of the Toxic Effects of Pollutants on Germination and Early Growth of Higher Plants. Test No. 208: Terrestrial Plant Test: Seedling Emergence and Seedling Growth Test. ISO: Geneva, Switzerland, 2016.
18. Tuan, P.A.; Sun, M.; Nguyen, T.-N.; Park, S.; Ayele, B.T. 1—Molecular Mechanisms of Seed Germination. In *Sprouted Grains*; Feng, H., Nemzer, B., DeVries, J.W., Eds.; AACC International Press: Washington, DC, USA, 2019; pp. 1–24. [CrossRef]
19. Makhaye, G.; Mofokeng, M.M.; Tesfay, S.; Aremu, A.O.; Van Staden, J.; Amoo, S.O. Chapter 5—Influence of plant biostimulant application on seed germination. In *Biostimulants for Crops from Seed Germination to Plant Development*; Gupta, S., Van Staden, J., Eds.; Academic Press: Cambridge, MA, USA, 2021; pp. 109–135. [CrossRef]
20. Reddy, Y.A.N.; Reddy, Y.N.P.; Ramya, V.; Suma, L.S.; Narayana Reddy, A.B.; Krishna, S.S. Chapter 8—Drought adaptation: Approaches for crop improvement. In *Millets and Pseudo Cereals, Woodhead Publishing Series in Food Science, Technology and Nutrition*; Singh, M., Sood, S., Eds.; Woodhead Publishing: Sawston, UK, 2021; pp. 143–158. [CrossRef]
21. Lozano, Y.M.; Rillig, M.C. Effects of microplastic fibers and drought on plant communities. *Environ. Sci. Technol.* **2020**, *54*, 6166–6173. [CrossRef]
22. Matras, E.; Gorczyca, A.; Pociecha, E.; Przemieniecki, S.W.; Oćwieja, M. Phytotoxicity of Silver Nanoparticles with Different Surface Properties on Monocots and Dicots Model Plants. *J. Soil Sci. Plant Nutr.* **2022**, *22*, 1647–1664. [CrossRef]
23. Piršelová, B.; Kuna, R.; Libantová, J.; Moravčíková, J.; Matušíková, I. Biochemical and physiological comparison of heavy metal-triggered defense responses in the monocot maize and dicot soybean roots. *Mol. Biol. Rep.* **2011**, *38*, 3437–3446. [CrossRef]
24. Małkowski, E.; Sitko, K.; Szopiński, M.; Gieroń, Ż.; Pogrzeba, M.; Kalaji, H.M.; Zieleźnik-Rusinowska, P. Hormesis in Plants: The Role of Oxidative Stress, Auxins and Photosynthesis in Corn Treated with Cd or Pb. *Int. J. Mol. Sci.* **2020**, *21*, 2099. [CrossRef]
25. Agathokleous, E.; Kitao, M.; Calabrese, E.J. Hormesis: A compelling platform for sophisticated plant science. *Trends Plant Sci.* **2019**, *24*, 318–327. [CrossRef]
26. Kendig, E.L.; Le, H.H.; Belcher, S.M. Defining hormesis: Evaluation of a complex concentration response phenomenon. *Int. J. Toxicol.* **2010**, *29*, 235–246. [CrossRef] [PubMed]

Disclaimer/Publisher's Note: The statements, opinions and data contained in all publications are solely those of the individual author(s) and contributor(s) and not of MDPI and/or the editor(s). MDPI and/or the editor(s) disclaim responsibility for any injury to people or property resulting from any ideas, methods, instructions or products referred to in the content.

Article

Evaluation of the Physical and Shape Memory Properties of Fully Biodegradable Poly(lactic acid) (PLA)/Poly(butylene adipate terephthalate) (PBAT) Blends

Marica Bianchi [1], Andrea Dorigato [1,*], Marco Morreale [2,*] and Alessandro Pegoretti [1]

[1] Department of Industrial Engineering and INSTM Research Unit, University of Trento, Via Sommarive 9, 38123 Trento, Italy
[2] Faculty of Engineering and Architecture, Kore University of Enna, Cittadella Universitaria, 94100 Enna, Italy
* Correspondence: andrea.dorigato@unitn.it (A.D.); marco.morreale@unikore.it (M.M.)

Abstract: Biodegradable polymers have recently become popular; in particular, blends of poly(lactic acid) (PLA) and poly(butylene adipate terephthalate) (PBAT) have recently attracted significant attention due to their potential application in the packaging field. However, there is little information about the thermomechanical properties of these blends and especially the effect induced by the addition of PBAT on the shape memory properties of PLA. This work, therefore, aims at producing and investigating the microstructural, thermomechanical and shape memory properties of PLA/PBAT blends prepared by melt compounding. More specifically, PLA and PBAT were melt-blended in a wide range of relative concentrations (from 85/15 to 25/75 wt%). A microstructural investigation was carried out, evidencing the immiscibility and the low interfacial adhesion between the PLA and PBAT phases. The immiscibility was also confirmed by differential scanning calorimetry (DSC). A thermogravimetric analysis (TGA) revealed that the addition of PBAT slightly improved the thermal stability of PLA. The stiffness and strength of the blends decreased with the PBAT amount, while the elongation at break remained comparable to that of neat PLA up to a PBAT content of 45 wt%, while a significant increment in ductility was observed only for higher PBAT concentrations. The shape memory performance of PLA was impaired by the addition of PBAT, probably due to the low interfacial adhesion observed in the blends. These results constitute a basis for future research on these innovative biodegradable polymer blends, and their physical properties might be further enhanced by adding suitable compatibilizers.

Keywords: PLA; PBAT; biodegradable polymers; blends; shape memory polymers

Citation: Bianchi, M.; Dorigato, A.; Morreale, M.; Pegoretti, A. Evaluation of the Physical and Shape Memory Properties of Fully Biodegradable Poly(lactic acid) (PLA)/Poly(butylene adipate terephthalate) (PBAT) Blends. *Polymers* 2023, 15, 881. https://doi.org/10.3390/polym15040881

Academic Editors: Jesús-María García-Martínez and Emilia P. Collar

Received: 19 December 2022
Revised: 31 January 2023
Accepted: 2 February 2023
Published: 10 February 2023

Copyright: © 2023 by the authors. Licensee MDPI, Basel, Switzerland. This article is an open access article distributed under the terms and conditions of the Creative Commons Attribution (CC BY) license (https://creativecommons.org/licenses/by/4.0/).

1. Introduction

In recent years, the increasing difficulties of plastic disposal have raised concerns worldwide. Most plastic waste ends up in landfills, oceans, soil and water, representing thus serious hazards for plants, animals and humans. According to recent statistics, every year up to 12.7 million tons of plastic enter the oceans, causing the death of several seabirds and aquatic animals. The severity of this problem will exponentially increase with the increase in global non-biodegradable plastic consumption [1–3]. Plastic waste is estimated to triple by 2060, as claimed by the latest forecast by the OECD's Global Plastic Outlook, rising from 353 million tons of waste in 2019 to 1014 million tons in the next decades [4]. Therefore, measures and solutions must be taken to avoid irreversible consequences on ecosystems and human beings. One of the major generators of plastic pollution is packaging, responsible for almost half of the global total due to its short lifespan [5]. This fact, highlighted in several statistics, has led to a growing demand for the use of biodegradable polymers for packaging materials since they could significantly minimize environmental pollution [5].

Nowadays, biodegradable polymers (BPs) have become a topic of great interest as an alternative to overcome the issues related to non-biodegradable polymeric waste. BPs can be decomposed by microbes (bacteria, some fungi and algae, etc.) existing in nature into CO_2, H_2O, CH_4 and biomass, which can be integrated into the natural ecosystem without any ecotoxic effects [1]. One of the most studied BPs is poly(lactic acid) (PLA), a linear aliphatic polyester derived from lactic acid and synthesized from renewable resources, such as sugarcane or corn starch [6–8]. Due to its good mechanical properties (similar to those of polystyrene), processability, reasonable price and commercial availability, it has gained interest in academia and industry. PLA shows a Young's modulus of around 3 GPa, a tensile strength between 50 and 70 MPa and an impact strength close to 2.5 kJ/m^2 [9]. It can be processed at the industrial level with the same processing technologies used for traditional non-biodegradable polymers, but with lower greenhouse gas emissions [9]. PLA is the most used biodegradable polymer in the food packaging industry and it is commercialized mainly for single-use disposal packaging applications, such as bottles, cold drink cups, blister packages and flexible films [9]. It presents several advantages over non-biodegradable plastics that are commonly used for packaging, including good transparency, degradation in biological environments (for example, in soil and compost), biocompatibility and FDA approval [5]. In addition, it exhibits a thermo-responsive shape memory behavior that makes it even more attractive for the development of active and smart packaging [10–16].

In more detail, the polymer's thermo-responsive shape memory behavior allows it to undergo large controllable shape changes in response to a change in temperature [17–19]. In the case of semi-crystalline PLA, the crystalline domains act as the net points and determine the initial permanent shape of the material, while the glass transition temperature of PLA (approx. 60 °C) acts as a molecular switch and is responsible for the fixability of the desired temporary shape. PLA macromolecules exhibit high mobility when the material is heated above its glass transition, allowing the polymer to be easily deformed into the appropriate temporary shape. Cooling PLA under loads below T_g strongly reduces its mobility, thereby fixing the imposed deformation. Finally, when the polymer is heated above T_g, the switching domains progressively regain their mobility, and the material returns to its original condition [17–19].

Although PLA possesses many advantages for packaging, it also has some drawbacks that prevent its industrial exploitation, including poor toughness and sensitivity to thermal degradation [20,21]. One possible strategy to find more PLA commercial applications in the sustainable packaging field, while maintaining its biodegradability, consists of blending PLA with other BPs that have complementary characteristics [21]. Melt-blending techniques are generating a lot of interest since they are a cost-effective, easy and readily available processing technology at the industrial level that allows us to obtain packaging formulations with a desired performance by varying the blend composition [21]. Poly(butylene adipate-co-terephthalate) (PBAT), an aliphatic-aromatic copolymer, is considered an interesting toughening candidate [22]. It is a fully biodegradable polymer based on fossil resources, produced by poly-condensation between butanediol (BDO), adipic acid (AA) and terephthalic acid (PTA) [23,24]. PBAT is characterized by high elasticity, good thermal stability at elevated temperatures, wear and fracture resistance, and compatibility with many other natural polymers. However, its stiffness and strength are rather poor [22–25]. PBAT's properties can be slightly adjusted by varying the relative ratio of the two structural units: the rigid butylene terephthalate (BT) segments, derived from 1,4-butanediol and terephthalic acid monomers, and the flexible butylene adipate (BA) segments, derived from 1,4-butanediol and adipic acid monomers. Similar to PLA, PBAT has found application in packaging (for example, in the production of plastic bags), as well as in the biomedical field, in hygiene products and in mulch film production [23–25].

In view of their complementary characteristics and common application fields, blending stiff PLA with flexible PBAT could represent a valuable technical solution to develop eco-sustainable packaging with tailorable properties [26–28]. PLA/PBAT blends have been thoroughly researched over the years [29–37]. Farsetti et al. [37] found that PLA/PBAT blends with different relative concentrations produced by melt compounding exhibited two-phase behavior and two separate glass transition temperatures. The blends maintained quite a high modulus and tensile strength compared to those of neat PLA, and small amounts of PBAT improved both the elongation at break and the impact resistance. Gu et al. [32] investigated the melt rheology of PLA/PBAT blends with PBAT contents lower than 30 wt% prepared by twin-screw extrusion. The incorporation of PBAT was found to improve the melt processability, since the PLA/PBAT blends were characterized by a more pronounced shear-thinning behavior compared to that of the neat PLA. Several studies have focused on the preparation and characterization of flexible PLA/PBAT biodegradable films for agricultural uses. For example, Weng et al. [33] prepared PLA/PBAT films and investigated their biodegradation behavior in a real soil environment. They observed that the blends presented a lower biodegradation rate compared to that of the neat polymers, and this experimental evidence was attributed to the poor compatibility between the PLA and PBAT phases, as highlighted by SEM micrographs. Palsikowski et al. [36] prepared flexible PBAT/PLA films compatibilized with a chain extender and investigated the effect of the addition of the chain extender on their biodegradation. The results highlighted that the presence of the chain extender led to a delay in the film biodegradation, and the blends showed an intermediate behavior compared to that of the neat polymers. However, few studies in the literature have investigated the use of these blends in the packaging field, and the majority of these studies have primarily dealt with food packaging. For example, Paulsen et al. [26] studied the suitability of PLA/PBAT films in the packaging of broccoli. They found that broccoli heads packaged in these films achieved a shelf life of 21 days at 2 °C, which is a more extended postharvest storage with respect to that of low-density polyethylene films. Weng et al. [27] discovered that cinnamaldehyde-loaded corn starch/PBAT/PLA blend film effectively preserved the quality of soy-protein-based meat analogues by inhibiting the growth of *Escherichia coli* and *Staphylococcus aureus* during 4 °C storage. Qui et al. [28] produced PLA/PBAT films incorporated with nano-polyhedral oligomeric silsesquioxane (POSS (epoxy)8) as a reactive compatibilizer via melt processing. They noticed that the addition of POSS to the PLA/PBAT films improved the storage capacity of the packaging films and that the excellent permeability of PBAT created a low-moisture environment capable of suppressing fungal growth.

Considering the recent interest in PLA/PBAT blends and in the shape memory capability of PLA for the development of smart packaging with a thermo-responsive shape, this work is focused on the investigation of the thermomechanical and shape memory properties of fully biodegradable PLA/PBAT blends, produced by melt compounding and hot pressing. Therefore, this study aims at investigating the effect of the blend composition on the processability and on the most important physical features of these materials, in view of their future application in the packaging field. To the authors' knowledge this is the first work dealing with the investigation of the shape memory behavior of PLA/PBAT blends.

2. Materials and Methods
2.1. Materials and Methods

Ingeo™ 2500HP, supplied by NatureWorks® LLC (Minnetonka, MN, USA) in pellet form, was the PLA used in this work. According to the producer's technical datasheet, it was characterized by a specific gravity of 1.24 g/cm^3, a melt flow index (MFI of 8 g/10 min (210 °C, 2.16 kg) and a melting temperature of 165–180 °C. Polymer granules of PBAT Technipol® Bio 1160 were purchased from Sipol Spa (Mortara, PV, Italy). According to the supplier datasheet, its specific gravity was 1.23 g/cm^3, the MFI was 20 g/10 min (160 °C, 2.16 kg) and melting temperature was about 115 °C.

2.2. Sample Preparation

PLA/PBAT blends at different relative ratios were prepared by melt compounding in a Rheomix 600 internal mixer (Thermo Haake®, Waltham, MA, USA), equipped with counter-rotating rotors. The compounding temperature was kept at 180 °C while the mixing time was 10 min. The rotor speed was set at 60 rpm. Prior to processing, PLA and PBAT pellets were dried at 50 °C overnight to remove moisture. After compounding, the blends were hot-pressed in a hydraulic press for 10 min at a pressure of 7 bar and a temperature of 180 °C, followed by water cooling. The processing temperature was selected after preliminary trials, and viscosimetric measurements on both virgin and processed samples (not reported for the sake of brevity) demonstrated that the molecular weight of the constituents was retained after melt-compounding and hot-pressing operations. In this way, square sheets with in-plane dimensions of 150 × 150 mm² and a thickness of 1 mm were prepared. The composition (in wt% and vv%) of the prepared samples along with their codes is reported in Table 1. The final number in the code refers to the PBAT weight concentration in the blend.

Table 1. List of the prepared samples and their nominal composition.

Sample	PLA Content (wt%/vv%)	PBAT Content (wt%/vv%)
PLA	100.0/100.0	-
PLA_PBAT_15	85.0/84.9	15.0/15.1
PLA_PBAT_30	70.0/69.8	30.0/30.2
PLA_PBAT_45	55.0/54.8	45.0/45.2
PLA_PBAT_60	40.0/39.8	60.0/60.2
PLA_PBAT_75	25.0/24.8	75.0/75.2
PBAT	-	100.0/100.0

2.3. Characterization

2.3.1. Rheological Properties

To examine how the blend's composition affects the viscosity in the molten state, dynamic rheological tests were performed on a DHR-2 rheometer (TA Instrument, New Castle, DE, USA) in parallel plate oscillatory mode. The measurements were conducted using 25 mm diameter plates at 180 °C and setting a gap distance of 1 mm. Disc samples (diameter 25 mm, thickness 1 mm) were selected for the tests. The measurements were conducted in frequency sweep mode, applying a maximum shear strain of 1%, which, according to the literature, is within the limit of viscoelasticity for these systems [38]. In this way, the trends of the complex viscosity (η^*), storage modulus (G') and loss modulus (G'') were evaluated in an angular frequency range of 0.1–1000 rad/s.

2.3.2. Microstructural Properties

In order to investigate the dispersion of the two constituents in the blends, optical microscope micrographs were acquired on polished samples, by using an upright incident-light optical CH-9435 Heerbrugg microscope (Heerbrugg, Switzerland).

A field emission scanning electron microscope (FESEM) AG-SUPRA 40 (Carl Zeiss, Ober-Kochen, Germany), operating at an accelerating voltage of 2.5 kV, was employed to investigate the interfacial adhesion between the PLA and PBAT phases. The samples were cryofractured after being submerged in liquid nitrogen for one hour. The fracture surface was analyzed after Pd/Pt sputtering to provide enhanced electrical conductivity.

2.3.3. Thermal Properties

In order to investigate the degradation resistance of the produced blends, thermogravimetric analysis (TGA) was performed by using a Mettler TG50 thermo-balance (Mettler-Toledo GmbH, Schwerzenbach, Switzerland). Samples with a mass of 20 mg were tested at a heating rate of 10 °C/min from 25 °C up to 700 °C under a nitrogen flow of 10 mL/min. The temperature that was associated with a mass loss of 5% ($T_{5\%}$) and the residual mass at 700 °C (m_{700}) were evaluated. In addition, the temperatures corresponding to the maximum mass loss rate of PLA (T_{peak1}) and PBAT (T_{peak2}) were determined from the first derivative of the thermogravimetric curve (DTG).

Differential scanning calorimetry (DSC) tests were performed on the prepared blends by means of a Mettler DSC30 calorimeter (Mettler Toledo GmbH, Schwerzenbach, Switzerland). Measurements were carried out under 100 mL/min nitrogen flow according to the following protocol: (I) the first heating scan was from −80 °C to 200 °C at 10 °C/min, (II) the cooling scan was from 200 °C to −80 °C at 10 °C/min, (III) and the final heating scan was from −80 °C to 200 °C at 10 °C/min. Specimens with a mass of 10 mg were tested. DSC analysis allowed the determination of the glass transition temperature (T_g), the melting and cold crystallization temperatures (T_m, T_{cc}) and specific enthalpy values (ΔH_m, ΔH_{cc}) of the blend constituents. Furthermore, the PLA and PBAT crystallinity degree (χ_c) was calculated according to Equation (1):

$$\chi_c\ (\%) = 100 \cdot \frac{\Delta H_m - \Delta H_{cc}}{\Delta H_m^0 \cdot w} \qquad (1)$$

where ΔH_{cc} is the cold crystallization enthalpy, ΔH_m^0 is the theoretical melting enthalpy of fully crystalline polymer and w is the weight fraction of the polymer phase in the blend. ΔH_m^0 was taken as 93 J/g for fully crystalline PLA and 114 J/g for fully crystalline PBAT [20,24].

2.3.4. Mechanical Properties

Dynamic-mechanical analysis was carried out through a DMA Q800 (TA instrument, New Castle, DE, USA) machine, in order to investigate the dependence of the storage modulus (E') and loss modulus (E") of the produced blends on temperature. The tests were performed on rectangular specimens (gauge length 10 mm, width 5 mm, thickness 1 mm) with a frequency of 1 Hz, a strain amplitude of 0.05% and a heating ramp from 25 °C to 130 °C at 3 °C/min. One specimen was tested for each composition. In this way, the values of the storage modulus at 25 °C (E'_{25}) and at 85 °C (E'_{85}) were determined to evaluate the storage modulus drop below and above the glass transition temperature of PLA.

An Instron 5969 (Instron®, Norwood, MA, USA) tensile testing machine, equipped with a load cell of 10 kN and operating at a cross-head speed of 5 mm/min, was used to determine the tensile properties of the blends under quasi-static conditions. The tests were performed on ISO 527 type 1BA specimens at 25 °C. The elastic modulus (E) was measured as the secant value between two strain levels (0.05 and 0.25 mm/mm) by means of a resistance extensometer. Moreover, the tensile stress at break (σ_b) and the strain at break (ε_b) were determined. Five specimens were tested for each composition.

2.3.5. Evaluation of the Shape Memory Behavior

The investigation of the shape memory behavior of the prepared PLA/PBAT blends was performed on prismatic rectangular specimens (width 5 mm, thickness 1 mm, length 20 mm) by means of a DMA Q800 (TA instrument, New Castle, DE, USA) apparatus, operating in single cantilever mode. Since at 25 °C and 85 °C the PLA phases in the blends are in the glassy and in the rubbery state, respectively, these temperatures were selected for exploring the shape memory performance of the samples. One specimen was tested for each formulation. The tests were carried out according to the thermomechanical cycle typically employed for SMPs [39–48], consisting of a programming step, followed by a recovery step. More specifically, the specimens were subjected to the following thermomechanical history: (I) heating of the specimen and equilibrium at 85 °C; (II) application of a ramp force of 0.8 N/min (the ramp force was reduced to 0.1 N/min for PLA_PBAT_75 blend); (III) drive-off of the motor when a displacement (d_{imp}) of 8000 μm was achieved; (IV) cooling and equilibrium of the sample at 25 °C (cooling rate of 20 °C/min); (V) isothermal step for 5 min; (VI) drive-on of the motor and application of a constant load equal to 0.001 N to continuously monitor the sample displacement as a function of the temperature; (VII) equilibrium at 85 °C; and (VIII) isothermal for 30 min at 85 °C. The graphical representation and the schematization of the thermomechanical cycle is illustrated in Figure 1a,b (the graphical representation of the thermomechanical cycles of all blends are reported in Figure S1). In Figure 1b, d_{imp} is the programmed displacement at step (III), d_{fix} is the effectively fixed displacement at step (VI) and d_{30} is the displacement reached by the movable clamp at the end of the recovery process at step (VIII).

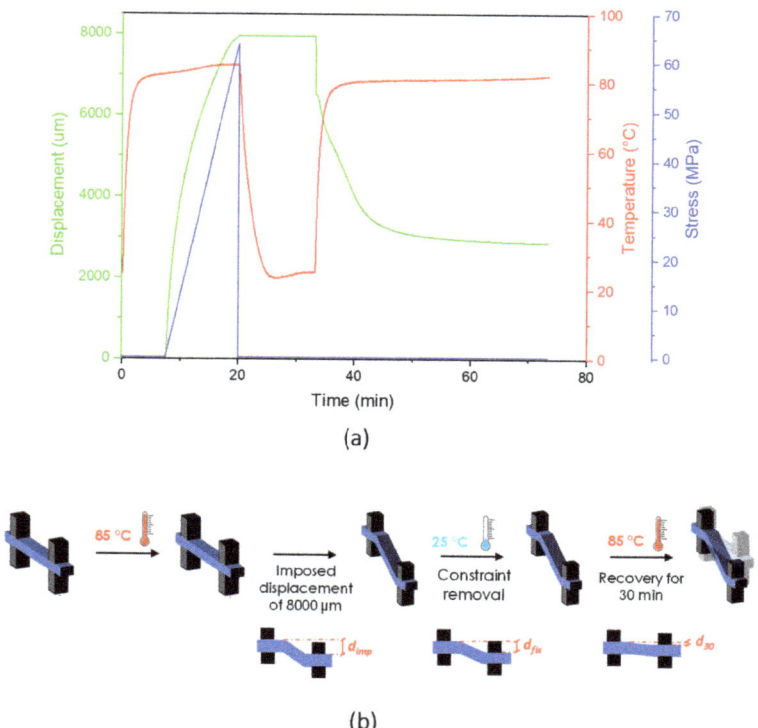

Figure 1. (a) Graphical representation, as an example, of PLA_PBAT_15 sample and (b) illustration of the thermomechanical cycle adopted to investigate the shape memory behavior of the PLA/PBAT blends.

The strain fixity (SF) and strain recovery (SR) parameters were then determined in order to describe the shape memory performance of the PLA/PBAT blends. The first parameter assesses the material's capacity to fix the mechanical deformation introduced during the programming process, while the second one indicates how much strain is recovered during the recovery step. Both SF and SR should be close to 100% in a well-performing SMP. For each specimen, the SF and SR were, respectively, calculated according to Equations (2) and (3):

$$SF(\%) = \frac{d_{fix}}{d_{imp}} \cdot 100 \qquad (2)$$

$$SR(\%) = \frac{d_{imp} - d_{30}}{d_{imp}} \cdot 100 \qquad (3)$$

3. Results and Discussion

3.1. Rheological Measurements

Dynamic rheological tests were carried out to investigate the effect of the blend composition on the rheological properties at the processing temperature (180 °C) in order to obtain information about the processability of these materials. Figure 2a–c show the results of the dynamic rheological tests performed on neat PLA, neat PBAT and all the prepared blends in terms of the dynamic moduli (G′, G″) and complex viscosity (η*).

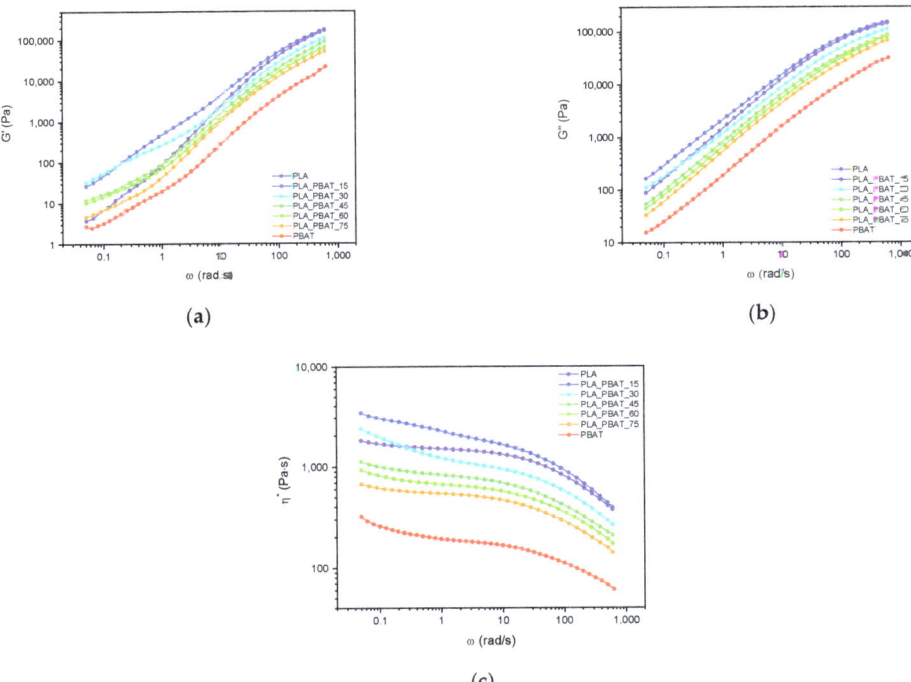

Figure 2. Results of the dynamic rheological tests at 180 °C on the neat matrices and on the prepared blends. Trends of (**a**) storage modulus; (**b**) loss modulus; and (**c**) complex viscosity as a function of the angular frequency.

Figure 2a,b show, respectively, the storage and loss modulus of the PLA/PBAT blends as a function of the frequency. All the samples exhibit a similar trend of the G′ in the high-frequency range, while differences can be seen at lower frequencies. The G′ of all blends in the low-frequency range was higher than that of neat PLA and neat PBAT, suggesting that the two polymers are probably immiscible [49]. Furthermore, significant differences in the slopes of the curves appear at low frequencies, i.e., with the appearance of a shoulder in the PLA_PBAT_15 blend. This behavior is very likely to be due to a droplet/matrix morphology [50–53] and it substantially disappears at higher compositions. This could be attributed to the formation of more complex morphologies showing slow relaxation dynamics [51–53] and will be further investigated and discussed in the following morphological characterization section.

Comparing the loss and the storage modulus values, it can be observed that the G″ is greater than the G′; this, combined with their overall trends, suggests that a liquid-like rheological behavior prevails over a solid-like one for all of the investigated blends. Furthermore, the frequency dependence of the G″ is almost the same for all the samples in the entire frequency range. Finally, when increasing the PBAT content in the blends, both the G′ and G″ decrease.

The complex viscosity of the neat PLA and PBAT (Figure 2c) is similar to the values reported in the literature, highlighting the negligible thermal degradation these materials undergo during the production process [54]. On a general level, all the samples exhibited a shear thinning behavior, more pronounced in the high-frequency range. The η* of PLA is approximately one order of magnitude higher than that of PBAT, as shown in Figure 2c. Therefore, this suggests that it is much easier for PBAT to disperse within the PLA matrix, once processed at the molten state. With the increasing PBAT concentration, the complex viscosity of the blends decreases, except for that of PLA_PBAT_15 and PLA_PBAT_30 in the low-frequency regime, attributable to yield stress, likely to be due to resistance to the flow of the PBAT domains in the PLA matrix.

3.2. Morphological Characterization

The optical microscope images of the prepared blends are shown in Figure 3a–e.

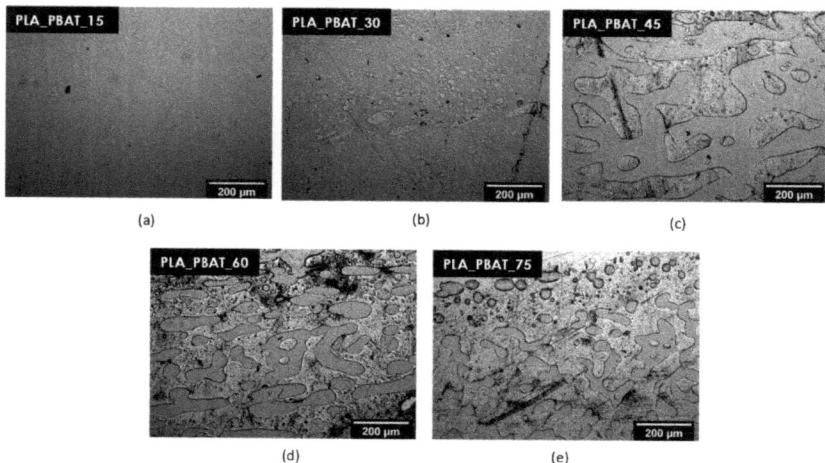

Figure 3. Optical microscope images of the prepared PLA/PBAT blends. (**a**) PLA_PBAT_15; (**b**) PLA_PBAT_30; (**c**) PLA_PBAT_45; (**d**) PLA_PBAT_60; and (**e**) PLA_PBAT_75.

The bright field micrographs shown in Figure 3a–e provide interesting information about the microstructural features of the produced PLA/PBAT blends. Firstly, it is possible to notice that PLA and PBAT are immiscible since all the analyzed compositions show an evident phase separation, which is consistent with the observation reported in other studies on these systems [22,35]. Secondly, depending on the PLA/PBAT ratio, the two phases exhibit either a sea-island morphology (Figure 3a,b,e), i.e., a discrete domain (regularly or irregularly shaped) embedded in the surrounding matrix, or an almost co-continuous structure (Figure 3c,d), i.e., large interconnected domains with irregular shapes. Focusing at first on blends with low PBAT contents, PLA_PBAT_15 (Figure 3a) presents a fine structure with small and spherical PBAT domains evenly dispersed in the PLA matrix. After a statistical analysis of the domain size distribution, it is possible to conclude that the size of PBAT domains increases with the PBAT concentration, from 3.6 ± 0.7 μm for the PLA_PBAT_15 blend up to 11.9 ± 8.9 μm for the PLA_PBAT_30 sample. Moreover, in the PLA_PBAT_30 blend (Figure 3b), there is the coexistence of small and large PBAT droplets, with either spherical or irregularly elongated shapes. The formation of these small PBAT domains in the PLA matrix could be responsible for the high complex viscosity values evidenced in the low-frequency range in the PLA_PBAT_15 and PLA_PBAT_30 samples (see Figure 2c). The dispersed PBAT phase probably hindered the mobility of the PLA chains, inducing a viscosity increase. A significant change in morphology is observable in the PLA_PBAT_45 blend (Figure 3c). This sample exhibits an almost co-continuous morphology, with a 3D network of large and elongated PBAT domains in the PLA matrix. Moreover, it is interesting to observe that small PLA droplets are immersed in some PBAT domains. This experimental evidence suggests that PLA/PBAT is a polymer blend characterized by high viscosity, since double-emulsion structures are generally present in blends with this feature [55]. As can be seen in Figure 3d, PLA_PBAT_60 also presents a co-continuous morphology. However, differently to PLA_PBAT_45, PBAT is the primary constituent in this case, highlighting that in the PBAT concentration interval between 45 wt% and 60 wt%, the phase inversion occurred. Future works will aim at identifying the exact composition associated to this inversion. By a more detailed analysis of the blend containing 60 wt% of PBAT, it is possible to observe that PLA domain size and shape are broadly distributed, since small PLA droplets are present in the PBAT matrix together with the coarser domains. Finally, by further increasing the PBAT content in the blends up to 75 wt% (Figure 3e), the size of the PLA domains, as well as the statistical dispersion of their size distribution, decreases. Their elongated shape probably derives from the coalescence of smaller domains. In conclusion, it is interesting to point out that, on a general level, when PBAT is in the dispersed phase of the blend, its domains are smaller than those formed by PLA when PBAT is the primary constituent. In fact, as it has been evidenced by rheological measurements, PBAT presents a significantly lower viscosity, which favors its breakup in the PLA matrix.

SEM observations were also performed on these blends in order to obtain further information about the interfacial adhesion conditions. The micrographs of the cryofracture surface of the prepared blends are presented in Figure 4a–g.

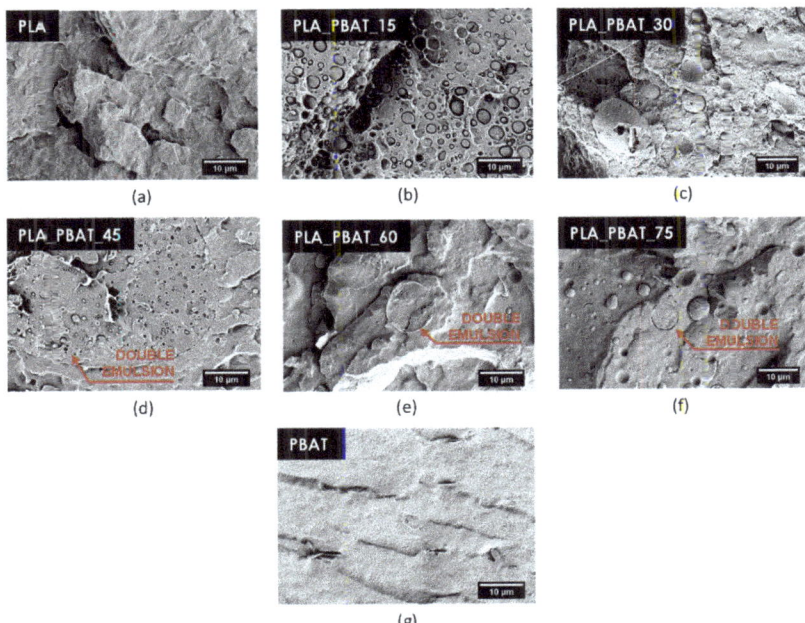

Figure 4. SEM micrographs of the cryofracture surface of neat PLA, neat PBAT and the prepared blends: (**a**) PLA; (**b**) PLA_PBAT_15; (**c**) PLA_PBAT_30; (**d**) PLA_PBAT_45; (**e**) PLA_PBAT_60; (**f**) PLA_PBAT_75; and (**g**) PBAT.

Neat PLA and PBAT (Figure 4a,g) present a different cryofracture surface appearance, which helps in identifying the two components in the blends (Figure 4b–f). In particular, the cryofracture surface of PLA appears rough (Figure 4a), while that of PBAT is smoother (Figure 4g). The micrographs of the blends reported in Figure 4b–f show that the interfacial adhesion between PLA and PBAT is rather poor, since voids are clearly visible at the interface. The neat separation between the two phases and the presence of detachment phenomena, mostly visible in the PLA_PBAT_15 and PLA_PBAT_45 samples (Figure 4b,d), are evidence of the immiscibility and the poor compatibility of these two polymers. In future works, the interfacial adhesion needs to be improved by using suitable compatibilizers.

The SEM images reveal the presence of very small domains in the order of 1–5 μm (of PLA in the PLA_PBAT_15 and PLA_PBAT_30 samples, and of PBAT in the PLA_PBAT_60 and PLA_PBAT_75 samples) dispersed in the major phase, which has not been possible to observe in the optical micrographs. This confirms the broad domain size distribution already detected in Figure 3a,b,d,e. Furthermore, as indicated by the red arrows in Figure 4d–g, double-emulsion structures can be detected at this magnification level.

3.3. Thermal Properties

Thermogravimetric tests were performed in order to investigate the thermal degradation behavior of the prepared blends and to explore the influence of blend composition on their degradation resistance. The thermogravimetric curves of the prepared samples along with the corresponding derivative curves are presented in Figure 5a,b, while the most significant results are reported in Table 2.

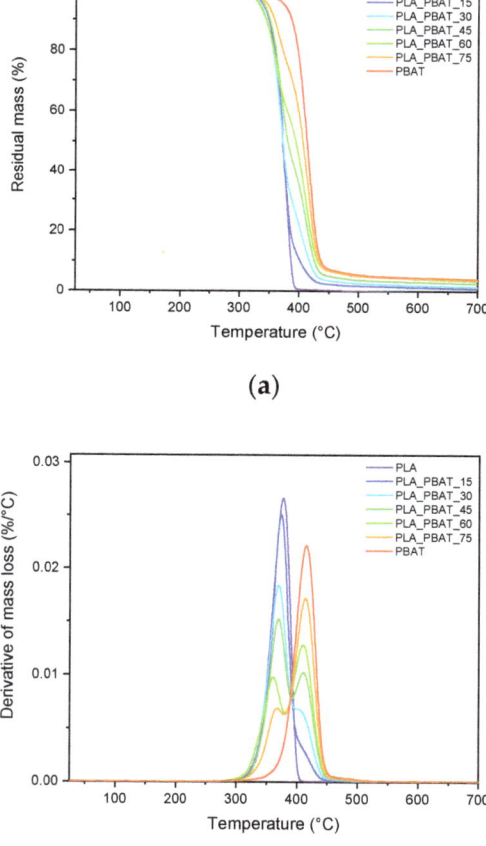

Figure 5. TGA thermograms of neat PLA, neat PBAT and their blends. Trends of (**a**) residual mass and (**b**) mass loss derivative as a function of temperature.

Table 2. Results of the TGA tests on neat PLA, neat PBAT and their blends.

Sample	$T_{5\%}$ (°C)	T_{peak1} (°C)	T_{peak2} (°C)	m_{700} (%)
PLA	339.7	376.1	-	0.0
PLA_PBAT_15	341.2	372.6	-	0.9
PLA_PBAT_30	341.8	368.8	406.0	1.5
PLA_PBAT_45	343.7	368.5	409.9	2.6
PLA_PBAT_60	345.3	360.0	409.4	3.4
PLA_PBAT_75	346.5	366.4	412.6	3.8
PBAT	374.3	-	414.4	4.2

$T_{5\%}$ = temperature corresponding to a mass loss of 5%; T_{peak1} = temperature of the maximum degradation rate of PLA; T_{peak2} = temperature of the maximum degradation rate of PBAT; m_{700} = mass residue at 700 °C.

As it is possible to observe from Figure 5a,b, the thermal degradation of neat polymers occurs in one single step. PLA and PBAT start degrading at temperatures close to 340 °C and 370 °C, respectively, and show maximum degradation rates at 376 and 414 °C, as reported in Table 2. The higher degradation resistance of PBAT is probably attributable to the presence of benzene rings in its molecular structure [56]. A plateau in the TGA curves above 400 °C for PLA and above 450 °C for PBAT can be noticed, highlighting that no significant mass loss in the nitrogen atmosphere occurs at higher temperatures. As for the residue at 700 °C (m_{700}), PLA completely degrades without leaving any solid residue, while PBAT presents an m_{700} value of 4.2%. This indicates that partial coking and carbonization occurs for PBAT. Consequently, it is possible to observe that the residual mass increases proportionally to the PBAT amount in the samples.

The decomposition process of all the PLA/PBAT blends takes place in two different steps, associated with the thermal degradation of the PLA and PBAT phases. This is further evidence of the immiscibility of these polymers. With the increasing PBAT concentration, the thermal stability of the blends slightly increases. Indeed, as can be seen from Table 2, the temperature corresponding to a 5% weight loss ($T_{5\%}$) shifts to higher temperatures, moving from the PLA_PBAT_15 to PLA_PBAT_75 sample.

In order to investigate the thermal properties of the PLA/PBAT blends, a DSC analysis has been carried out. The DSC thermograms collected in the first and second heating scans of the neat PLA, neat PBAT and PLA/PBAT blends are shown in Figure 6a,b, while the most significant results are reported in Table 3. The DSC tests collected in the cooling scan did not provide noteworthy results and were therefore not reported for the sake of brevity.

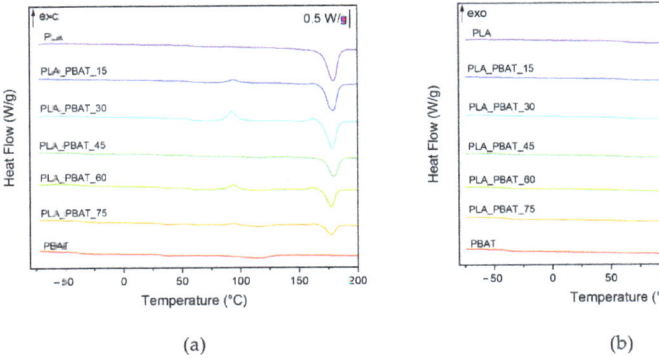

Figure 6. DSC thermograms of neat PLA, neat PBAT and PLA/PBAT blends. (a) First heating scan, (b) second heating scan.

Table 3. Results of the DSC tests on neat PLA, neat PBAT and PLA/PBAT blends (first and second heating scans).

Sample	T_{gPBAT} (°C)	T_{gPLA} (°C)	T_{ccPLA} (°C)	ΔH_{ccPLA} (J/g)	T_{mPBAT} (°C)	ΔH_{mPBAT} (J/g)	T_{mPLA} (°C)	ΔH_{mPLA} (J/g)	χ_{PLA} (%)	χ_{PBAT} (%)
1st heating										
PLA	-	57.6	-	-	-	-	178.0	58.3	62.2	-
PLA_PBAT_15		58.9	93.3	20.3				48.6	35.8	
PLA_PBAT_30	−36.9	56.5	93.0	15.2			177.1	39.8	37.7	
PLA_PBAT_45	−39.1	57.3	-	-	114.1	2.7	178.5	32.1	62.7	6.3
PLA_PBAT_60	−39.6	56.5	93.8	6.9	115.2	3.3	176.5	22.2	41.2	4.9
PLA_PBAT_75	−39.1	58.5	93.7	1.8	114.7	5.9	176.1	13.0	47.9	6.9
PBAT	−39.7	-	-	-	113.7	11.9	-	-	-	10.4
2nd heating										
PLA	-	62.4	-	-	-	-	179.1	54.5	58.6	-
PLA_PBAT_15	−44.1	61.4	-	-			178.0	43.0	54.4	
PLA_PBAT_30	−41.9	61.2	-	-			177.3	38.6	59.3	
PLA_PBAT_45	−39.5	59.2	-	-	114.0	3.2	177.6	31.4	61.5	6.3
PLA_PBAT_60	−39.9	59.0	-	-	114.0	4.7	177.4	22.1	59.4	6.9
PLA_PBAT_75	−38.5	59.1	-	-	113.4	6.3	176.5	14.0	56.6	7.5
PBAT	−39.0	-	-	-	114.2	11.2	-	-	-	9.8

T_{gPBAT} = glass transition temperature of PBAT; T_{gPLA} = glass transition temperature of PLA; T_{ccPLA} = cold crystallization temperature of PLA; T_{mPBAT}: melting temperature of PBAT; ΔH_{mPBAT} = melting enthalpy of PBAT; T_{mPLA}: melting temperature of PLA; ΔH_{mPLA} = melting enthalpy of PLA; χ_{PLA} = crystallinity degree of PLA; χ_{PBAT} = crystallinity degree of PBAT.

By analyzing the results reported in Table 3 and in the DSC thermograms in Figure 6a,b, it is possible to observe that the DSC traces of neat PLA and PBAT present the typical profile of semi-crystalline materials. In the first heating scan, the melting point of neat PLA is 178.0 °C, while the glass transition temperature is 57.6 °C. The crystallinity degree of neat PLA is very high (62.1%) and no cold crystallization peaks are observable, probably because the preliminary drying promoted the crystallization of the material. As it is possible to observe from Figure 6a, the endothermic peak of PLA is preceded by a small exothermic peak. According to the literature, this signal corresponds to the reorganization of the α'-crystal of PLA into α-crystals [57]. The α-phase is an ordered structure produced at a high crystallization temperature, while the α'-phase is a metastable disordered phase that develops at low temperatures. Above 150 °C, the α'-crystals become unstable and the phase transformation into α-crystals takes place. In regards to PBAT, it shows a T_g at −39.7 °C and a weak and broad endothermic melting peak at 113.7 °C, indicating its low crystallization rate and limited crystallinity degree (10.4%).

Focusing on the first heating DSC scan of the PLA/PBAT blends, it is interesting to observe that the T_m and T_g values of the PLA and PBAT phases in the blends correspond to those detected for neat polymers, and are thus unaltered by the addition of another phase in the blends. This experimental evidence, together with the well-distinct melting phenomena of the two constituents, confirms that the PLA and PBAT phases are immiscible. In the PLA_PBAT_15 sample, the weak signals associated with the glass transition and melting of the PBAT phase do not permit the determination of the T_g and T_m for this polymer. In all the blends except for PLA_PBAT_45, an exothermic peak at a temperature close to 93 °C is noted, attributable to the cold-crystallization of PLA macromolecules, which, after the glass transition, acquire sufficient energy to rearrange and crystallize. The crystallinity degree of PLA is significantly reduced in the blends, especially in the PLA_PBAT_15 and PLA_PBAT_30 samples. This suggests that the PLA crystallization process of PLA is hindered by the presence of a second phase in the blend. On the other hand, the peculiar morphology of the PLA_PBAT_45 blend, characterized by the presence of large PBAT domains, has probably favored the crystallization of PLA. Indeed, the crystallization degree of this sample is 62.7%, comparable of that of neat PLA. The crystallization rate of PBAT is also impaired by the presence of PLA in the blends, since the χ_{PBAT} in the blends is lower than that of the neat PBAT.

In the second heating scan, the cold crystallization peak of PLA disappears. This confirms that water cooling after hot pressing partially hindered the crystallization of PLA within the blends. In comparison to the first heating stage, the crystallization degree of both PLA and PBAT is higher in almost all the samples, close to the values of the neat constituents. Finally, no significant differences in the T_m and T_g of PLA and PBAT determined by the first and second heating scan can be detected.

3.4. Mechanical Properties

This work, as explained above, aims to evaluate the shape memory performance of PLA/PBAT blends using the glass transition of PLA as the switching temperature. In order to identify the temperature range in which the shape memory behavior of the prepared materials should be evaluated, a DMA analysis was carried out. The trends of the storage modulus and loss modulus as a function of temperature are presented in Figure 7a,b, while the results of the DMA tests on the prepared samples are reported in Table 4.

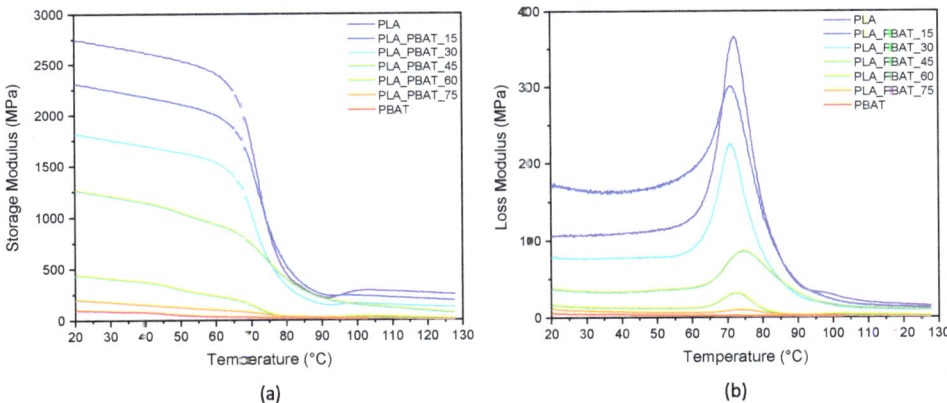

Figure 7. Storage modulus (a) and loss modulus (b) of neat PLA, neat PBAT and PLA/PBAT blends.

Table 4. Results of the DMA tests on the prepared blends.

Sample	Storage Modulus at 25 °C (MPa)	Storage Modulus at 85 °C (MPa)
PLA	2725.0	278.2
PLA_PBAT_15	2279.3	336.5
PLA_PBAT_30	1775.0	208.4
PLA_PBAT_45	1221.6	276.9
PLA_PBAT_60	420.2	19.5
PLA_PBAT_75	188.2	24.5
PBAT	87.9	15.3

As is visible from Figure 7a,b, the glass transition temperature of PLA covers a temperature range between 55 °C and 85 °C.. The intensity of the loss modulus peak, associated with this transition, progressively decreases and broadens with the increasing PBAT concentration in the blends, while the corresponding peak remains for all the samples close to a temperature of 70–80 °C. The storage modulus of the blends decreases significantly in the range of 65–80 °C and the drop becomes more pronounced with the increasing PLA content in the samples. Above 90 °C, the cold crystallization of PLA occurs and the modulus rises because of the presence of newly formed crystals. This is particularly evident for the PLA, PLA_PBAT_15, PLA_PBAT_30 and PLA_PBAT_60 samples. It is important to underline that the DSC thermograms of the neat PLA do not reveal any cold crystallization peaks. It can be hypothesized that the lower heating rate used and the orientation of the PLA macromolecules in the stress direction obtained in the DMA tests have probably promoted the cold crystallization of the neat PLA.

Since at 25 °C and at 85 °C the PLA phases in the blends are, respectively, in the glassy and rubbery states, these temperatures were selected for exploring the shape memory performance of the samples. More specifically, 85 °C was chosen as the temperature at which the samples were deformed to the temporary shape. On the other hand, 25 °C was chosen as the temperature at which the deformed samples were cooled down to fix the temporary shape. To avoid any interference with the cold crystallization phenomenon of PLA, temperatures higher than 85 °C were not considered to explore the shape memory behavior. The storage moduli of the samples at 25 °C and at 85 °C are reported in Table 4. By passing from 25 to 85 °C, all the blends experience a storage modulus drop of about one order of magnitude. However, it is possible to notice that the higher the PLA content in the sample, the higher the E' difference at the two considered temperatures. Since the greater the drop in the storage modulus, the better the shape memory performances of the material, higher shape memory effects are expected in blends with low PBAT contents.

Quasi-static tensile tests were carried out in order to investigate the influence of the PLA/PBAT relative ratio on the mechanical properties of the blends. The representative stress–strain curves of the neat PLA, neat PBAT and of PLA/PBAT blends are shown in Figure 8, while the main results are reported in Table 5.

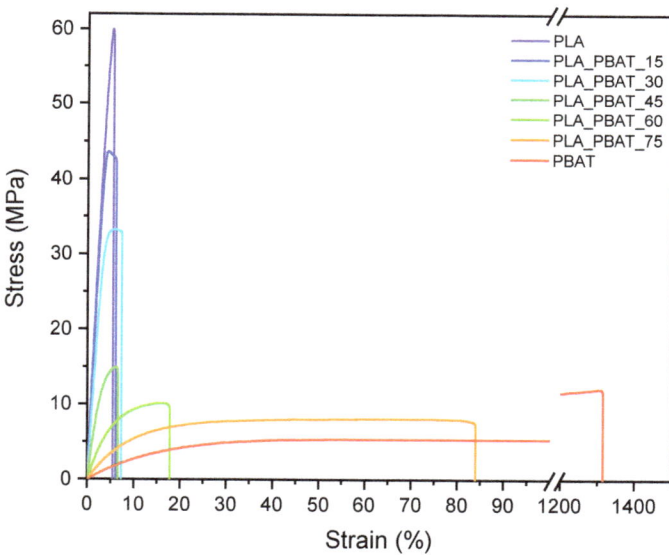

Figure 8. Representative stress–strain curves of neat PLA, neat PBAT and PLA/PBAT blends.

Neat PLA is characterized by an average tensile strength of 59.2 MPa, a Young's modulus of 4119.4 MPa and a strain at break of 5.8%. On the other hand, neat PBAT presents a much lower tensile strength (10.7 MPa) and elastic modulus (87.7 MPa), but a significantly greater strain at break (1237.8%). As can be seen from Table 5, the addition of

PBAT to PLA leads to a general decrease in stiffness and strength, while the elongation at break remains almost comparable to that of PLA up to a PBAT concentration of 45 wt%. This means that as long as PLA is the matrix constituent of the blend, negligible toughening effects can be obtained, probably due to the low interfacial adhesion between the two phases and the intrinsic, relative brittleness of PLA. The limited interfacial adhesion also explains the considerable σ_b drop with limited PBAT contents. A significant increase in blend ductility is noticed for PBAT contents higher than 60 wt%, i.e., when phase inversion has occurred and PBAT becomes the major constituent. In particular, PLA_PBAT_60 and PLA_PBAT_75 are characterized by a strain at break 3.5 and 14.4 times higher than that of neat PLA, respectively. However, it is interesting to note that, even though PBAT constitutes the matrix of these blends, ε_b is decisively lower than that of neat PBAT. The presence of large PLA domains and, once again, the low adhesion between the two phases are probably responsible for this result. Regarding the elastic modulus and tensile strength, the prepared blends exhibit high values of E and σ_b up to a PBAT content of 30 wt%, while experiencing a drop for higher PBAT concentrations. Once more, this can be explained by recalling the blend morphology. As it was observed in Figure 3a,b, in the PLA_PBAT_15 and PLA_PBAT_30 samples, the PBAT phase is evenly dispersed in the PLA matrix, and the dimensions of the domains are smaller compared to those of the other samples. Moving to the PLA_PBAT_30 to PLA_PBAT_45 sample, the morphology completely changes. Large and irregularly shaped PBAT domains appear, and the interfacial area increases, causing a reduction in blend stiffness and strength. In the future, it will be interesting to enhance the interfacial adhesion between the two phases by using suitable compatibilizers. In conclusion, it is possible to say that when PBAT is added to the PLA matrix as a dispersed component, the tensile strength and elastic modulus decrease but the elongation at break remains the same. In contrast, when PBAT is the major constituent, i.e., PLA is the dispersed component, the stiffness and strength increase with the increasing PLA content, but the elongation at break is significantly reduced.

Table 5. Result of quasi-static tensile test at 25 °C on the prepared samples.

Sample	E (MPa)	σ_b (MPa)	ε_b (%)
PLA	4119.4 ± 673.3	59.2 ± 2.4	5.8 ± 0.7
PLA_PBAT_15	3030.9 ± 219.2	42.2 ± 0.7	6.0 ± 0.3
PLA_PBAT_30	2295.4 ± 66.2	32.3 ± 1.1	7.9 ± 1.3
PLA_PBAT_45	1694.3 ± 78.2	11.8 ± 3.4	5.5 ± 1.4
PLA_PBAT_60	221.9 ± 17.9	9.2 ± 0.3	20.3 ± 2.9
PLA_PBAT_75	87.7 ± 17.9	7.7 ± 0.6	83.8 ± 12.2
PBAT	87.7 ± 2.0	10.7 ± 1.2	1237.8 ± 136.1

E = elastic modulus; σ_b = stress at break; ε_b = strain at break.

The elastic modulus of the blends was theoretically predicted by using three models, i.e., the Series, Parallel and Equivalent Box models. More specifically, the Series and Parallel models were used to predict the upper and lower boundaries of blend stiffness [58], according to the expressions reported in Equations (4) and (5), respectively.

$$E_b = E_{PLA} \cdot V_{PLA} + E_{PBAT} \cdot V_{PBAT} \tag{4}$$

$$E_b = \frac{E_{PLA} \cdot E_{PBAT}}{(V_{PLA} \cdot E_{PBAT} + V_{PBAT} \cdot E_{PLA})} \tag{5}$$

where E_b is the elastic modulus of the blend, E_{PLA} and E_{PBAT} are the elastic moduli of neat PLA and PBAT, respectively, while V_{PLA} and V_{PBAT} are the volume fractions of PLA and PBAT in the samples. The Parallel model assumes that the continuous phase consists of the higher modulus polymer and therefore represents the upper boundary. On the other hand, the lower boundary is represented by the Series model, which assumes that the lower modulus component is the continuous phase. The Equivalent Box model (EBM) combined with the phase continuity percolation approach was also used to predict the modulus of these polymer blends [59]. The EBM operates with partly parallel (subscript p) and partly series (subscript s) couplings of two components. It is a two-parameter model in that of the four volume fractions, only two are independent variables. The volume fractions of the constituents are correlated as reported in Equations (6) and (7):

$$V_{PLA} = (V_{PLAp} + V_{PLAs}) \qquad (6)$$

$$V_{PBAT} = (V_{PBATp} + V_{PBATs}) \qquad (7)$$

where $V_{PLA} + V_{PBAT} = V_{PLAs} + V_{PLAp} + V_{PBATs} + V_{PBATp} = 1$. The tensile modulus of the blend (E_b) is given as reported in Equation (8):

$$E_b = E_p \cdot V_p + E_s \cdot V_s = E_{PLA} \cdot V_{PLAp} + E_{PBAT} \cdot V_{PBATp} + \frac{V_s^2}{\frac{V_{PLAs}}{E_{PLA}} + \frac{V_{PBATs}}{E_{PBAT}}} \qquad (8)$$

where E_p and E_s are, respectively, the elastic moduli of the parallel and series branch of the system, V_p is the sum of V_{PLAp} and V_{PBATp}, and V_s is the sum of V_{PLAs} and V_{PBATs}. Using the universal formula provided by percolation theory for the elastic modulus of binary systems, Kolarik and coworkers derived equations for the determination of the volume fractions in parallel, as reported in Equations (9) and (10):

$$V_{PLAp} = \left(\frac{V_{PLA} - V_{PLAcr}}{1 - V_{PLAcr}}\right)^T \qquad (9)$$

$$V_{PBATp} = \left(\frac{V_{PBAT} - V_{PBATcr}}{1 - V_{PBATcr}}\right)^T \qquad (10)$$

where V_{PLAcr} and V_{PBATcr} are the critical volume fractions at which the component PLA and PBAT become partially continuous and T is the critical exponent. The best fit of the experimental results was obtained with $V_{PLAcr} = V_{PBATcr} = 0.1$ and $T = 1.6$. For further details about the EBM, the reader can refer to the work of Kolarik et al. [59].

The experimental and theoretical values of the elastic modulus of the prepared blends as a function of the volume fraction of PBAT are shown in Figure 9.

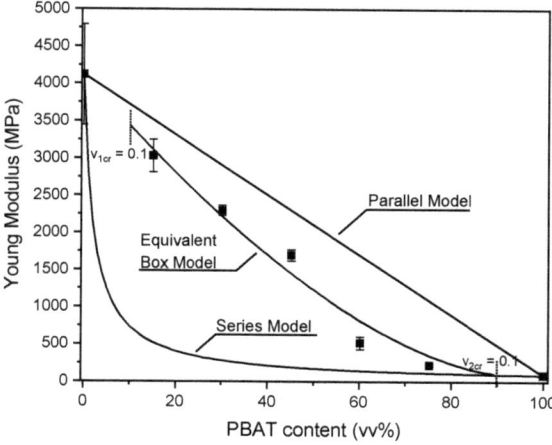

Figure 9. Experimental and theoretical values of the elastic modulus of the produced blends as a function of the PBAT content.

All the elastic modulus values fall into the range between the Parallel and Series models, suggesting that PLA and PBAT are partially compatible, even though they are not miscible. It is interesting to notice that the experimental results are in good agreement with the theoretical predictions based on the Equivalent Box model. Finally, the elastic modulus of the PLA_PBAT_75 sample predicted by the Series model practically coincides with the experimental value. This indicates that for this composition, PBAT is the continuous phase with PLA dispersed within it.

3.5. Evaluation of the Shape Memory Behavior

Testing the shape memory properties of polymers using a DMA apparatus in single cantilever mode is quite useful when one needs to compare the shape memory performance of materials showing different elongation at break values. Table 6 shows the strain fixity and strain recovery parameters of the neat PLA, neat PBAT and PLA/PBAT blends.

Table 6. Strain fixity and strain recovery parameters of neat PLA, neat PBAT and PLA/PBAT blends.

Sample	SF (%)	SR (%)
PLA	83.8	56.9
PLA_PBAT_15	81.9	63.5
PLA_PBAT_30	77.3	73.6
PLA_PBAT_45	73.1	64.3
PLA_PBAT_60	80.2	54.7
PLA_PBAT_75	91.4	14.3
PBAT	91.3	8.7

As expected, neat PLA is characterized by a good strain fixity capability, reaching an SF value of 83.8%. It is worth noting that the strain recovery of neat PLA is quite limited, considering the values reported in the literature [10,11,14,15,60–66]. Only slightly more than half of the imposed deformation (56.7%) is recovered during the recovery step. This experimental result can probably be attributed to its high crystallinity degree. However, to confirm this hypothesis, future works will need to focus on varying the processing conditions to obtain a PLA with lower crystallinity and investigating its shape memory properties. In regards to neat PBAT, an excellent strain fixability (SF value of 91.3%) and negligible recovery capabilities (SR value of 8.7%) have been found. Due to the absence of a transition temperature in the explored temperature range, neat PBAT cannot exhibit a shape memory behavior.

With the addition of PBAT, the shape memory performance of PLA undergoes a particular trend. As can be seen from Table 6, while the fixing capability tends to slightly decrease up to a PBAT concentration of 45% and then to increase again for higher PBAT amounts, the strain recovery parameter exhibits almost the opposite trend. It increases up to a PBAT content of 30 wt% and then decreases. The higher SR parameters of PLA_PBAT_15 and PLA_PBAT_30 compared to neat PLA could be probably related to the lower crystallinity degree of PLA in these blends, as evidenced by the DSC analyses. The samples constituted by a PBAT matrix (PLA_PBAT_60 and PLA_PBAT_75) fix the deformation well but show inferior recovery capabilities. In particular, in the PLA_PBAT_75 blend, the SR is almost complementary to the SF parameter, indicating that just a very small deformation is recovered during the recovery step. The higher fixability and the lower recovery capabilities of these blends are probably attributable to two aspects, as illustrated in Figure 10a,b. The first is related to the fact that PBAT is a flexible thermoplastic polymer characterized by very low crystallinity. The crystalline domains should act as net points, avoiding the irreversible sliding of the macromolecules when the polymer is mechanically stressed. However, since the crystallinity degree of PBAT is very low, part of the polymer chains slides irreversibly and, when the stress is released, a great deal of the deformation imposed is maintained.

The second aspect is related to the poor adhesion between the PLA and PBAT phases, which probably causes a non-optimal stress transfer between the two constituents. PLA provides at the same time both the transition temperature that guarantees the fixability of the temporary shape and the retractive forces that allow the material to recover the original permanent shape. In more detail, the net points, or the crystalline domains, determine the permanent shape of the polymer. The glassy area, instead, acts as a molecular switch and is responsible for the fixability of the temporary shape. Consequently, the PLA domains should act as a spring in the PBAT matrix and guarantee recovery to the initial condition. However, the poor adhesion between the phases does not allow a good transfer of these retractive forces from the PLA domains to the PBAT matrix and, as a result, the blend exhibits poor recovery capabilities (Figure 10b). To confirm this hypothesis, it will be useful in future works to perform the same shape memory tests on compatibilized blends. Furthermore, the difference in SR parameters among PLA_PBAT_60 and PLA_PBAT_75 (54.7% and 14.3%, respectively) is probably due to the morphologies these blends show. Both PLA_PBAT_60 and PLA_PBAT_75 are high-PBAT-content blends; however, while PLA_PBAT_60 presents a co-continuous structure, PLA_PBAT_75 exhibits a sea-island morphology (Figure 3d,e). The elongated and interconnected PLA domains in PLA_PBAT_60 probably promote a higher recovery capability with respect to that of PLA_PBAT_75. It has already been mentioned that only PLA presents a shape memory behavior, and only PLA can provide the retractive forces to allow the blend to return to the original conditions. Consequently, the presence of larger PLA domains in the blend is beneficial from the point of view of the recovery performances.

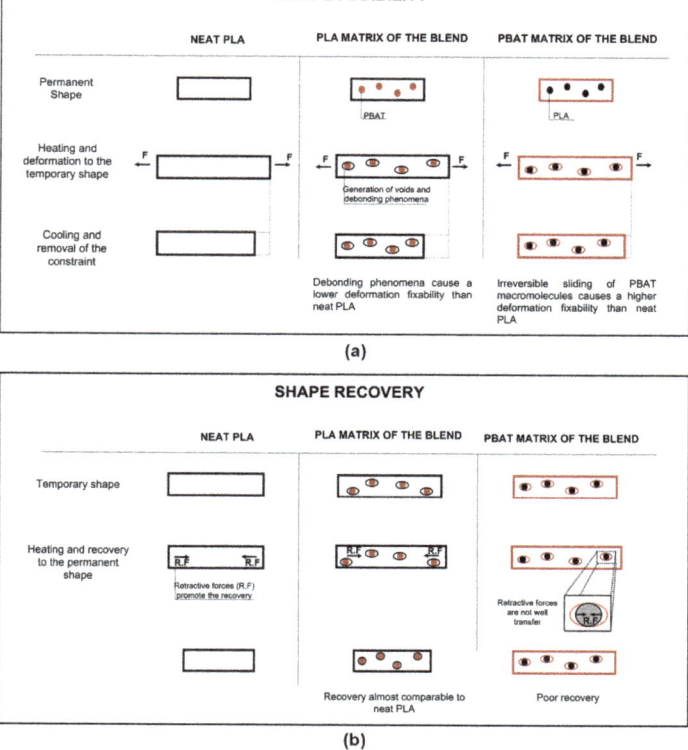

Figure 10. Schematization of the proposed mechanism for the shape memory behavior of the produced blends. Model of (**a**) strain fixity and (**b**) strain recovery behavior.

The best compromise between shape fixability and recovery capability was obtained in the PLA_PBAT_30 sample, for which SF and SR values of 77.3% and 73.6% were obtained, respectively. However, it should be noticed that the compositions exhibiting the more interesting shape memory properties do not coincide with the blends showing a significant improvement in PLA ductility. Once again, the addition of a proper compatibilizer could give a satisfactory answer to this issue.

4. Conclusions

The present work investigated the microstructural, thermomechanical and shape memory properties of biodegradable PLA/PBAT blends produced by melt compounding at different relative amounts. Rheological measurements, optical microscopy and SEM micrographs evidenced the immiscibility and the low interfacial adhesion between the PLA and PBAT phases. Depending on the PLA/PBAT ratio, the two constituents exhibited either a sea-island morphology or an almost co-continuous structure, and it was shown that phase inversion occurred in a PBAT concentration interval between 45 wt% and 60 wt%. The immiscibility of these blends was also confirmed by a DSC analysis, as the glass transition temperatures of PLA and PBAT corresponded to the values of neat polymers and were unaffected by the blend composition. The TGA analysis revealed that the addition of PBAT slightly improved the thermal stability of PLA, probably because of the presence of rigid benzene rings in the molecular structure of PBAT.

Tensile tests evidenced that the stiffness and strength of the blends decreased with the increasing PBAT concentration, while the elongation at break remained comparable to that of neat PLA (5.8%) up to a PBAT content of 45 wt%. A significant increment in blend ductility was registered only for higher PBAT concentrations (elongation at break up to 83.8% with a PBAT amount of 75 wt%).

The shape memory performance of PLA was impaired by the addition of PBAT. In particular, a reduction in the strain recovery parameter was registered with the increasing PBAT content, probably due to the low adhesion between the constituents. The best compromise between shape fixability and recovery capability was obtained with a PBAT content of 30 wt% and SF and SR values of 77.3% and 73.6%, respectively.

These results constitute a basis for future research on innovative multifunctional biodegradable polymer blends with promising applications in the packaging field and suggest that their properties could be enhanced by applying suitable compatibilizers. Future works should focus on the selection of the most appropriate compatibilizer, the effect of blend compatibilization on the physical and shape memory properties of these systems and a deep investigation of the gas permeability of the blends.

Supplementary Materials: The following supporting information can be downloaded at: https://www.mdpi.com/article/10.3390/polym15040881/s1, Figure S1: Thermo-mechanical cycle of (a) PLA_PBAT_15, (b) PLA_PBAT_30, (c) PLA_PBAT_45, (d) PLA_PBAT_60, (e) PLA_PBAT_75.

Author Contributions: Conceptualization, A.D. and A.P.; Data curation, M.B.; Funding acquisition, A.D. and A.P.; Investigation, M.B.; Methodology, M.B.; Project administration, A.P.; Visualization, M.B.; Writing—original draft, M.B. and M.M.; Writing—review and editing, A.D., A.P. and M.M. All authors have read and agreed to the published version of the manuscript.

Funding: This research received no external funding.

Institutional Review Board Statement: Not applicable.

Data Availability Statement: Data available on request.

Conflicts of Interest: The authors declare no conflict of interest.

References

1. Shen, M.; Song, B.; Zeng, G.; Zhang, Y.; Huang, W.; Wen, X.; Tang, W. Are biodegradable plastics a promising solution to solve the global plastic pollution? *Environ. Pollut.* **2020**, *263*, 114469. [CrossRef]
2. Nanda, S.; Patra, B.R.; Patel, R.; Bakos, J.; Dalai, A.K. Innovations in applications and prospects of bioplastics and biopolymers: A review.Environ. *Chem. Lett.* **2022**, *20*, 379–395. [CrossRef]
3. Luckachan, G.E.; Pillai, C.K.S. Biodegradable Polymers- A Review on Recent Trends and Emerging Perspectives. *J. Polym. Environ.* **2011**, *19*, 637–676. [CrossRef]
4. Anna, F. Plastic Waste: Here's What It Could Look Like by 2060. Statista. Available online: https://www.weforum.org/agenda/2022/07/recycling-efforts-not-enough-to-solve-plastic-waste-problem/ (accessed on 29 September 2022).
5. Armentano, I.; Bitinis, N.; Fortunati, E.; Mattioli, S.; Rescignano, N.; Verdejo, R.; Lopez-Manchado, M.A.; Kenny, J.M. Multifunctional nanostructured PLA materials for packaging and tissue engineering. *Prog. Polym. Sci.* **2013**, *38*, 1720–1747. [CrossRef]
6. Pang, X.; Zhuang, X.; Tang, Z.; Chen, X. Polylactic acid (PLA): Research, development and industrialization. *Biotechnol. J.* **2010**, *5*, 1125–1136. [CrossRef] [PubMed]
7. Fredi, G.; Dorigato, A.; Dussin, A.; Xanthopoulou, E.; Bikiaris, D.N.; Botta, L. Compatibilization of Polylactide/Poly(ethylene 2,5-furanoate) (PLA/PEF) Blends for Sustainable and Bioderived Packaging. *Molecules* **2022**, *27*, 6371. [CrossRef]
8. Rigotti, D.; Soccio, M.; Dorigato, A.; Gazzano, M.; Siracusa, V.; Fredi, G.; Lotti, N. Novel Biobased Polylactic Acid/Poly(pentamethylene 2,5-furanoate) Blends for Sustainable Food Packaging. *ACS Sustain. Chem. Eng.* **2021**, *9*, 13742–13750. [CrossRef]
9. Bax, B.; Müssig, J. Impact and tensile properties of PLA/Cordenka and PLA/flax composites. *Compos. Sci. Technol.* **2008**, *68*, 1601–1607. [CrossRef]
10. Leonés, A.; Sonseca, A.; López, D.; Fiori, S.; Peponi, L. Shape memory effect on electrospun PLA-based fibers tailoring their thermal response. *Eur. Polym. J.* **2019**, *117*, 217–226. [CrossRef]
11. Peponi, L.; Sessini, V.; Arrieta, M.P.; Navarro-baena, I.; Sonseca, A.; Dominici, F.; Gimenez, E.; Torre, L.; Tercjak, A.; Lopez, D.; et al. Thermally-activated shape memory effect on biodegradable nanocomposites based on PLA/PCL blend reinforced with hydroxyapatite. *Polym. Degrad. Stab.* **2018**, *151*, 36–51. [CrossRef]
12. Arrieta, M.P.; Sessini, V.; Peponi, L. Biodegradable poly (ester-urethane) incorporated with catechin with shape memory and antioxidant activity for food packaging. *Eur. Polym. J.* **2017**, *94*, 111–124. [CrossRef]
13. Mehrpouya, M.; Vahabi, H.; Janbaz, S.; Darafsheh, A.; Mazur, T.R.; Ramakrishna, S. 4D printing of shape memory polylactic acid (PLA). *Polymer* **2021**, *230*, 124080. [CrossRef]
14. Zhao, W.; Huang, Z.; Liu, L.; Wang, W.; Leng, J.; Liu, Y. Porous bone tissue scaffold concept based on shape memory PLA/Fe$_3$O$_4$. *Compos. Sci. Technol.* **2021**, *203*, 108563. [CrossRef]
15. Ehrmann, G.; Ehrmann, A. Investigation of the shape-memory properties of 3D printed PLA structures with different infills. *Polymers* **2021**, *13*, 164. [CrossRef] [PubMed]
16. Liu, C.; Qin, H.; Mather, P.T. Review of progress in shape-memory polymers. *J. Mater. Chem.* **2007**, *17*, 1543–1558. [CrossRef]
17. Pandini, S.; Baldi, F.; Paderni, K.; Messori, M.; Toselli, M.; Pilati, F.; Gianoncelli, A.; Brisotto, M.; Bontempi, E.; Riccò, T. One-way and two-way shape memory behaviour of semi-crystalline networks based on sol-gel cross-linked poly(ε-caprolactone). *Polymer* **2013**, *54*, 4253–4265. [CrossRef]
18. Dorigato, A.; Pegoretti, A. Shape memory epoxy nanocomposites with carbonaceous fillers and in-situ generated silver nanoparticles. *Polym. Eng. Sci.* **2019**, *59*, 694–703. [CrossRef]
19. Dorigato, A.; Pegoretti, A. Evaluation of the shape memory behavior of a poly(cyclooctene) based nanocomposite device. *Polym. Eng. Sci.* **2018**, *58*, 430–437. [CrossRef]
20. Ludwiczak, J.; Frąckowiak, S.; Leluk, K. Study of thermal, mechanical and barrier properties of biodegradable pla/pbat films with highly oriented mmt. *Materials* **2021**, *14*, 7189. [CrossRef]
21. Arrieta, M.P.; Samper, M.D.; Aldas, M.; Lopez, J. On the Use of PLA-PHB Blends for Sustainable Food Packaging Applications. *Materials* **2017**, *10*, 1008. [CrossRef]
22. Al-Itry, R.; Lamnawar, K.; Maazouz, A. Rheological, morphological, and interfacial properties of compatibilized PLA/PBAT blends. *Rheol. Acta* **2014**, *53*, 501–517. [CrossRef]
23. Ferreira, F.V.; Cividanes, L.S.; Gouveia, R.F.; Lona, L.M.F. An overview on properties and applications of poly(butylene adipate-co-terephthalate)–PBAT based composites. *Polym. Eng. Sci.* **2019**, *59*, 7–15. [CrossRef]
24. Fukushima, K.; Wu, M.; Bocchini, S.; Rasyida, A.; Yang, M. PBAT based nanocomposites for medical and industrial applications. *Mat. Sci. Eng.* **2021**, *32*, 1331–1351. [CrossRef]
25. Jian, J.; Xiangbin, Z.; Xianbo, H. An overview on synthesis, properties and applications of poly(butylene-adipate-co-terephthalate)–PBAT. *Adv. Ind. Eng. Polym. Res.* **2020**, *3*, 19–26. [CrossRef]
26. Paulsen, E.; Lema, P.; Martinèz-Romero, D.; Garcia-Viguera, C. Use of PLA/PBAT stretch-cling film as an ecofriendly alternative for individual wrapping of broccoli heads. *Sci. Hortic.* **2022**, *304*, 111260. [CrossRef]
27. Wang, L.; Xu, J.; Zhang, M.; Zheng, H.; Li, L. Preservation of soy protein-based meat analogues by using PLA/PBAT antimicrobial packaging film. *Food Chem.* **2022**, *380*, 132022. [CrossRef] [PubMed]
28. Qiu, S.; Zhou, Y.; Waterhouse, G.I.N.; Gong, R.; Xie, J.; Zhang, K.; Xu, J. Optimizing interfacial adhesion in PBAT/PLA nanocomposite for biodegradable packaging films. *Food Chem.* **2021**, *334*, 127487. [CrossRef]

29. Steen, K.V.; Lange, C. PBAT: A comprehensive software package for genome-wide association analysis of complex family-based studies. *Hum. Genom.* **2005**, *2*, 67–69. [CrossRef]
30. Weng, Y.; Jin, Y.; Meng, Q.; Wang, L.; Zhang, M.; Wang, Y. Biodegradation behavior of poly(butylene adipate-co-terephthalate) (PBAT), poly(lactic acid) (PLA), and their blend under soil conditions. *Polym. Test.* **2013**, *32*, 918–926. [CrossRef]
31. Dammak, M.; Fourati, Y.; Tarres, Q.; Delgado-Aguilar, M.; Mutjè, P.; Boufi, S. Blends of PBAT with plasticized starch for packaging applications: Mechanical properties, rheological behaviour and biodegradability. *Ind. Crops Prod.* **2020**, *144*, 112061. [CrossRef]
32. Gu, S.Y.; Zhang, K.; Zhan, H. Melt rheology of polylactide/poly(butylene adipate-co-terephthalate) blends. *Carbohyd. Polym.* **2008**, *74*, 79–85. [CrossRef]
33. Nayak, S.K.; Mohanty, S. Biodegradable Nanocomposites of Poly(butylene adipate-co-terephthalate) (PBAT) and Organically Modified Layered Silicates. *J. Polym. Environ.* **2012**, *20*, 195–207.
34. Zhang, T.; Han, W.; Zhang, C.; Weng, Y. Effect of chain extender and light stabilizer on the weathering resistance of PBAT/PLA blend films prepared by extrusion blowing. *Polym. Degrad. Stab.* **2021**, *183*, 109455. [CrossRef]
35. Su, S.; Duhme, M.; Kopitzky, R. Uncompatibilized pbat/pla blends: Manufacturability, miscibility and properties. *Materials* **2020**, *13*, 4897. [CrossRef] [PubMed]
36. Palsikowski, P.A.; Kuchnier, C.N.; Pinheiro, I.F. Biodegradation in Soil of PLA/PBAT Blends Compatibilized with Chain Extender. *J. Polym. Environ.* **2018**, *26*, 330–341. [CrossRef]
37. Farsetti, S.; Cioni, B.; Lazzeri, A. Physico-mechanical properties of biodegradable rubber toughened polymers. *Macromol. Symp.* **2011**, *301*, 82–89. [CrossRef]
38. Mohammadi, M.; Heuzey, M.C.; Carreau, P.J.; Taguet, A. Morphological and Rheological Properties of PLA, PBAT, and PLA/PBAT Blend Nanocomposites Containing CNCs. *Nanomaterials* **2021**, *11*, 357. [CrossRef]
39. Kalita, H. *Shape Memory Polymers: Theory and Application*; De Gruyter: Berlin, Germany; Boston, MA, USA, 2018.
40. Guo, J.; Wang, Z.; Tong, L.; Lv, H.; Liang, W. Shape memory and thermo-mechanical properties of shape memory polymer/carbon fiber composites. *Composites* **2015**, *76*, 162–171. [CrossRef]
41. Scalet, G.; Pandini, S.; Messori, M.; Toselli, M.; Auricchio, F. A one-dimensional phenomenological model for the two-way shape-memory effect in semi-crystalline networks. *Polymer* **2018**, *158*, 130–148. [CrossRef]
42. Pandini, S.; Inverardi, N.; Scalet, G.; Battini, D.; Bignotti, F.; Marconi, S.; Auricchio, F. Shape memory response and hierarchical motion capabilities of 4D printed auxetic structures. *Mech. Res. Commun.* **2020**, *103*, 103463. [CrossRef]
43. Atli, B.; Gandhi, F.; Karst, G. Thermomechanical Characterization of Shape Memory Polymers. *J. Intell. Mater. Syst. Struct.* **2009**, *20*, 87–95. [CrossRef]
44. Rahmatabadi, D.; Aberoumand, M.; Soltanmohammadi, K.; Soleyman, E.; Ghasemi, I.; Baniassadi, M.; Abrinia, K.; Zolfagharian, A.; Bodaghi, M.; Baghani, M. A New Strategy for Achieving Shape Memory Effects in 4D Printed Two-Layer Composite Structures. *Polymers* **2022**, *14*, 5446. [CrossRef] [PubMed]
45. Yarali, E.; Taheri, A.; Baghani, M. A comprehensive review on thermomechanical constitutive models for shape memory polymers. *J. Intell. Mater. Syst. Struct.* **2020**, *31*, 1243–1283. [CrossRef]
46. Inverardi, N.; Pandini, S.; Gemmo, G.; Toselli, M.; Messori, M.; Scalet, G.; Auricchio, F. Reversible Stress-Driven and Stress-Free Two-Way Shape Memory Effect in a Sol-Gel Crosslinked Polycaprolactone. *Macromol Symp.* **2022**, *405*, 2100254. [CrossRef]
47. Merlettini, A.; Pandini, S.; Agnelli, S.; Gualandi, C.; Paderni, K.; Messori, M.; Toselli, M.; Focarete, M.L. Facile fabrication of shape memory poly(3-caprolactone) non-woven mat by combining electrospinning and sol–gel reaction. *RSC Adv.* **2016**, *6*, 43964. [CrossRef]
48. Liu, Y.; Gall, K.; Dunn, M.L.; McCluskey, P. Thermomechanical recovery couplings of shape memory polymers in flexure. *Smart Mater. Struct.* **2003**, *12*, 947. [CrossRef]
49. Nofar, M.; Salehiyan, R.; Ray, S.S. Rheology of poly (lactic acid)-based systems. *Polym. Rev.* **2019**, *59*, 465–509. [CrossRef]
50. Palierne, J.F. Linear rheology of viscoelastic emulsions with interfacial tension. *Rheol. Acta* **1990**, *29*, 204–214. [CrossRef]
51. Castro, M.; Carrot, C.; Prochazka, F. Experimental and theoretical description of low frequency viscoelastic behaviour in immiscible polymer blends. *Polymer* **2004**, *45*, 4095–4104. [CrossRef]
52. Casamento, F.; D'Anna, A.; Arrigo, R.; Frache, A. Rheological behavior and morphology of poly(lactic acid)/low-density polyethylene blends based on virgin and recycled polymers: Compatibilization with natural surfactants. *J. Appl. Polym. Sci.* **2021**, *138*, 50590. [CrossRef]
53. D'Anna, A.; Arrigo, R.; Frache, A. PLA/PHB Blends: Biocompatibilizer Effects. *Polymers* **2019**, *11*, 1416. [CrossRef] [PubMed]
54. Arias, A.; Sojoudiasli, H.; Heuzey, M.C.; Huneault, M.A.; Wood-Adams, P. Rheological study of crystallization behavior of polylactide and its flax fiber composites. *J. Polym. Res.* **2017**, *24*, 46. [CrossRef]
55. Qiao, Z.; Wang, Z.; Zhang, C.; Yuan, S.; Zhu, Y.; Wang, J. PVAm–PIP/PS composite membrane with high performance for CO_2/N_2 separation. *AIChE J.* **2012**, *59*, 215–228. [CrossRef]
56. Bo, L.; Guan, T.; Wu, G.; Ye, F.; Weng, Y. Biodegradation Behavior of Degradable Mulch with Poly (Butylene Adipate-co-Terephthalate) (PBAT) and Poly (Butylene Succinate) (PBS) in Simulation Marine Environment. *Polymers* **2022**, *14*, 1515.
57. Di Lorenzo, M.L.; Androsch, R. Influence of $α'$-/$α$-crystal polymorphism on properties of poly(L-lactic acid). *Polym. Int.* **2019**, *68*, 320–334. [CrossRef]
58. Ebadi-Dehaghani, H.; Khonakdar, H.A.; Barikani, M.; Jafari, S.H. Experimental and theoretical analyses of mechanical properties of PP/PLA/clay nanocomposites. *Composites* **2015**, *69*, 133–144. [CrossRef]

59. Kolarik, J. Simultaneous Prediction f the Modulus and Yield Strenght of Bianry Polymer Blends.Tensile moduli of co-continuous polymer blends. *Polym. Eng. Sci.* **1996**, *36*, 2518–2524. [CrossRef]
60. Zhang, W.; Zhang, F.; Lan, X.; Leng, J.; Wu, A.S.; Bryson, T.M.; Cotton, C.; Gu, B.; Sun, B.; Chou, T. Shape memory behavior and recovery force of 4D printed textile functional composites. *Compos. Sci. Technol.* **2018**, *160*, 224–230. [CrossRef]
61. Barletta, M.; Gisario, A.; Mehrpouya, M. 4D printing of shape memory polylactic acid (PLA) components: Investigating the role of the operational parameters in fused deposition modelling (FDM). *J. Manuf. Process.* **2021**, *61*, 473–480. [CrossRef]
62. Senatov, F.S.; Niaza, K.V.; Zadorozhnyy, M.Y.; Maksimkin, A.V.; Kaloshkin, S.D.; Estrin, Y.Z. Mechanical properties and shape memory effect of 3D-printed PLA-based porous scaffolds. *J. Mech. Behav. Biomed. Mat.* **2016**, *57*, 139–148. [CrossRef]
63. Jia, H.; Gu, S.; Chang, K. 3D printed self-expandable vascular stents from biodegradable shape memory polymer. *Adv. Polym. Technol.* **2018**, *37*, 3222–3228. [CrossRef]
64. Ehrmann, G.; Ehrmann, A. 3D printing of shape memory polymers. *J. Appl. Polym. Sci.* **2021**, *138*, 50847. [CrossRef]
65. Li, Z.; Liu, Y.; Lu, H.; Shu, D. Tunable hyperbolic out-of-plane deformation of 3D-printed auxetic PLA shape memory arrays. *Smart Mat. Struct.* **2022**, *31*, 75025. [CrossRef]
66. Ma, S.; Jiang, Z.; Wang, M.; Zhang, L.; Liang, Y.; Zhang, Z.; Ren, L.; Ren, L. 4D printing of PLA/PCL shape memory composites with controllable sequential deformation. *Bio-Des. Manuf.* **2021**, *4*, 867–878. [CrossRef]

Disclaimer/Publisher's Note: The statements, opinions and data contained in all publications are solely those of the individual author(s) and contributor(s) and not of MDPI and/or the editor(s). MDPI and/or the editor(s) disclaim responsibility for any injury to people or property resulting from any ideas, methods, instructions or products referred to in the content.

MDPI
St. Alban-Anlage 66
4052 Basel
Switzerland
Tel. +41 61 683 77 34
Fax +41 61 302 89 18
www.mdpi.com

Polymers Editorial Office
E-mail: polymers@mdpi.com
www.mdpi.com/journal/polymers

www.ingramcontent.com/pod-product-compliance
Lightning Source LLC
LaVergne TN
LVHW070718100526
838202LV00013B/1119

MDPI
St. Alban-Anlage 66
4052 Basel
Switzerland
Tel. +41 61 683 77 34
Fax +41 61 302 89 18
www.mdpi.com

Polymers Editorial Office
E-mail: polymers@mdpi.com
www.mdpi.com/journal/polymers